Fundamentals of Computational Fluid Dynamics

Tapan K. Sengupta
Department of Aerospace Engineering
IIT Kanpur

Universities Press

Universities Press (India) Private Limited

Registered Office
3-6-747/1/A & 3-6-754/1, Himayatnagar,
Hyderabad 500 029 (A.P.), India
e-mail: info@universitiespress.com

Distributed by
Orient Blackswan Private Limited

Registered Office
3-6-752, Himayatnagar, Hyderabad 500 029 (A.P.), India

Other Offices
Bangalore / Bhopal / Bhubaneshwar / Chennai
Ernakulam / Guwahati / Hyderabad / Jaipur / Kolkata
Lucknow / Mumbai / New Delhi / Patna

First Published 2004
Reprinted 2005, 2009, 2011

Cover and book design

ISBN: 978 81 7371 478 8

Set in Times New Roman 10 on 12 by
Krishtel eMaging Solutions Pvt.Ltd., Chennai 600 017

Printed in India at
Graphica Printers
Hyderabad 500 013

Published by
Universities Press (India) Private Limited
3-6-747/1/A & 3-6-754/1, Himayatnagar
Hyderabad 500 029 (A.P), India

Front cover: Stream function (top three panels) and vorticity contours at the indicated time for the problem of instability of shear layer caused by a translating vortex in the free stream. The arrowheads indicate the streamwise location of the free-stream vortex Navier-Stokes equation computed by compact schemes discussed in Chapter 11. [For details, see: Vortex-induced instability of incompressible wall-bounded shear layer, T.K. Sengupta, S.De & S. Sarkar, *J. Fluid Mech.*, **493**, 277–286 (2003).]

This work is dedicated to
my parents
Shri R.N. Sengupta and the late Smt. Leela Sengupta.

Contents

Preface

Computational Fluid Dynamics (CFD), despite its very appellation, has diverse applications in applied physics, mathematics and various branches of engineering. Although the physical descriptions of various phenomena in these fields vary, the unity comes from the fact that the physical laws governing them essentially stem from the same fundamental conservation laws. From the mathematical point of view, the governing partial differential equations can be classified into one of the three types, viz. parabolic, elliptic and hyperbolic partial differential equation. The considerations and techniques of solving these equations are similar if the types of equations are similar.

However, the situation is not so simple for practical engineering systems. Firstly, the governing differential equations are of *mixed type*, i.e., different parts of the spatial domain have different types of predominant characteristics – this can also happen with variation in time. It can be further complicated by the complexities of the geometries involved. Nonetheless, a systematic knowledge of the techniques for different equation types will help in solving complex practical problems. It is with this idea that this book is presented to the reader, with emphasis on description and analyses of various techniques of CFD.

In the presence of so many titles on the same topic, appearance of another book may appear as an anachronism. However, I wish to assure the readers that this book would not have been written if I did not feel the need for this book. One finds there are two classes of books/ monograph available on CFD. In the first category are books which contain descriptions of large number of numerical methods or schemes with very little detail, and which end with the familiar refrain that *CFD is more of an art than science*. In the second category are books devoted solely to a single topic, with an in-depth treatment of the matter. In writing this book I have tried to provide a balance between the two. Here, a few methods have been discussed concurrently to show the fundamental principles of computing involved in them. I have found from experience that students find it easier to learn by such method of association, and in fact prefer it. This approach would, additionally, encourage readers to grasp other methods equally well.

In this book I have purposely added the spectral framework for the analysis of CFD techniques. It fulfils two purposes. Firstly, it provides a uniform framework from where one can analyse all the methods and make a comparison of them. Thus, this provides a scale by which all methods can be calibrated. This is especially useful when disparate techniques are viewed and their performances compared. For example, in Chapter 12, readers are familiarized with popular finite volume (FV) and finite element (FE) techniques. Starting from the fundamentals, these techniques are compared using the spectral yardstick so that anyone planning to use them can see the relative merits and demerits of each technique. The chapter ends with a comparative study of FEM, FVM and finite difference methods (FDM).

Secondly, from a theoretical standpoint, one can see the utility of spectral description of any dynamical system if proper orthogonal decomposition (POD) is adopted as a descriptive tool. This tool has been in use in the recent times for studying the dynamical system models of turbulence. As direct numerical simulation (DNS) is appearing more and more feasible, it is perfectly natural to invoke such a tool. In POD, a *random* field ν_i is represented by a deterministic vector ϕ_i such that the projection of the latter to the former is maximized, i.e. the quantity $\langle |\nu\phi_i|^2 \rangle$ is maximized subject to given initial and boundary conditions. Without going into details, it can be shown that the deterministic vector comes out as a solution of an integral eigenvalue relation given by,

$$\int_{\text{region}} R_{i,j}(r, r')\phi_j(r')dr' = \lambda\phi_i,$$

where the kernel of the above is the two-point correlation function. For a system with finite energy content, among the many possible solutions, there are countably-infinite orthogonal eigen functions, $\phi_i' s$, in each one of them. An optimal description (in the mean-square error sense) is obtained by Galerkin projection of the governing equation, and the eigen functions belong to Fourier spectral modes. Hence a spectral mode of description will always provide an optimal description, while other methods will always be trying to reach this ideal goal.

With the growing importance of resolving various length and time scales for practical problems, we see a requirement for the reader to be familiar with the rudiments of the concept of length and time scales. Thus, there is a very elementary description of wave mechanics in Chapter 4. At this stage when DNS and large eddy simulation (LES) is becoming commonplace for high Reynolds numbers, this is a basic minimum that the reader must be aware of. CFD has matured from its initial preoccupation of finding *stable* algorithms alone! Now, instead, it is important that a correct numerical method be neutrally stable while the energy of the system propagates at the correct speed. The latter is denoted by what is known as dispersion relation property (DRP). It is now realized by the serious practitioners of CFD that to track complex flows or weak signals (as in computational aero-acoustics (CAA) and computational electromagnetics (CEM)), we must take into account these properties. In addition, one must avoid spurious conditions being generated at the boundaries and interfaces—a topic covered under the rubric of non-reflecting boundary condition, buffer layer techniques, and phase matched layer (PML) techniques. We do not intend to cover these aspects; these require an understanding of compressible flow techniques and are not discussed here. As DRP is an important requirement for all incompressible turbulent flows, readers are made familiar with it.

In the above discussion it is implicit that there are newer methods emerging which are capable of solving problems using DNS and LES. At this point I would like to alert the reader that there is a *hope* among some CFD practitioners using very low accuracy schemes, that with the advances in computer hardware, in a none-too-distant future, DNS will be possible with such methods. This book will alert the newcomers in the field against such complacency. It is with this idea that in Chapter 11 the readers have been introduced to near-spectral accuracy compact schemes –

those that have been proven over the last decade or so – as a vital weapon in the armoury of CFD practitioners. Also, in this chapter I have shown the readers how to choose among different options by properly analyzing the schemes for the above mentioned properties. Special attention has to be paid to the boundaries that seldom create numerical instabilities by violating the property of the characteristics and spuriously *exciting* the system. It is quite common to note that many methods having this problem and/or spurious dispersion, produces results having the superficial attributes of turbulence and have been termed as DNS results.

Finally, with the advent of higher accuracy methods, CFD practitioners are being confronted with a new danger in the form of aliasing error. Earlier, such problems used to bother spectral method practitioners for the nonlinear terms. New high-accuracy methods approaching near-spectral accuracy faces the same problem today, even while evaluating linear terms in the transformed plane! This always perplexes a newcomer to the field. This problem, arising out of solving equations in the transformed plane, has plagued computing people for decades – the cause was never diagnosed as due to aliasing. In Chapter 10 the source of this problem has been discussed, and a few available techniques have been listed to circumvent it.

The core material of this book was developed while I was teaching graduate level courses on CFD for one semester. I have been informed by students that initially they find newer concepts, as discussed in Chapters 4 and 10, difficult to grasp in the absence of a single book that comprehensively covers all the aspects. This prompted me to write this book. If readers find that the treatment of topics are not uniform, it is due to the fact that some of the topics covered belong to the realm of classical methods (with well developed theories), while other topics are emerging ones as a result of ongoing contemporary research.It is hoped that readers would find the approach I have adopted beneficial in comprehending the techniques of CFD in an analytical manner.

The preparation of this manuscript has been facilitated by a grant from CDTE, IIT Kanpur. Several students of IIT Kanpur and National University of Singapore have carried out the various computational work reported here. I would like to particularly mention Umendra Singh Chauhan, Manoj Nair, Gaurav Ganeriwal, Manojit Chattopadhyay, L.Yi Pin and Anurag Dipankar. I gratefully acknowledge these contributions.

I would appreciate receiving feedback from readers on any errors and misprints so that they can be corrected in future editions.

Tapan K. Sengupta
Professor
Department of Aerospace Engineering
Indian Institute of Technology Kanpur
Kanpur 208 016
tksen@iitk.ac.in

1 Basic Ideas of Computational Fluid Mechanics

1.1 Computational Fluid Mechanics

Since the publication of *Principia* by Newton in 1687 till the end of the second world war, advancements in fluid mechanics were brought about by the complimentary approaches of theoretical and experimental investigations. This we may term as the period of classical fluid mechanics.

The physical aspects of fluid flow are governed by the conservation principles of mass, momentum and energy. These principles, when expressed in terms of mathematical equations while making very few assumptions lead to a set of nonlinear partial differential equations (PDE). Till today, there are no generic theories available that allow general solution of these PDEs. Even when the PDEs can be linearised, they do not yield to analytical methods for practical problems due to complications of geometries. It is only during the second half of the last century, following the advent of digital computers as a tool to solve the linear and nonlinear PDEs, that we have witnessed a virtual revolution in many branches of physics including fluid mechanics. One such discipline is computational fluid dynamics (CFD).

The basic theoretical advances in numerical techniques were made over the previous centuries so much so that L.F. Richardson (1922) sought to solve the system of equations for weather forecasting using a desk calculator. The results of this pioneering effort were not particularly encouraging but the seeds were sown for what was to become a major initiative in mathematical physics! The use of a desk calculator in solving a viscous fluid dynamical problems was initiated by Thom (1928, 1933). Similar efforts were replicated by Kawaguti (1953) and Allen & Southwell (1955) for the solution of Navier-Stokes equation for flow past a circular cylinder. While Kawaguti (1953) solved the problem for a low Reynolds number of 40, the effort by Allen & Southwell was particularly the harbinger of later activities in CFD using digital computers. Their solution was obtained for a Reynolds number of 1000, in which they solved the problem in the transformed plane using residual relaxation methods.

Using digital computers for the first time, Charney et al. (1950) produced the first successful results for weather forecasting at 500 mb with an equivalent barotropic model for nonlinear vorticity transport equation. Several important milestones were achieved by researchers working in the area of numerical weather prediction—including in the areas of large eddy simulation (LES) and direct numerical simulation (DNS).

Kopal (1947) was the first to use a digital computer to solve the Taylor-Maccoll equation (an ODE) numerically. A similar result was presented earlier by Maccoll (1938) who did not use a digital computer. This solves the problem of conical flow past a cone at supersonic speed and consists of fitting an isentropic conical flow with a conical shock. While a desk calculator allows one to solve a problem, the use of digital computers allows one to systematically solve the same problem repeatedly for different parameter values.

For compressible flow problems that require solving PDEs, the beginning was made with the important paper of Von Neumann (1944) that contained his proposal on numerical methods for the hydrodynamic shock problem. The analysis provided in this paper was further refined in a subsequent work by Neumann & Richtmeyer (1950) and is considered as a classic work. In this the authors introduced the concept of adding explicit numerical dissipation essentially to capture the shock that is adequately given by the Rankine–Hugoniot relation. The added dissipation smears the shock, but at the same time provides numerical stability to the algorithm for one-dimensional inviscid calculations of a propagating shock wave using non-conservative Lagrangian equations. Ludford et al. (1953) advocated the use of arbitrarily large values of artificial "viscosity coefficient" in the viscous equation given in a Lagrangian framework instead. The major theoretical advances in CFD for hyperbolic PDE were made by Courant et al. (1928,1967) in which not only was the method of characteristics discussed, but also contained the important idea of numerical stability, which is given by the CFL criterion in terms of the non-dimensional number bearing their initials. The other notable theoretical contributions for the solution of compressible flow equations came from Lax (1954) and Godunov (1959).

1.2 Role of Computational Fluid Mechanics

It is needless to delve into the supremacy of one method of understanding flow physics over the other. When CFD as a subject began, it was with the modest goal of being a complementary tool to experiment and theory. Over the last five decades CFD has matured to a point where it now has the capability to supplant both experimental and theoretical approaches for some specific problems. This is specifically true for laminar flows and hence this assertion leaves out a major area of investigation related to turbulent flows. However, the progress made by researches in the areas of DNS and LES in recent times hold the promise of even crossing this last frontier of classical physics! Additionally, CFD is also helping to unravel newer theoretical phenomenon supplanting experimental fluid mechanics which used to play the same role earlier. For example, in some recent investigations conducted by Sengupta et al. (1997), a new class of upstream propagating waves have been shown to exist. This is owing to a combination of CFD and theoretical studies undertaken simultaneously. Such exercises promise to explain diverse phenomena in fluid mechanics.

In Fig. 1.1 the relationship among the triad of CFD, theory and experiments are shown on the faces of a pyramid. The apex is the knowledge of fluid mechanics that practitioners aspire. Face-1 is the classical approach followed over the last three centuries. Face-2 shows the synergy of experiments and CFD in adding to the body of existing knowledge and is a useful tool in engineering analysis. The CFD tools used here are usually not very sophisticated and the major goal is to obtain results with acceptable accuracy. In the early stages of development of CFD, benchmark experiments were used to develop and fine-tune the CFD codes. The tuning of bells and whistles in CFD codes was called the *art of CFD*. In fact, this stage of CFD activities could be termed as "post-diction" as opposed to "prediction". One of the most important aspects of modern CFD is experienced in the aircraft industry—the use of CFD in preliminary design stages has significantly reduced the number of costly wind tunnel tests. This also cuts down the time taken for a design cycle. The wind tunnel tests are done to fine-tune the final design.

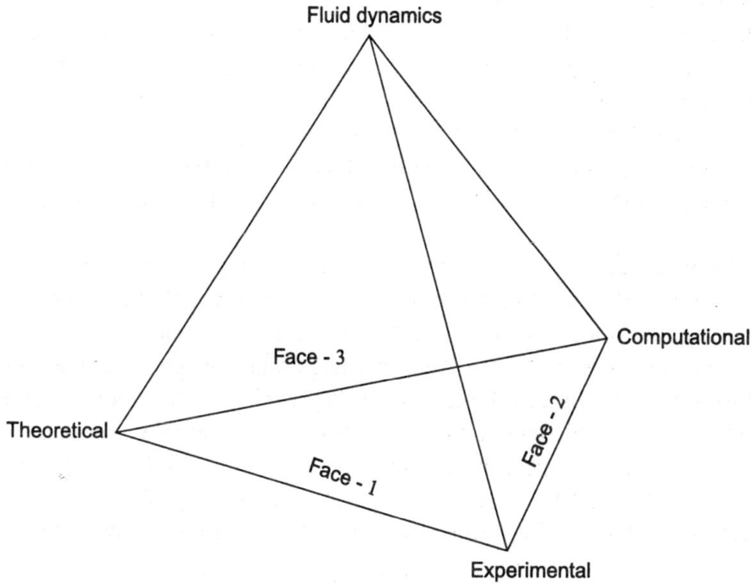

Fig. 1.1 *Triad in fluid mechanics.*

Face-3 of the pyramid shows the complementary approaches to fluid mechanics by CFD and theoretical studies. And this interaction holds the maximum promise of unraveling many new phenomena and explaining past unsolved problems. However, this requires the mathematical models used in CFD to be sophisticated enough to include relevant physical mechanisms (as in DNS and LES, for example) for resolving a wide range of spatial and temporal scales. The result obtained by such CFD simulations can supplant experiments in terms of the obtained details and difficult regions of the flow. This is one of the future directions of CFD. CFD as a tool is proving to be prime mover in our understanding of the basic fluid mechanics following this route. The attendant high accuracy methods developed for DNS has encouraged researches in associated fields like computational aero-acoustics (CAA) and computational electromagnetics (CEM). Such activities demand resolving very small scales and hence brings with it practical problems of coping with and interpreting massive outputs. This opens up new fields of study like data mining. The present purpose of this text is to emphasize this last aspect of CFD—as a tool to interpret fluid mechanics.

The growth of CFD is due to the advent and subsequent growth of capabilities of digital computers. It has been noted that the relative cost of computation comes down by an order of magnitude every eight years. The exponential growth of relative computer performance (defined by computer speed and memory) is propelled mostly by the introduction of newer technologies. However, the growth rate due to a component innovation saturates in a period of about fifteen years, and by then another new technology becomes the pacing item. It is conjectured that this type of exponential growth can be sustained at the current level of understanding for another decade or so. Currently, most of the speed of supercomputers is due to interprocessor communications, memory structure and the computer architecture. Otherwise a desktop computer and a supercomputer uses almost similar chipset for computing and has the same speed per processor. Apart from parallel computing, it is expected that new advances in algorithm should contribute to significant speed-up in solving complex problems for some more time to come. For example, finite difference methods used in CFD are using several compact schemes that allow solving problems with fewer grid points while yielding same spectral accuracy. Hence they effectively provide higher speed-up of computations.

Finally, it must be noted that, with all the accruing benefits of computer architecture, newer algorithms etc., we will also see another aspect of computing which has not received sufficient attention so far becoming important. This is the aspect of high accuracy computing in CFD, CAA and CEM. Since it is now possible to resolve a very large range of spatial and temporal scales, the traditional methods must be subjected to checks for their inherent accuracy to solve the problem of DNS for both incompressible and compressible flows. For example, one must ensure that the algorithm used must preserve dispersion relation when used for DNS, CAA and CEM, along with very high spectral accuracy. This is seen to be compromised very often by even the so-called higher order accurate methods. This will be the focus of our attention in one of the later chapter.

References

[1] ALLEN, D.N., DE G., & SOUTHWELL, R.V., Relaxation methods applied to determine the motion, in two dimensions, of a viscous fluid past a fixed cylinder, *Quarterly J. of Mechanics and Applied Mathematics*, **8**, 129–145, (1955).

[2] CHARNEY, J.G., FJORTOFT, R., & VON NEUMANN, J., Numerical integration of the barotropic vorticity equation, *Tellus*, **2**, 237–254, (1950).

[3] COURANT, R., FRIEDRICHS, K.O., & LEWY, H., Uber die partiellen differenzengleichurgen der mathematischen physik, *Mathematische Annalen*, **100**, 32–74, (1928).

[4] COURANT, R., FRIEDRICHS, K.O., & LEWY, H., On the partial differential equations of mathematical physics, *IBM Journal*, 215–234, (1967).

[5] GODUNOV, S.K., Finite difference method for numerical computation of discontinuous solutions of the equations of fluid dynamics, *Matematicheskii Sbornik*, **47**, 271–306, (1959).

[6] KAWAGUTI, M., Numerical solution of the Navier-Stokes equations for the flow around a circular cylinder at Reynolds number 40, *J. Physical Society of Japan*, **8**, 747–757, (1953).

[7] KOPAL, Z., *Tables of Supersonic Flows about Cones*, MIT, (1947).

[8] LAX, P.D., Weak solutions of nonlinear hyperbolic equations and their numerical computation, *Communication on Pure and Applied Mathematics*, **7**, 159–193, (1954).

[9] LUDFORD, G., POLACHEK, H., & SEEGER, R.J., On unsteady flow of compressible viscous fluids, *J. of Applied Physics,* **24**, 490–495, (1953).

[10] MACCOLL, J.W.,The conical shock wave formed by a cone moving at high speed, *Proc. Roy. Soc. A*, 459, (1937).

[11] RICHARDSON, L.F., *Weather Prediction by Numerical Process*, Cambridge University Press, reprinted Dover, (1965), (1922).

[12] SENGUPTA, T.K., RANA, V.& NAIR, M.T., Boundary layer excited by low frequency disturbances - Klebanoff mode, *Journal of Fluid and Structures,* **11**, 845–853, (1997).

[13] THOM, A., An investigation of fluid flow in two dimensions, Aerospace Research Center, *R. and M.*, **No. 1194**, United Kingdom, (1928).

[14] THOM, A., The flow past circular cylinders at low speed, *Proceedings of the Royal Society of London*, **A 141**, 651–666, (1933).

[15] VON NEUMANN, J., Proposal and analysis of a numerical method for the treatment of hydrodynamical shock problems, *Nat. Def. and Res. Com. Report,* **AM–551**, (1944).

[16] VON NEUMANN, J., & RICHTMYER, R.D., A method for the numerical calculation of hydrodynamic shocks, *J. Applied Physics*, **21**, 232–257, (1950).

2 Governing Equations of Fluid Mechanics

2.1 Basic Equations of Fluid Mechanics

The basic equations of fluid mechanics are nothing but the statement of conservation principles for mass, momentum and energy. The governing equations can be stated in different forms. From a theoretical perspective, it is immaterial as to which form of equations are used, but in CFD this is of paramount importance. In one form, the solutions would be correct, in another form it may lead to numerical instability, and in yet another form it may give totally unphysical results. Thus, it is important that one learns the various forms of governing equations. We will state them in this chapter in order to point out their differences and similarities, and talk of the possible implications of their usage in CFD. The following steps are followed in CFD:

(a) Choose the major physical principles that are to be satisfied, i.e. whether mass is conserved or created. How is the momentum balance achieved? And whether energy is conserved in the field of interest or created there by external agencies.

(b) Based on the stated physical principles, select an appropriate model for the flow.

(c) Finally, from combined application of (a) and (b) extract the mathematical equations stating the same. Use of CFD comes in while solving these mathematical equations.

Firstly, we will concentrate on the second aspect, i.e. choosing a suitable model that incorporates all the important physical events. Unlike in solid mechanics problems note that fluid mechanics problems are characterized by "infinite" or very large number of degrees of freedom. This is because fluid mechanics is characterized by very low physical diffusion. However, this diffusion cannot be completely ignored from the whole flow domain. For example, it is known that for streamlined bodies, dissipation is important within a thin shear layer, whereas in the outer part of the flow it is negligible and often considered so. The smooth blending of these two parts of the flow is to be performed consistently, which is not a trivial exercise. To study fluid flow we must adopt a framework for analysis. This could be done by considering either of the following.

2.1.1 Finite control volume

In this framework we consider a fixed closed volume encompassing a finite region of the flow, with the volume defining the *control volume* demarcated from the surrounding fluids by the *control surface*. This *control volume* can be fixed or can move in space. In either case, the fundamental physical principles (as in (a) above) are applied to the fluid inside the *control volume* and to the fluid crossing the *control surface*. This procedure is similar to the one in deriving the governing differential equations, except that the control volume is finite here rather than being infinitesimally small.

The fluid flow equations that we obtain directly by applying the conservation principles to a finite *control volume* are in integral form. From this integral form of the equation one can derive the differential form by using Taylor series expansion. The equations so obtained – either in differential or integral form – are called the *conservation form* of the governing equations if the control volume is fixed in space. If the control volume moves with the fluid, the governing equations in either the integral or differential form are in the *non-conservation or convective form*.

2.1.2 Infinitesimal fluid element

In this framework the considered fluid element is small in the mathematical sense, but in the physical sense big enough to contain a large number of fluid molecules so that the continuum assumption holds. One can apply the physical principles to this element and directly obtain differential equations in the *conservation from* when the element is fixed in space, and in the *non-conservative* or *convective form* when the fluid element convects with the surrounding fluid.

In both these frameworks, we need to define the variation of physical quantities that vary with both space and time by introducing *substantive derivative*.

2.1.3 Substantive derivative

Before deriving the governing equations we will introduce this concept commonly used in fluid mechanics. This is the acceleration experienced by a fluid elements that moves with the flow. The flow field is described by the velocity vector in the Cartesian frame by

$$\vec{V} = u\hat{i} + v\hat{j} + w\hat{k}, \tag{2.1}$$

where u, v and w are the Cartesian components of the velocity vector in x, y, and z-direction respectively.

The other unknowns are the density (ρ) and the pressure (p) that completely defines the fluid motion in primitive variables.

Consider an element of fluid at $t = t_1$ to be located at (x_1, y_1, z_1) in a Cartesian frame of reference. At a later time, $t = t_2$, let this element be found at (x_2, y_2, z_2). If the fluid density is ρ_1 and ρ_2 respectively at these two time instants, then one can represent ρ_2 in terms of ρ_1 by using Taylor series

$$\rho_2 = \rho_1 + \left(\frac{\partial \rho}{\partial x}\right)_1 (x_2 - x_1) + \left(\frac{\partial \rho}{\partial y}\right)_1 (y_2 - y_1) + \left(\frac{\partial \rho}{\partial z}\right)_1 (z_2 - z_1) + \left(\frac{\partial \rho}{\partial t}\right)_1 (t_2 - t_1) + \text{higher order terms.} \tag{2.2}$$

In the limit $t_2 \to t_1$, Equation (2.2) can be rewritten as

$$\lim_{t_2 \to t_1} \left(\frac{\rho_2 - \rho_1}{t_2 - t_1}\right) \equiv \frac{D\rho}{Dt} = \frac{\partial \rho}{\partial t} + u\frac{\partial \rho}{\partial x} + v\frac{\partial \rho}{\partial y} + w\frac{\partial \rho}{\partial z} \tag{2.3}$$

because

$$\lim_{t_2 \to t_1} \frac{x_2 - x_1}{t_2 - t_1} = \left(\frac{dx}{dt}\right) = u \text{ and so on.}$$

The first term on the right hand side represents the *local change* while the rest of the terms on the right hand side are due to *convective change*. Physically, the *local change* represents the time rate of change at a fixed point and the *convective change* is the time rate of change of the property because of the movement of the element and is due to spatial inhomogeneity of the flow field. Thus the substantive derivative is given by

$$\frac{D}{Dt} = \frac{\partial}{\partial t} + u\frac{\partial}{\partial x} + v\frac{\partial}{\partial y} + w\frac{\partial}{\partial z}. \tag{2.4}$$

This can also be represented in a coordinate frame-independent context by

$$\frac{D}{Dt} = \frac{\partial}{\partial t} + \vec{V}.\nabla, \tag{2.5}$$

where $\vec{V} = u\hat{i} + v\hat{j} + w\hat{k}$
and $\nabla = \hat{i}\frac{\partial}{\partial x} + \hat{j}\frac{\partial}{\partial y} + \hat{k}\frac{\partial}{\partial z}$.

The governing equations of fluid motion that are to be derived here are nothing but the conservation principles. We begin by stating the conservation of mass.

2.2 Equation of Continuity

This physical principle is simple to comprehend and provides a balance between mass entering and leaving the *control volume* through the control surfaces with the mass that may be created inside the *control volume*. For unsteady compressible flows, this leads to

$$\frac{D\rho}{Dt} + \rho\nabla.\vec{V} = \frac{\partial\rho}{\partial t} + \nabla.(\rho\vec{V}) = 0. \tag{2.6}$$

For incompressible flows ρ = const. and Equation (2.6) simplifies to

$$\nabla.(\vec{V}) = 0. \tag{2.7}$$

Any vector field that satisfies the equivalent condition given by Equation (2.7) is called the *divergence-free* or *solenoidal field*.

2.3 The Momentum Equation

In this section we will derive the important physical principles for conservation of translation momentum for flows which is nothing but Newton's second law of motion applied to a control volume system. We choose for our flow model, the moving fluid element shown below in Fig. 2.1. We have indicated the forces acting in the x-direction only. Newton's second law states that the force vector acting on a control volume is equal to the rate of change of momentum. As we have decided to follow the control volume in its movement, then in effect we are considering a *control mass system* for which the mass does not change with time. Hence the rate of change of momentum is nothing but the mass times the acceleration experienced by the fluid in the control volume. If we decide to write only the x-component of this relation we get,

$$F_x = ma_x, \tag{2.8}$$

with the notation carrying their conventional meaning. Let us look little closely at the left hand side of Equation (2.8).

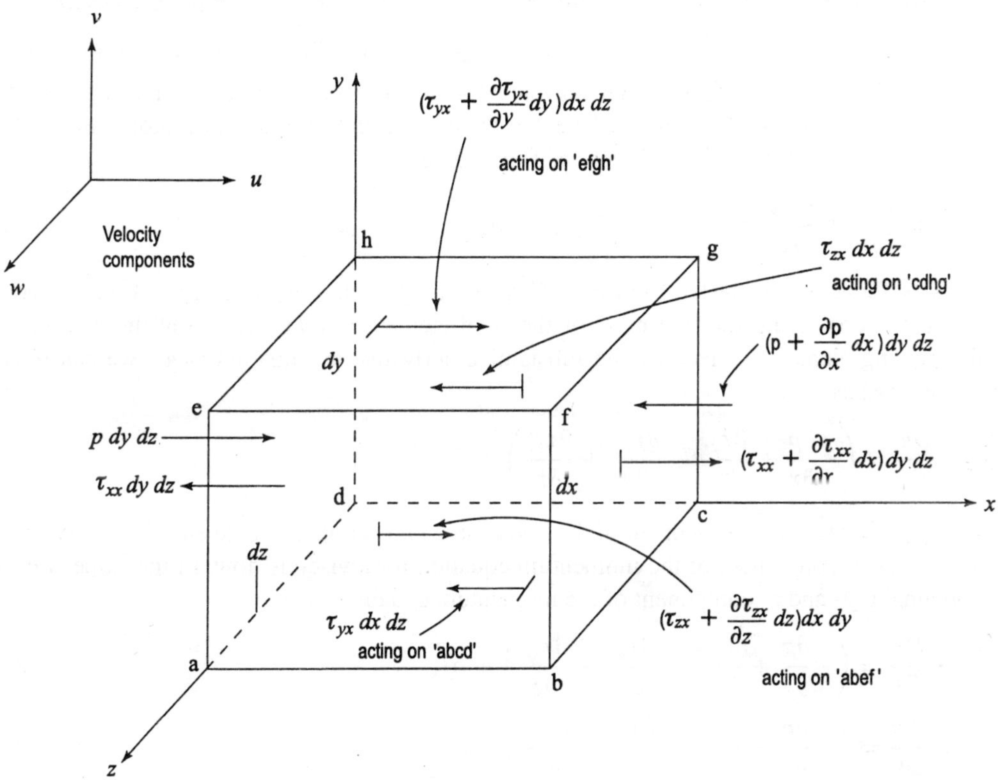

Fig. 2.1 *Infinitesimally small, moving fluid element with forces only in the x-direction shown.*

What are the forces that act on the fluid in the control volume? Firstly, we know that there are *body forces* that act on the volume *from a distance* and do not depend on the geometry of the body. Examples are the gravitational and the electromagnetic forces. Additionally, there are the *surfaces forces* that act directly on the control surfaces of the fluid element. Since it acts on the surface elements, the geometry is an important parameter in deciding the forces acting on it. These *surfaces forces* can be further subdivided into *normal* and *shear forces* caused solely by normal and shear stresses.

In Fig. 2.1, the volume of the indicated element is $(dx\, dy\, dz)$ and if the local density is ρ, then the body force acting in the x-direction can be simply written as

$$= \rho f_x (dx\, dy\, dz), \tag{2.9}$$

where f_x is the associated acceleration in the x-direction.

The surface forces are as indicated in Fig. 2.1, for the x-component momentum balance with τ_{ij} denoting the stress tensor acting in the j-direction on a plane whose normal vector is in the i-direction. For example, on face *abcd* the only force acting in the x-direction is due to the shear stress τ_{yx} and on the face *efgh* the corresponding force is created due to the shear stress $\tau_{yx} + \frac{\partial \tau_{yx}}{\partial y} dy$. The net force in the x-direction due to this pair of shear stresses on the surfaces *abcd* and *efgh* is $\tau_{yx} dx\, dz - \left(\tau_{yx} + \frac{\partial \tau_{yx}}{\partial y} dy\right) dx\, dz$, i.e., $-\frac{\partial \tau_{yx}}{\partial y} dx\, dy\, dz$. The direction of the forces are as indicated in Fig. 2.1. The adopted sign convention is that u, v and w are the increasing functions of their spatial arguments x, y and z. One can therefore count all the contributory stresses that give rise to surface force in the x-direction. This is given by

$$= \left(-\frac{\partial p}{\partial x} + \frac{\partial \tau_{xx}}{\partial x} + \frac{\partial \tau_{yx}}{\partial y} + \frac{\partial \tau_{zx}}{\partial z}\right) dx\, dy\, dz. \tag{2.10}$$

Thus the force term on the left hand side of Equation (2.8) is the algebraic sum of Equations (2.9) and (2.10). Considering that the mass of the fluid element is $\rho dx.dy.dz$ and the acceleration of the moving element is given by the substantive derivative in Equation (2.4), we can rewrite Equation (2.8) as

$$\rho \frac{Du}{Dt} = \left(-\frac{\partial p}{\partial x} + \frac{\partial \tau_{xx}}{\partial x} + \frac{\partial \tau_{yx}}{\partial y} + \frac{\partial \tau_{zx}}{\partial z}\right) + \rho f_x, \tag{2.11}$$

where $\rho dxdydz f_x$ is the component of the body force $(\vec{F})$ acting upon the element $dxdydz$.

This is the x-component of the momentum equation for a viscous flow. In the same way one can obtain the y- and z- component of the momentum equation as

$$\rho \frac{Dv}{Dt} = \left(-\frac{\partial p}{\partial y} + \frac{\partial \tau_{xy}}{\partial x} + \frac{\partial \tau_{yy}}{\partial y} + \frac{\partial \tau_{zy}}{\partial z}\right) + \rho f_y; \tag{2.12}$$

$$\rho \frac{Dw}{Dt} = \left(-\frac{\partial p}{\partial z} + \frac{\partial \tau_{xz}}{\partial x} + \frac{\partial \tau_{yz}}{\partial y} + \frac{\partial \tau_{zz}}{\partial z}\right) + \rho f_z. \tag{2.13}$$

Equations (2.11) to (2.13) are the Navier-Stokes equation, and are written in *non-conservative* or *convective form*. It is also customary to write the above equations in vector form that is coordinate system independent, and is given by

$$\rho \frac{D\vec{V}}{Dt} = -\nabla p + \nabla.\partial_j \tau_{ji} + \rho \vec{F}, \tag{2.14}$$

where δ_{ij} is the Kronecker delta function.

It is also possible to express the Navier-Stokes equation in *conservation form.* Using the vector form of continuity equation as given in Equation (2.6), one can express the substantive derivative in Equation (2.11) as

$$\rho\frac{Du}{Dt} = \frac{\partial(\rho u)}{\partial t} + \nabla.(\rho u\vec{V}). \tag{2.15}$$

Then Equation (2.11) can be written in the following conservative form

$$\frac{\partial(\rho u)}{\partial t} + \nabla.(\rho u\vec{V}) = -\frac{\partial p}{\partial x} + \frac{\partial \tau_{xx}}{\partial x} + \frac{\partial \tau_{yx}}{\partial y} + \frac{\partial \tau_{zx}}{\partial z} + \rho f_x. \tag{2.16}$$

Similarly, one can write the other two components of the momentum equation in *conservative form.* These equations are still not usable as one does not have informations about the general stress system in fluid. It was Newton, who suggested that the "stress is proportional to the strain rate." In fluids the strain rates are related to the velocity gradient, as given in many textbooks (see e.g. Schlichting, 1979). These types of functional relationships between flow quantities are also called constitutive relations because they are related to the property of the fluid. For example, the perfect gas law is a constitutive relation for a particular type of gas. The fluids that abide by the above relation between stress and strain rates are called the Newtonian fluids. For many applications in fluid mechanics this is a satisfactory relation. However, there are also many applications where the fluid cannot be termed as Newtonian. For example blood is a non-Newtonian fluid and so are many rheological fluids, plastics and thixotropic fluids. For Newtonian fluids, Stokes (1845) has shown

$$\tau_{ij} = \lambda\delta_{ij}\frac{\partial v_k}{\partial x_k} + \mu\left(\frac{\partial v_i}{\partial x_j} + \frac{\partial v_j}{\partial x_i}\right) - p\delta_{ij}, \tag{2.17}$$

where p is the thermodynamic pressure or hydrostatic stress. If p_m is the mechanical pressure, then by definition

$$p_m = \frac{\tau_{ii}}{3},$$

where μ is the molecular viscosity and λ is the bulk viscosity coefficient. In writing Equation (2.17) we have used the Einstein's summation notation, i.e., any repeated index implies summation over that index. In general, from the first principle in thermodynamics, one does not have a relationship between the two viscosity coefficients. It was Stokes, who again hypothesized a relationship between these two by equating thermodynamic and mechanical pressure. According to this,

$$3\lambda + 2\mu = 0. \tag{2.18}$$

This relationship is obtained by equating the thermodynamic pressure with the mechanical pressure since,

$$p - p_m = \left(\lambda + \frac{2}{3}\mu\right)\frac{\partial v_k}{\partial x_k} = -\left(\lambda + \frac{2}{3}\mu\right)\frac{1}{\rho}\frac{D\rho}{Dt}.$$

Using the constitutive relation as given in Equation (2.17) we can write down the final form of Navier-Stokes equation for compressible flow as

$$\frac{\partial(\rho u)}{\partial t}+\frac{\partial(\rho u^2)}{\partial x}+\frac{\partial(\rho uv)}{\partial y}+\frac{\partial(\rho uw)}{\partial z}=-\frac{\partial p}{\partial x}+\frac{\partial}{\partial x}\left(\lambda\nabla.\vec{V}+2\mu\frac{\partial u}{\partial x}\right)$$
$$+\frac{\partial}{\partial y}\left(\mu\left[\frac{\partial v}{\partial x}+\frac{\partial u}{\partial y}\right]\right)+\frac{\partial}{\partial z}\left(\mu\left[\frac{\partial u}{\partial z}+\frac{\partial w}{\partial x}\right]\right)+\rho f_x; \tag{2.19}$$

$$\frac{\partial(\rho v)}{\partial t}+\frac{\partial(\rho uv)}{\partial x}+\frac{\partial(\rho v^2)}{\partial y}+\frac{\partial(\rho vw)}{\partial z}=-\frac{\partial p}{\partial y}+\frac{\partial}{\partial y}\left(\lambda\nabla.\vec{V}+2\mu\frac{\partial v}{\partial y}\right)$$
$$+\frac{\partial}{\partial x}\left(\mu\left[\frac{\partial v}{\partial x}+\frac{\partial u}{\partial y}\right]\right)+\frac{\partial}{\partial z}\left(\mu\left[\frac{\partial v}{\partial z}+\frac{\partial w}{\partial y}\right]\right)+\rho f_y; \tag{2.20}$$

$$\frac{\partial(\rho w)}{\partial t}+\frac{\partial(\rho uw)}{\partial x}+\frac{\partial(\rho wv)}{\partial y}+\frac{\partial(\rho w^2)}{\partial z}=-\frac{\partial p}{\partial z}+\frac{\partial}{\partial z}\left(\lambda\nabla.\vec{V}+2\mu\frac{\partial w}{\partial z}\right)$$
$$+\frac{\partial}{\partial x}\left(\mu\left[\frac{\partial w}{\partial x}+\frac{\partial u}{\partial z}\right]\right)+\frac{\partial}{\partial y}\left(\mu\left[\frac{\partial v}{\partial z}+\frac{\partial w}{\partial y}\right]\right)+\rho f_z. \tag{2.21}$$

One can simplify the equations further for incompressible flows. For incompressible flow the equation of continuity is given by

$$\nabla.\vec{V}=0.$$

This equation when used in Equations (2.19)–(2.21), one readily sees that the terms associated with the bulk viscosity coefficient terms drop out. Furthermore, if one considers the incompressible flow without any heat addition, then the coefficient of molecular viscosity will not be function of space and can be treated as a constant. Instead of writing the component form of the momentum equations, we write the vector form of the Navier-Stokes equation for incompressible flow as

$$\frac{\partial\vec{V}}{\partial t}+(\vec{V}.\nabla)\vec{V}=-\frac{\nabla p}{\rho}+\nu\nabla^2\vec{V}+\vec{F}. \tag{2.22}$$

Equations written so far have been in terms of the physical variables, namely pressure and velocity. This kind of equations are customarily called the *equations in primitive variables.*

2.4 The Energy Equation

The energy equation is nothing but the first law of thermodynamics stated for a control volume system. In Fig. 2.2 the control volume element has been shown with only the stress system acting in the x-direction.

The first law of thermodynamics for a control system simply states that the rate of change of energy inside the control volume must be due to the heat interaction associated with the control volume plus the work done reversibly due to the boundary displacement by the body and the surface forces. Each of these constituents are obtained next.

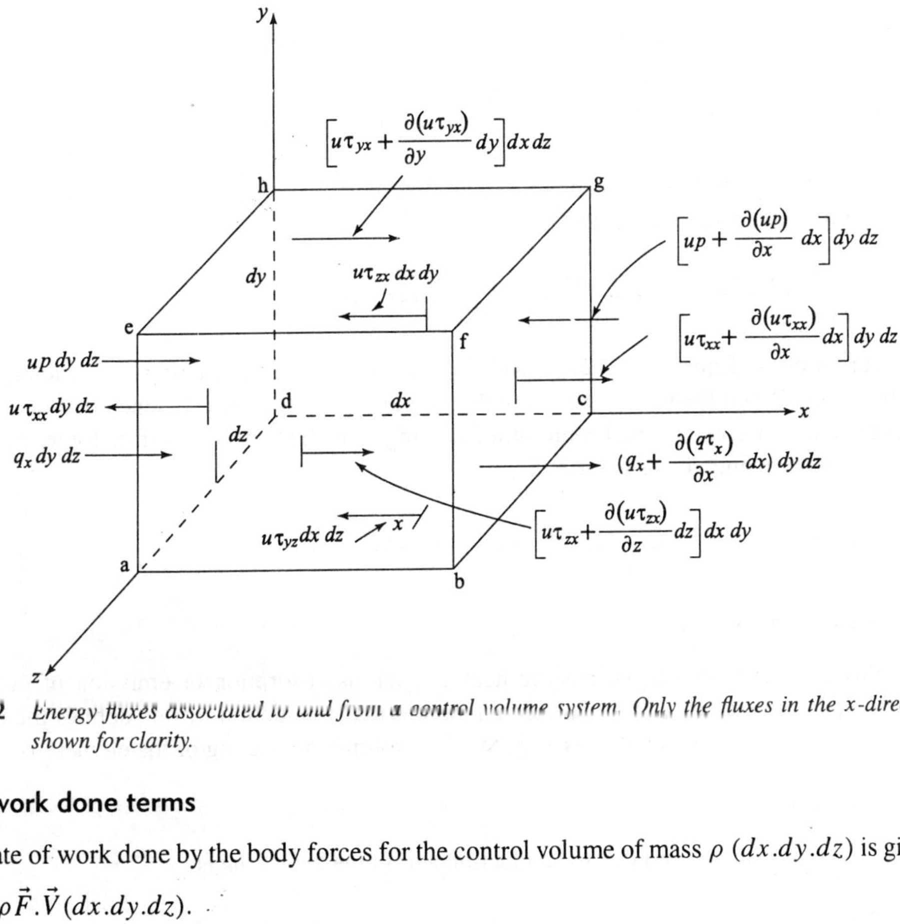

Fig. 2.2 *Energy fluxes associated to and from a control volume system. Only the fluxes in the x-direction are shown for clarity.*

The work done terms

The rate of work done by the body forces for the control volume of mass $\rho\ (dx.dy.dz)$ is given by

$$\rho \vec{F}.\vec{V}(dx.dy.dz). \tag{2.23}$$

The contribution of the surface forces acting on the element are due to normal and shear stresses. Only the x-component of forces are shown in Fig. 2.2 and the rate of work done will be the product of the forces in the x-direction times the x-component of velocity (u). On the face *abcd* the work done term is

$$\tau_{yx} dx\, dz.u.$$

Similarly the work done term due to the shear stress acting on the face *efgh* is given by,

$$\left[u\tau_{yx} + \frac{\partial(u\tau_{yx})}{\partial y} dy\right] dxdz.$$

Since the two shear forces are acting in the opposite direction, the net energy flux in the x-direction due to the shear stresses in the x-direction on faces *abcd* and *efgh* is

$$\frac{\partial(u\tau_{yx})}{\partial y}dxdydz.$$

Similarly, the normal stresses acting on the faces *adhe* and *bcgf* gives rise to the work done term

$$\left[up - \left(up + \frac{\partial(up)}{\partial x}dx\right)\right]dydz = -\frac{\partial(up)}{\partial x}dxdydz.$$

Now, considering all the forces shown in Fig. 2.2, the net energy flux for the control volume due to these forces is simply obtained as

$$\left[-\frac{\partial(up)}{\partial x} + \frac{\partial(u\tau_{xx})}{\partial x} + \frac{\partial(u\tau_{yx})}{\partial y} + \frac{\partial(u\tau_{zx})}{\partial z}\right]dxdydz. \tag{2.24}$$

The expressions in Equations (2.23) and (2.24) are obtained by considering forces acting in the x-direction only. When the surface forces in the y- and z-directions are also included in deriving the work done term we get the total contribution coming from the body and surface forces as (using a mix of summation and vector notation),

$$\left[-\nabla.(p\vec{V}) + \frac{\partial}{\partial x_j}(u_i\tau_{ji})\right]dxdydz + \rho\vec{F}.\vec{V}dxdydz. \tag{2.25}$$

The heat transfer terms

The net flux of heat is due to volumetric heating such as absorption or emission of radiation and heat transfer across the control surface due to thermal conduction. If one defines the rate of volumetric heat addition per unit mass as $\dot{q}$, then the volumetric heating of the element is

$$= \rho\dot{q}dxdydz. \tag{2.26}$$

In Fig. 2.2, the heat transferred due to conduction across the face *adhe* is

$$\dot{q}_x dydz,$$

where $\dot{q}_x$ is the heat flux in the x-direction per unit time per unit area by thermal conduction. Similarly the heat flux out of element across face *bcfg* is

$$\left[\dot{q}_x + \frac{\partial\dot{q}_x}{\partial x}dx\right]dydz.$$

The net heat flux in the x-direction into the fluid element by thermal conduction is

$$= -\frac{\partial\dot{q}_x}{\partial x}dxdydz. \tag{2.27}$$

Thus the heating of the fluid element due to thermal conduction by considering all directional conduction is given by,

$$= -\frac{\partial(\dot{q}_j)}{\partial x_j}dv. \tag{2.28}$$

Oncè again, summation convection applies. Using Newton's law for the conductive heat transfer one can relate the directional heat transfer with temperature gradient by

$$\dot{q}_j = -k\frac{\partial T}{\partial x_j},$$

where k is the thermal conductivity.

Hence the total heat interaction term is obtained by using Newton's law in Equation (2.28) and adding Equation (2.26) to this to give

$$= \left[\rho\dot{q} + \frac{\partial}{\partial x_j}\left(k\frac{\partial T}{\partial x_j}\right)\right]dxdydz. \tag{2.29}$$

The rate of change of energy terms

The total energy of a moving fluid per unit mass is the sum of its internal energy per unit mass (e), and its kinetic energy per unit mass $\left(\frac{V^2}{2}\right)$. Since we are following a moving fluid element, the time rate of change of energy per unit mass is given by the substantial derivative

$$= \rho\frac{D}{Dt}\left(e + \frac{V^2}{2}\right)dxdydz. \tag{2.30}$$

The final form

The final form of the energy equation is obtained by collating the terms given by Equations (2.24), (2.29) and (2.30) to give the *non-conservation* or *convective* form of the energy equation as

$$\rho\frac{D}{Dt}\left(e + \frac{V^2}{2}\right) = \left[\rho\dot{q} + \frac{\partial}{\partial x_j}\left(k\frac{\partial T}{\partial x_j}\right)\right] - \nabla.(p\vec{V}) + \frac{\partial}{\partial x_j}(u_i\tau_{ji}) + \rho\vec{F}.\vec{V}. \tag{2.31}$$

2.5 Alternate Form of Energy Equation

Occasionally, the energy equation is written in terms of internal energy only. This is achieved by rewriting the momentum Equation (2.16) after multiplying by u and multiplying the y- and z-momentum equations by v and w respectively. The component equations are obtained as,

$$\rho \frac{D}{Dt}\left(\frac{u^2}{2}\right) = -u\frac{\partial p}{\partial x} + u\left(\frac{\partial \tau_{xx}}{\partial x} + \frac{\partial \tau_{yx}}{\partial y} + \frac{\partial \tau_{zx}}{\partial z}\right) + \rho u f_x; \tag{2.32a}$$

$$\rho \frac{D}{Dt}\left(\frac{v^2}{2}\right) = -v\frac{\partial p}{\partial y} + v\left(\frac{\partial \tau_{xy}}{\partial x} + \frac{\partial \tau_{yy}}{\partial y} + \frac{\partial \tau_{zy}}{\partial z}\right) + \rho v f_y; \tag{2.32b}$$

$$\rho \frac{D}{Dt}\left(\frac{w^2}{2}\right) = -w\frac{\partial p}{\partial z} + w\left(\frac{\partial \tau_{xz}}{\partial x} + \frac{\partial \tau_{yz}}{\partial y} + \frac{\partial \tau_{zz}}{\partial z}\right) + \rho w f_z. \tag{2.32c}$$

Adding these three equations one gets

$$\rho \frac{D}{Dt}\left(\frac{V^2}{2}\right) = -\vec{V}.\nabla p + \rho \vec{F}.\vec{V} + v_j \frac{\partial \tau_{ij}}{\partial x_i}. \tag{2.33}$$

Subtracting Equation (2.33) from Equation (2.31) one gets,

$$\rho\frac{De}{Dt} = \rho\dot{q} + \frac{\partial}{\partial x_j}\left(k\frac{\partial T}{\partial x_j}\right) - p\nabla.\vec{V} + \tau_{xx}\frac{\partial u}{\partial x} + \tau_{yx}\frac{\partial u}{\partial y} + \tau_{zx}\frac{\partial u}{\partial z}$$
$$+ \tau_{xy}\frac{\partial v}{\partial x} + \tau_{yy}\frac{\partial v}{\partial y} + \tau_{zy}\frac{\partial v}{\partial z} + \tau_{xz}\frac{\partial w}{\partial x} + \tau_{yz}\frac{\partial w}{\partial y} + \tau_{zz}\frac{\partial w}{\partial z}. \tag{2.34}$$

Equations (2.31) and (2.34) can be written in terms of the flow variables by using the stress strain relation (2.17) and noting that the stress tensors are symmetric as

$$\rho\frac{De}{Dt} = \rho\dot{q} + \frac{\partial}{\partial x_j}\left(k\frac{\partial T}{\partial x_j}\right) - p\nabla.\vec{V} + \lambda(\nabla.\vec{V})^2$$
$$+ \mu\left[2\left(\frac{\partial u}{\partial x}\right)^2 + 2\left(\frac{\partial v}{\partial y}\right)^2 + 2\left(\frac{\partial w}{\partial z}\right)^2\right]$$
$$+ \mu\left[\left(\frac{\partial u}{\partial y} + \frac{\partial v}{\partial x}\right)^2 + \left(\frac{\partial u}{\partial z} + \frac{\partial w}{\partial x}\right)^2 + \left(\frac{\partial v}{\partial z} + \frac{\partial w}{\partial y}\right)^2\right]. \tag{2.35}$$

2.6 The Energy Equation in Conservation Form

This is easily obtained from Equation (2.35) by noting that

$$\rho\frac{De}{Dt} = \rho\frac{\partial e}{\partial t} + \rho\vec{V}.\nabla e$$

and

$$\nabla.(\rho e\vec{V}) = e\nabla.(\rho\vec{V}) + \rho\vec{V}.\nabla e.$$

Therefore,

$$\rho \frac{De}{Dt} = \frac{\partial(\rho e)}{\partial t} - e\left[\frac{\partial \rho}{\partial t} + \nabla.(\rho \vec{V})\right] + \nabla.(\rho e \vec{V}).$$

The quantity in the square bracket is zero by Equation (2.6), and thus the left hand side of Equation (2.35) is now expressed in the *conservation form*. The right hand side of equation (2.35) is already in the conservation form.

2.7 Notes on Governing Equations

(*a*) The governing equations are in general coupled partial differential equations whose general solutions are not known *analytically*.

(*b*) The difference between the *conservation* and *non-conservation forms* arises solely due to the terms on left hand side of the equations. In the conservation form the terms mostly appear as divergence of some fluxes. Thus this form of equation is also known as the *divergence form*.

(*c*) We have listed five equations for six unknowns (ρ, p, u, v, w, e). In fluid mechanics the additional equation is provided by using the equation of state. For gases often one uses

$$p = \rho RT,$$

which however brings in a new variable T, which can be augmented by introducing the seventh equation which is nothing but a thermodynamic relation between state variables,

$$e = e(p, T).$$

For a calorically perfect gas, this is further simplified to

$$e = c_v T$$

with c_v as the specific heat at constant volume.

2.8 Strong Conservation Form and Weak Conservation Form

The form of equations derived so far involves taking divergence of various flux terms (mass, momentum and energy flux components). These equations are said to be in weak conservation form. In a physical sense the flux terms are the important contributors to the governing equations and hence the divergence form is found suitable for application in CFD often. All the governing equations can also be written in the generic form,

$$\frac{\partial U}{\partial t} + \frac{\partial F}{\partial x} + \frac{\partial G}{\partial y} + \frac{\partial H}{\partial z} = J. \tag{2.36}$$

This equation can be used to represent all the equations together (in CFD these are often referred together as Navier-Stokes equation) in conservation form, where F, G, H and U are column vectors given by,

$$U = \left[\rho \quad \rho u \quad \rho v \quad \rho w \quad \rho\left(e + \frac{V^2}{2}\right)\right]^T;$$

$$F = \Bigg[\rho u \quad (\rho u^2 + p - \tau_{xx}) \quad (\rho uv - \tau_{xy}) \quad (\rho uw - \tau_{xz})$$

$$\rho u\left(e + \frac{V^2}{2}\right) + pu - k\frac{\partial T}{\partial x} - u\tau_{xx} - v\tau_{xy} - w\tau_{xz}\Bigg]^T;$$

$$G = \Bigg[\rho v \quad (\rho uv - \tau_{yx}) \quad (\rho v^2 + p - \tau_{yy}) \quad (\rho vw - \tau_{yz})$$

$$\rho v\left(e + \frac{V^2}{2}\right) + pv - k\frac{\partial T}{\partial y} - u\tau_{yx} - v\tau_{yy} - w\tau_{yz}\Bigg]^T;$$

$$H = \Bigg[\rho w \quad (\rho uw - \tau_{zx}) \quad (\rho vw - \tau_{zy}) \quad (\rho w^2 + p - \tau_{zz})$$

$$\rho w\left(e + \frac{V^2}{2}\right) + pw - k\frac{\partial T}{\partial z} - u\tau_{zx} - v\tau_{zy} - w\tau_{zz}\Bigg]^T;$$

and

$$J = [0 \quad \rho f_x \quad \rho f_y \quad \rho f_z \quad \rho\vec{F}.\vec{V} + \rho\dot{q}]^T.$$

In Equation (2.36) the column vectors F, G and H are the flux terms and J is the source term. For an unsteady problem, U is called the state vector because these are obtained numerically as a function of time.

When the conservation laws are written in the form given by Equation (2.36) we regard the equation to be in strong conservation form as opposed to the weak conservation form written down earlier in Equation (2.35). The reason for this nomenclature is simple because in the strong conservation form the discretization method does not affect the solution in the interior as the neighbouring cell terms "telescope" into each other, leaving only the boundary terms as it should be, for a true boundary value problem. This is not the case with the weak conservation form where the numerical solution depends on the process of discretization. This is one of the reasons for the popularity of the strong conservation form. It is also possible to show that for compressible flows in the presence of discontinuities for the physical variables, the fluxes still remain continuous. It is to be expected that the continuous functions are easier to compute than the discontinuous functions. This is the second reason for using governing equations in the strong conservation form. The third reason is that this form remains invariant even when the equations of motions are transformed in an arbitrary curvilinear coordinate system. This will be discussed in Chapter 9.

2.9 Types of Boundary and Initial Conditions

It is clear that the vast multitude of flow phenomena that we observe in nature are governed by conservation laws which could be stated by five mathematical equations! The different answers that we perceive for different flow situations are thus not due to different governing principles—the differences arise mainly due to boundary and initial conditions. The boundary and initial conditions taken together are called the auxiliary conditions. Auxiliary conditions are specified in three ways:

(*a*) **Dirichlet condition** when unknown values are prescribed on some boundary or as some initial condition. This is the easiest of all possible auxiliary conditions prescription to be applied numerically.

(*b*) **Neumann (Derivative) condition** when some gradient of an unknown is prescribed on some part of the boundary or as initial condition. (Numerical implementation of Neumann boundary condition for finite difference application is provided in Section 6.4).

(*c*) **Robin or mixed condition:** Here the function and its derivative are prescribed in a functional form for auxiliary condition.

In general one comes across Dirichlet and Neumann conditions mostly. In convective heat transfer problems one encounters Robin or mixed conditions. Also note that derivative boundary conditions are a major source of numerical errors, as computationally representing a derivative is always associated with approximation and leads to loss of information. This is also the reason why we will see that Neumann condition makes convergence a slow process for iterative solution of elliptic PDEs.

The boundary conditions and forcing terms dictate a particular solution to be obtained from the governing equations. Boundary conditions, as well as initial conditions, also determine the relative importance of different components of the general solution.

For viscous flows, the boundary conditions on a surface do not allow relative velocity between the surface and the fluid at the surface itself, if the flow is assumed to be a continuum. The so-called no-slip condition which is accepted unconditionally today, was the central issue of many research activities before it was accepted. Today, with interest on free molecular flows and flow at micro-scales increasing, the no-slip boundary conditions are being reconsidered for these classes of problems. For a body that is moving, the general boundary condition for viscous flow satisfying no-slip condition is given by

$$\vec{V}_{\text{fluid}} = \vec{V}_b,$$

where the right hand side is the prescribed velocity and in many physical situations a complicated function of space and time.

When we investigate the inviscid flow as a model of the actual flow, the flow slips over the surface and hence it must be tangential at the surface. This is represented by requiring that the fluid particles on the surface have the same normal velocity component as the normal component of body velocity at that point, i.c.

$$\vec{V}.\vec{n} = \vec{V}_b|_{\text{normal}}.$$

2.10 Equations of Motion in Non-Inertial Frames

The above discussions about boundary and initial conditions indicate the need to work in non-inertial frame of reference at times. For computing the flow field over complex geometries, limitations of Cartesian frames are well understood and a body-conforming generalized reference system is preferred. Furthermore, for unsteady problems involving moving boundaries, a body-fixed moving coordinate system makes the formulation simple. Otherwise, if the body's motion is complicated, it would necessitate prescribing boundary conditions which are complex functions of time. However, if we decide to fix a coordinate system that moves with the body, then in this frame of reference the boundary conditions are time-invariant.

Let us limit our discussions to incompressible flows for which the governing equations of motions are given by Equations (2.7) and (2.22). These equations are written for an inertial frame. In some problems of CFD, it is convenient to work in a moving frame of reference. This could be, for example, due to ease of applying boundary conditions, e.g. for flow past a wind turbine, turbo-machineries or a helicopter rotor blade. Such an approach is also convenient while investigating flow-induced vibrations or problems of aeroelasticity.

Let us consider a moving frame whose origin moves with a translational acceleration given by $\dot{W}_0$ with respect to the inertial frame. Let this moving frame be rotating at a rate $\Omega(t)$ with respect to the inertial frame. If the velocity vector is represented by $W(t)$ in the non-inertial frame, then the velocity vectors in these two frames are related by

$$\vec{V} = \vec{W} + \vec{\Omega}(t) \times \vec{r}, \tag{2.37}$$

where $\vec{r}$ is the displacement vector measured with respect to the non-inertial frame. It is straight forward to show that the equation of continuity remains invariant. The momentum equations that is given by Equation (2.22) can now be expressed in the moving frame by

$$\frac{\partial \vec{W}}{\partial t} + \vec{W}.\nabla\vec{W} + \dot{\vec{\Omega}} \times \vec{r} + \vec{\Omega} \times (\Omega \times \vec{r}) + 2\Omega \times \vec{W} + \dot{W}_0 = -\frac{\nabla p}{\rho} + \nu\nabla^2\vec{W}. \tag{2.38}$$

Note that the third term on the left hand side is the *Eulerian acceleration term;* the fourth term is the *centrifugal acceleration term* and the fifth term is the *Coriolis term*. One of the advantages of solving the flow problem in moving frame of reference is the clear view it affords to all the physical acceleration terms and their effects. However, the centrifugal and translational acceleration terms can be absorbed by redefining the pressure as

$$P = p + \frac{1}{2}(\vec{\Omega}.\vec{r})^2 - \frac{1}{2}\Omega^2 r^2 + \dot{\vec{W}}_0.\vec{r}. \tag{2.39}$$

In terms of this new pressure Equation (2.38) can be written as

$$\frac{\partial \vec{W}}{\partial t} + \vec{W}.\nabla\vec{W} + \dot{\vec{\Omega}} \times \vec{r} + 2\vec{\Omega} \times \vec{W} = -\nabla P + \nu\nabla^2\vec{W}. \tag{2.40}$$

Note that in solving Equation (2.40) the application of the boundary and initial conditions is simpler.

2.11 Equations of Motion in Terms of Derived Variables

We will illustrate this for the non-inertial frame of reference equations. For an inertial frame the equations can be derived simply by putting angular velocity and translational acceleration of the reference frame equal to zero. One of the most useful derived variable in fluid mechanics is the vorticity.

The kinematic definition of vorticity is given by

$$\vec{\omega} = \nabla \times \vec{W}. \tag{2.41}$$

Refer to the various useful vector identities and calculus relations provided in Appendix A in the following. To derive an equation for the vorticity one takes a curl of Equation (2.40), noting that

$$\nabla^2 \vec{W} = -\nabla \times \vec{\omega}. \tag{2.42}$$

In deriving Equation (2.42), the following identity has been used:

$$\nabla \times \nabla \times \vec{W} = \nabla(\nabla.\vec{W}) - \nabla^2 \vec{W}. \tag{2.43}$$

The first term on the right hand side of Equation (2.43) is identically zero from the equation of continuity for incompressible flows. The curl operation on Equation (2.40) yields

$$\frac{\partial \vec{\omega}}{\partial t} + \vec{W}.\nabla\vec{\omega} = \vec{\omega}.\nabla\vec{W} + \nu\nabla^2\vec{\omega} + 2\vec{\Omega}.\nabla\vec{W} - 2\dot{\vec{\Omega}}. \tag{2.44}$$

The left hand terms together constitute the substantive derivative of vorticity expressed in the moving frame. The first term on the right hand side is the *vortex-stretching* term and is identically zero for two dimensional flows. One can relate the vorticity in the two frames of references by taking curl of Equation (2.37) and using identities given in Appendix A.

$$\vec{\omega}_I = \vec{\omega} + 2\Omega. \tag{2.45}$$

The subscript I represents the inertial frame quantity for which the vorticity transport equation is given by

$$\frac{\partial \omega_I}{\partial t} + \vec{V}.\nabla\omega_I = \omega_I.\nabla\vec{V} + \nu\nabla^2\omega_I, \tag{2.46}$$

and the auxiliary relation between the vorticity and velocity is given by

$$\nabla^2 \vec{V} = -\nabla \times \omega_I. \tag{2.47}$$

Equations (2.46) and (2.47) are often solved in the so-called *vorticity–velocity formulation*. This is a derived variable formulation and often preferred for more meaningful and accurate solution over primitive variable solutions. The corresponding equations for the moving frame are given by Equations (2.42) and (2.44). The velocity–vorticity formulation as given above reveals that the non-inertial effects affect the solution through the initial and boundary conditions and the last two terms on the right hand side of Equation (2.44). These two terms require very minor modification of the numerical code. Thus the numerical algorithm will be almost identical for these frames of references. Secondly, it is well known that the application of pressure boundary condition gives rise to more problem than vorticity boundary condition. Finally, despite the requirements of handling more number of unknowns in derived variable formulations, the vorticity information is direct and more accurate. For high Reynolds number flows, it is the vorticity dynamics that controls flow evolution.

For two-dimensional flows, the derived variable equations are particularly significant because of the following reasons. Let us explore it in terms of stream function–vorticity formulation. Rewriting Equations (2.41) and (2.44) for incompressible flows in inertial frame of reference (for the sake of brevity we will not write the subscript 'I' indicating inertial frames quantities), one gets,

$$\frac{\partial \omega}{\partial t} + \vec{V}.\nabla\omega = \omega.\nabla V + \nu\nabla^2\omega, \tag{2.48}$$

$$\omega = \nabla \times \vec{V}. \tag{2.49}$$

Equations (2.44) and (2.48) are called the Vorticity Transport Equation (VTE). Also notice that for two-dimensional flows there is only one component of vorticity that is perpendicular to the plane of the flow and hence the first term on the right hand side of Equation (2.46) (the *vortex-stretching* term) will be identically zero. Instead of using Equation (2.49), one introduces the stream function (2 D) or *Vector Potential* (in 3D) via,

$$\vec{V} = \nabla \times \vec{\Psi}. \tag{2.50}$$

For 3D flows there are three components of $\vec{\Psi}$, thus the number of dependent variables becomes six (three for the vorticity and three for the vector potential). Contrast this with four unknowns for the primitive variable formulation for incompressible flows. The price one pays for this is due to difficulties encountered for the primitive variable formulation – where not only the pressure gradient term causes numerical problems, but satisfying continuity equation is not automatic and gives rise to large inaccuracies. The last problem is totally avoided in stream function/vector potential – vorticity formulations and is the main reason for using them in scientific calculations. Using Equation (2.50) in Equation (2.49) one gets

$$\vec{\omega} = \nabla \times \nabla \times \vec{\Psi}. \tag{2.51}$$

Using the vector identity given by Equation (2.43) and noting that the vector potential is solenoidal (i.e. $\nabla.\vec{\Psi} \equiv 0$), the above equations simplifies to

$$\nabla^2\vec{\Psi} = -\vec{\omega}. \tag{2.52}$$

Equations (2.46) and (2.52) provide six equations for the six unknowns in this vorticity–vector potential formulation.

For two-dimensional flows, furthermore, $\vec{\Psi}$ has only one non-zero out-of-plane, component, i.e.,

$$\vec{\Psi} = \lfloor 0 \ \ 0 \ \ \psi \rfloor^T.$$

Thus Equation (2.52) is a single equation for 2 D flows, since both the quantities (ψ and ω) are in out-of-plane direction. This has prompted some to identify this equation erroneously as scalar equation (see Grescho & Sani (1998) and Oran & Boris (2001)). The equations that are solved in ($\psi - \omega$) formulation of two dimensional flows are given by,

$$\frac{\partial \omega}{\partial t} + \vec{V}.\nabla_{2D}\omega = \upsilon \nabla^2_{2D}\omega, \tag{2.53}$$

$$\nabla^2_{2D}\psi = -\omega. \tag{2.54}$$

Equations (2.53) and (2.54) are two equations for the two unknowns ψ and ω. In these equations ∇_{2D} is the corresponding two-dimensional nabla operator.

There is thus an added benefit for (Ψ–ω) formulation in 2 D, for which the number of unknowns reduces to two as compared to three unknowns for primitive variable formulation.

Pressure Poisson Equation

For incompressible flows, whether the Navier-Stokes equation is solved using primitive or derived variables, often the pressure in the flow field is obtained by solving the pressure Poisson equation. For derived variable formulations, the reason for solving this equation is obviously to obtain the loads and moments acting on any body placed in the flow which is not available otherwise. For primitive variable formulation this is necessitated because neither the continuity nor the momentum equation can be identified as governing equation for pressure. Chorin (1967) suggested adding the time derivative of pressure to the continuity equation in his Artificial Compressibility Method for steady state flow calculation. This augmented equation allows linking the mass and momentum conservation.

However, this is inapplicable for unsteady flows. Similarly Patankar & Splading (1972) derived a pressure correction equation at each time step that satisfies the mass conservation correctly, as applied in a staggered grid.

Harlow & Welch (1965) in their Marker and Cell (MAC) method suggested solving the pressure Poisson equation to bypass the difficulties associated with the continuity equation. The pressure Poisson equation is implicit in the momentum equation and is derived from it by taking its divergence. However, for its application in CFD we cannot use the incompressibility condition directly. This is explained below.

The Navier-Stokes equation is rewritten as

$$\frac{\partial \vec{V}}{\partial t} + (\vec{V}.\nabla)\vec{V} = -\frac{1}{\rho}\nabla p + \nu\nabla^2 V + \vec{F}, \tag{2.55}$$

and

$$\nabla.\vec{V} = 0, \tag{2.56}$$

where $\vec{F}$ is the forcing or the body force term. We will consider this to be a conservative field (i.e. $\nabla.\vec{F} \equiv 0$), without any loss of generality. Taking the divergence of Equation (2.55) and using Equation (2.56) one gets the following equation:

$$-\frac{1}{\rho}\nabla^2 p = \nabla.\{(\vec{V}.\vec{\nabla})\vec{V}\}. \tag{2.57}$$

As

$$(\vec{V}.\nabla)\vec{V} = \frac{1}{2}\nabla(\vec{V}.\vec{V}) - \vec{V} \times \nabla \times \vec{V},$$

so Equation (2.57) can be written as

$$-\frac{1}{\rho}\nabla^2 p = \nabla^2\left(\frac{|\vec{V}|^2}{2}\right) - \nabla.(\vec{V} \times \vec{\omega}). \tag{2.58}$$

This is the pressure Poisson equation. But this equation is not solved in CFD (in MAC and its derivative methods), because numerically Equation (2.56) is not satisfied at every point in the field. The equation that is used instead is given by

$$-\frac{1}{\rho}\nabla^2 p = \nabla^2\left(\frac{|\vec{V}|^2}{2}\right) - \nabla.(\vec{V} \times \vec{\omega}) + \frac{\partial D}{\partial t} - \nu\nabla^2 D, \tag{2.59}$$

where $D = \nabla.\vec{V}$ and at any field point $D_{ij} \neq 0$. In solving Equation (2.59) one represents

$$\left.\frac{\partial D}{\partial t}\right|_{ij} = \frac{D_{ij}^{n+1} - D_{ij}^{n}}{\Delta t}, \tag{2.60}$$

with $D_{ij}^{n+1} \equiv 0$ satisfied numerically at each time step.

In many applications, neither Equation (2.58) nor Equation (2.59) are solved. Instead the following equations are used:

$$-\frac{1}{\rho}\nabla^2 p = \nabla.(\vec{V}.\nabla\vec{V}), \tag{2.61}$$

and

$$-\frac{1}{\rho}\nabla^2 p = \nabla.(\vec{V}.\nabla\vec{V}) + \frac{\partial D}{\partial t} - \nu\nabla^2 D. \tag{2.62}$$

Once again, in solving Equation (2.62) the condition given by Equation (2.60) is used. In some finite element applications, Equation (2.61) is called as *simplified pressure Poisson equation* while Equation (2.62) is termed as the *consistent pressure Poisson equation*. For example, Greshcho & Sani (1998) claim that for solving Equation (2.61) if D or $\frac{\partial D}{\partial \eta}$ (the normal derivative) is not zero at the boundary of the domain, then "Murphy's Law would tell us that divergence might sneak in!" This conclusion was drawn for the simpler case of straight boundaries.

In another variation (See Nair & Sengupta (1997)), the pressure Poisson equation (2.58) is written in terms of the total pressure $P_{tot} = \frac{p}{\rho} + \frac{1}{2}|\vec{V}|^2$ as

$$\nabla^2 P_{tot} = \nabla.(\vec{V} \times \vec{\omega}). \tag{2.63}$$

This equation is solved without the divergence terms for calculating the pressure alongside the solution of Navier-Stokes equation in ($\psi - \omega$) formulation. Since the velocity field is obtained from the stream function equation, it is truly divergence free. Also, the boundary condition for the pressure is obtained from the normal component of the Navier-Stokes equation given by

$$\frac{\partial p}{\partial n} = n.\left[-\nu\nabla \times \omega + \vec{F} + \vec{V} \times \vec{\omega} - \frac{\partial \vec{V}}{\partial t}\right]. \tag{2.64}$$

Note that the viscous diffusion term is written as the first term on the right hand side of Equation (2.64). Also on the no-slip wall $p = P_{tot}$, hence generating the boundary condition for Equation (2.63) is straightforward.

2.12 Advantages of Different Formulation

There are distinct physical and numerical advantages of solving Navier-Stokes equation using derived variables as opposed to primitive variables.

Charney et al. (1950) reported these advantages following the pioneering work done at Los Alamos Scientific Laboratory under the leadership of Von Neumann during world war II. This paper reports the first large scale weather forecasting simulation using vorticity transport equation. According to Roach (1976): "The authors explained the advantage of stability of the vorticity equations over the primitive equations." Similarly Speziale (1987) has also noted the advantages of the vorticity–velocity formulation for solving Navier-Stokes equation. This is related to incorporating easily the non-inertial effects for a moving body problem through initial and boundary conditions only when it is solved in a non-inertial frame of reference. The author also stated that, "a particular numerical algorithm which may be optimal for a given class of flows in an inertial frame of reference may not be so far the same class of flows in a non-inertial frame."

However, the logic of the school of thought that advocates usage of derived variables as opposed to primitive variables for incompressible flows is strengthened with the observation of Chorin (1968) that the difficulties in satisfying the continuity equation in the primitive variable formulation can trigger numerical instabilities. Also, it is advocated by Chorin (1968) and Williams (1969) that deriving pressure boundary condition is more cumbersome than deriving vorticity boundary conditions at the no-slip wall.

The debate on superiority of one equation set over the other to solve Navier-Stokes equation numerically is a long-standing one. The reader's attention is also drawn to Gresho & Sani (1998) who are critical in claiming: "we give little more than lip service to the 2 D $\psi - \omega$ formulation and that with snarling lips." However, if one shows any preference for derived variable formulation, it is solely due to the inability of primitive variable formulation to satisfy continuity from the stand point of high accuracy calculations of transitional flows. The incorrect satisfaction of continuity (more near the no-slip boundaries) constitutes an artificial distributed singularity in the flow that can trigger premature transition.

2.13 Other Forms of Navier-Stokes Equation

In the following, different forms of Navier-Stokes equations are reported based on primarily the nonlinear convection term of the primitive variable form, given by Equation (2.22).

$$\frac{\partial \vec{V}}{\partial t} + (V.\nabla)\vec{V} = -\frac{\nabla p}{\rho} + \nu\nabla^2\vec{V} + \vec{F}.$$

We will primarily use some of the vector identities given in Appendix A. Using the vector identity

$$\vec{V} \times \nabla \times \vec{V} = \frac{1}{2}\nabla(\vec{V}.\vec{V}) - \vec{V}.\nabla V \tag{2.65}$$

in Equation (2.22), we obtain the following equation:

$$\frac{\partial \vec{V}}{\partial t} - \vec{V} \times \vec{\omega} = -\nabla\left(\frac{p}{\rho} + \frac{|\vec{V}|^2}{2}\right) + \nu\nabla^2\vec{V} + \vec{F}. \tag{2.66}$$

This is the *rotational form* of Navier-Stokes equation. This is a very useful form in highlighting some physical events. For example if the flow is irrotational ($\vec{\omega} \equiv 0$), then the nonlinearity of the flow is not important. The nonlinear term merely redefines the pressure.

Furthermore, using the identity

$$\nabla^2\vec{V} = \nabla(\nabla.\vec{V}) - \nabla \times \nabla \times \vec{V} = -\nabla \times \omega,$$

one can rewrite Equation (2.66) as

$$\frac{\partial V}{\partial t} + \vec{\omega} \times \vec{V} = -\nabla(P_{\text{tot}}) - \nu\nabla \times \omega + \vec{F}. \tag{2.67}$$

This equation is particularly attractive as it does not require evaluation of derivatives higher than first order.

However, Kravchenko & Moin (1997) claim that when Equation (2.67) is solved by spectral method in DNS applications, the flow relaminarizes if no specific *dealising* technique is employed. However, second order finite difference formulation does not suffer from this problem.

The following identity

$$\nabla.(\vec{V}\vec{V}) = \vec{V}.\nabla\vec{V} + \vec{V}(\nabla.\vec{V})$$

where $\nabla.(\vec{V}\vec{V}) = \frac{\partial}{\partial x_j}(u_i u_j)$ is the dyadic product, can be used in Equation (2.22) to obtain

$$\frac{\partial \vec{V}}{\partial t} + \nabla.(\vec{V}\vec{V}) = -\frac{\nabla p}{\rho} + \nu\nabla^2\vec{V} + \vec{F}. \tag{2.68}$$

This is the *divergence form* of the Navier-Stokes equation. In CFD applications divergence form is always preferred, especially for compressible flows due to its superior conservation principles which are independent of the grid type. Such problems are truly driven by boundary and initial conditions. However, Kravchenko and Moin (1997) report that this form of equation is unstable for spectral, finite difference or compact difference methods if no specific *dealising* is adopted. When the nonlinear terms of the NS equation are *dealiased*, spectral method works. (But the compact method (discussed in Chapter 11) diverges; finite difference calculations with second order central differencing work only with staggered grids).

Finally, let us discuss the skew-symmetric form which was originally used by Arakawa (1966) in solving inviscid vorticity transport equation for weather prediction purposes. This is a higher order method which also exhibits a very desirable amplification factor related to numerical stability of the algorithm and analytically conserved vorticity, enstrophy and kinetic energy. This will be discussed in greater detail in Chapter 10. Subsequently, Zang (1991), Horiuti (1987), Blaisdell et al. (1996) and Kravchenko & Moin (1997) also used variations of this method. In these latter efforts the following form of Navier-Stokes equation is solved:

$$\frac{\partial \vec{V}}{\partial t} + \left[\vec{V}.\nabla\vec{V} + \frac{1}{2}V(\nabla.\vec{V})\right] = -\frac{\nabla p}{\rho} + \nu\nabla^2 V + \vec{F}. \tag{2.69}$$

The quantities within square brackets on the left hand side of Equation (2.69) can also be writen in the indicial notation by

$$\frac{1}{2}\left(\frac{\partial}{\partial x_j}(V_i V_j) + V_j\frac{\partial V_i}{\partial x_j}\right). \tag{2.70}$$

This is called the skew-symmetric formulation for Navier-Stokes equations. Skew-symmetric formulation is particularly recommended by Kravchenko and Moin (1997) for its specific property by which it reduces aliasing error.

Finally, we must mention that the above results, as reported in Kravchenko & Moin (1997), relate to computing a channel flow with very few grid points and non-dissipative schemes. Subsequently, when we discuss upwind schemes, we will note that the implicit numerical dissipation of higher order upwind schemes remove problems of aliasing that are mostly present at high wave numbers. Hence the actual choice of forms of Navier-Stokes equation depends mostly on the numerical method used.

Exercise

1. (a) What happens to the continuity equation for incompressible flows if there is a distributed mass source S at (x_0, y_0, z_0)?

(b) What is the physical significance of Equation (2.7)?

(c) Which is the conservation form and which is the non-conservation form shown in Equation (2.6)?

2. Consider a steady compressible flow and write down the continuity equation. Discretize the governing equations in both conservative and non-conservative forms and show the consequence of it by considering 1 D flow.
3. Start with the total energy $\left(e + \frac{V^2}{2}\right)$ and the following the same chain of operations show that

$$\rho \frac{D}{Dt}\left(e + \frac{V^2}{2}\right) = \frac{\partial}{\partial t}\left[\rho\left(e + \frac{V^2}{2}\right)\right] + \nabla\left[\rho\left(e + \frac{V^2}{2}\right)\vec{V}\right].$$

Finally write down the conservative form of total energy equation.
4. Write down the energy equation in terms of the enthalpy and total enthalpy.
5. Show that on the solid surface of the body

$$\nabla p = -\mu \nabla \times \omega.$$

This expression is often used in CFD to arrive at an integral constraint to obtain a single valued surface pressure.

References

[1] ARAKAWA, A., Computational design of long-time numerical integration of the equations of fluid motion : I Two-dimensional incompressible flow, *J. Comp. Phys.*, **1(1)**, pp. 119–143 (1966).

[2] BLAISDELL, G. A., SPYROPOULOS, E. T., & QIN, J. H., The effect of formulation of nonlinear terms on aliasing errors in spectral methods, *Appl. Numer. Math.*, **20**, 1–12, (1996).

[3] CHARNEY, J. G., FJORTOFT, R., & VON NEUMANN, J., Numerical integration of the barotropic vorticity equation. *Tellus*, **2(4)**, 237–254, (1950).

[4] CHORIN, A. J., A numerical method for solving incompressible viscous flow problem. *J. Comp. Phys.*, **2**, 12, (1967).

[5] CHORIN, A. J., Numerical solution of the Navier-Stokes equation, *Math. Comp.*, **22**, 745–762, (1968).

[6] GRESCHO, P. M., & SANI, R.L., *Incompressible Flow and the Finite Element Method,* John Wiley & Sons, Chichester, U.K., (1998).

[7] HARLOW, F.H., & WELCH, J.E., Numerical calculations of time dependent incompressible flow fluid with free surface, *Phys. Fluids*, **8(12)**, 2182–2189, (1965).

[8] HORIUTI, K., Comparison of conservative and rotational forms in large eddy simulation of turbulent channel flow, *J. Comp. Phys.*, **71**, 343, (1987).

[9] KRAVCHENKO, A. G., & MOIN, P., On the effect of numerical errors in large eddy simulations of turbulent flows, *J. Comp. Phys.*, **131**, 310–322, (1997).

[10] NAIR, M. T., & SENGUPTA, T. K., Unsteady flow past elliptic cylinders, *J. Fluids. and Structures*, **11**, 555–595, (1997).

[11] ORAN, E. S., & BORIS, J. P., *Numerical Simulation of Reactive Flow*, Cambridge University Press, U.K., 2nd ed., (2001).

[12] PATANKAR, S. V., & SPALDING, D. B., A calculation procedure for heat, mass and momentum transfer in three-dimensional parabolic flows, *J. Heat & Mass Transfer*, **15**, 1787–1806, (1972).

[13] ROACH, P. J., *Computational Fluid Dynamics*, Hermoza Publisher, Albuquerque, New Mexico, USA, (1976).

[14] SCHLICHTING, H., *Boundary Layer Theory*, 7th ed., McGraw Hill, New York, USA, (1979).

[15] SPEZIALE, C. G., On the advantages of the vorticity- velocity formulation of the equations of fluid dynamics, *J. Comp. Phys.*, **73**, 476–480, (1987).

[16] STOKES, G. G., On the theories of inertial friction of fluids in motion, *Trans. Cambr. Phil. Soc.*, **8**, 287–305, (1845).

[17] VICHNEVETSKY, R., & BOWLES, J.B., *Fourier Analysis of Numerical Approximations of Hyperbolic Equations.* Society for Industrial and Applied Mathematics, Philadelphia, USA, (1982).

[18] WILLIAMS, G. P., Numerical integration of the three dimensional Navier-Stokes equations for incompressible flow, *J. Fluid. Mech.*, **37**, 727–750, (1969).

[19] ZANG, T. A., On the rotation and skew-symmetric forms for incompressible flow simulations, *Appl. Numer. Math.*, **7**, 27, (1991).

3 Classification of Quasi-Linear PDEs

In the previous chapter we derived the most general form of conservation laws that are solved in CFD. For continuum flows, these are nothing but the Navier-Stokes equations. A quick look at them will show that they are nonlinear partial differential equations (PDEs). To be more precise, they are linear with respect to the highest derivative terms and such equations are called quasi-linear PDEs. It is perfectly possible to classify such equations based on the behaviour of the solution. In this chapter we will classify the quasi-linear PDEs so that we can derive specific numerical methods for each equation type.

Consider the moving boundary problem as shown in Fig. 3.1. This class of problems is also known as the propagation problem.

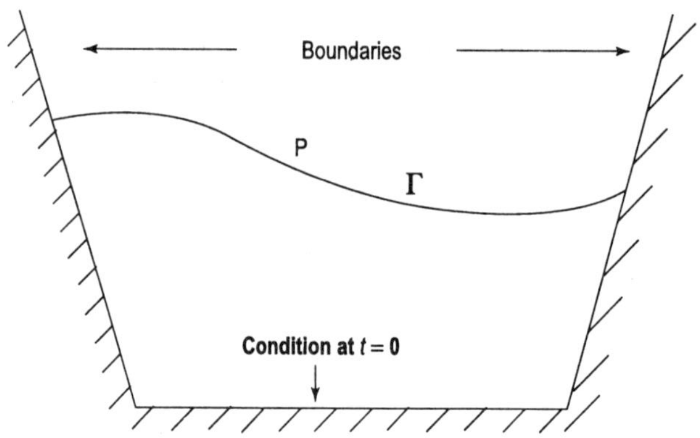

Fig. 3.1 *Representation of initial and boundary conditions for a general propagation problem.*

Let the system be described by following PDEs governing the two dependent variables u and v in two dimensions.

$$a_1u_x + b_1u_y + c_1v_x + d_1v_y = f_1, \tag{3.1}$$

$$a_2u_x + b_2u_y + c_2v_x + d_2v_y = f_2, \tag{3.2}$$

where a_1, a_2, b_1, b_2, ... f_2 are functions of x, y, v and u only. Let the solution be known from the initial condition up to Γ. Suppose we know the solution at Γ, is it sufficient to advance the solution beyond Γ?

The answer to this question is intimately related to whether u_x, u_y, v_x and v_y are uniquely determined at any arbitrary point P by values of u and v on Γ.

At P we have four relations: Equations (3.1) and (3.2) and

$$du = u_x dx + u_y dy, \tag{3.3}$$

$$dv = v_x dx + v_y dy, \tag{3.4}$$

for the evaluation of these four partial derivatives. These equations could be written in matrix form as

$$\begin{bmatrix} a_1 & b_1 & c_1 & d_1 \\ a_2 & b_2 & c_2 & d_2 \\ dx & dy & 0 & 0 \\ 0 & 0 & dx & dy \end{bmatrix} \begin{Bmatrix} u_x \\ u_y \\ v_x \\ v_y \end{Bmatrix} = \begin{Bmatrix} f_1 \\ f_2 \\ du \\ dv \end{Bmatrix}. \tag{3.5}$$

With u and v, $a_1, a_2, b_1, b_2 \ldots f_1, f_2$ are known at P. With the direction of Γ known, dx and dy are also known. As u and v are known along Γ, so du and dv are also known. Thus unique solutions of the partial derivatives are known if the (4×4) matrix in equation (3.5) is non-singular. However, the solution to the above problem becomes non-unique if the above matrix is singular. Consequently, discontinuities in the partial derivatives may occur as we cross Γ. Consider now, any point in the $x-y$ plane. The purpose of the classification is to seek lines or directions through this point along which the *derivatives* are discontinuous. Such lines, when they do exist, are called *characteristic lines*. It is to be noted that along such lines u and v are continuous. Thus the distinction between *characteristic lines* and *shock waves* is straightforward.

The determinant of the (4×4) matrix in equation (3.5) determines whether *characteristic lines* exist or not and if they do, what their properties are. Thus, we obtain the characteristic equation by equating the determinant to zero to get

$$(a_2c_1 - a_1c_2)(dy)^2 - (a_1d_2 - a_2d_1 + b_1c_2 - b_2c_1)dxdy + (b_1d_2 - b_2d_1)(dx)^2 = 0, \tag{3.6}$$

which is a quadratic in (dy/dx). If at P on the curve Γ has a slope (dy/dx) such that Equation (3.6) is satisfied, then the derivatives u_x, u_y, v_x and v_y are not uniquely determined along lines whose slopes are given by the calculated $\left(\frac{dy}{dx}\right)$. The directions specified by Equation (3.6) are the characteristic directions along which the discontinuities in the derivative propagate. The characteristic directions are furthermore determined by the property of the quadratic given by Equation (3.6), i.e., whether the characteristic determinant of Equation (3.6)

$$D = (a_1d_2 - a_2d_1 + b_1c_2 - b_2c_1)^2 - 4(a_1c_2 - a_2c_1)(b_1d_2 - b_2d_1) \tag{3.7}$$

is positive, zero or negative. Thus, the following possibilities exist:

(*a*) If $D > 0$, then $(dy/dx)_1$ and $(dy/dx)_2$ are real and distinct. Thus, through each and every point in the $x-y$ plane there exist two distinct directions along which the discontinuities emanate. When this is the case, then the system of Equations (3.1) and (3.2) are called the *hyperbolic equations*.

(*b*) If $D = 0$ then $(dy/dx)_1$ and $(dy/dx)_2$ are real but identical. Thus through each point one can get discontinuities in derivatives along this line. Such equations are called parabolic equations.

(*c*) If $D < 0$ then the characteristics are complex and such PDEs are called the *elliptic PDEs*.

The physical implications of the solution of PDEs are now apparent. For a hyperbolic system, the information propagates along the characteristic directions and it is possible to simplify the governing partial differential equations to ordinary differential equations along the direction of the characteristic (known as *compatibility equations*). Consult a good textbook on PDEs for elaboration. Similarly for the parabolic equation the information propagates only in one direction. The lack of any physical characteristics for elliptic partial differential equations implies the absence of any specific direction along which discontinuity or information propagates. This implies then that each and every point in the flow domain depends on every other point of the domain. Thus, every point eventually depends on every points on the boundaries. It is for this reason that elliptic problems are also known as *jury problems* (originally this nomenclature was given by Richardson (1922) who actually performed such calculations using desk calculators operated by human (jury!)) or *boundary value* problems while the parabolic problems are known as *evolution problems* with the flow variables evolving along the parabolic direction. Some additional illustrative examples follow, explaining the equations types.

Example 1: Consider the following quasi-linear equation:

$$au_{xx} + bu_{xy} + cu_{yy} = f \tag{3.8}$$

where a, b, c and f are function of x, y, u, u_x and u_y only.

We classify this equation by asking whether the knowledge of u, u_x and u_y on some curve Γ serves to determine u_{xx}, u_{xy}, u_{yy} uniquely beyond Γ? If these possibilities exist, we must have

$$d(u_x) = u_{xx}dx + u_{xy}dy \tag{3.9}$$

$$d(u_y) = u_{xy}dx + u_{yy}dy \tag{3.10}$$

along Γ.

Following the same chain of arguments, we can obtain the characteristic equation as

$$a(dy)^2 - b(dx)(dy) + c(dx)^2 = 0,$$

and the classification of the equation follows by looking at $D = b^2 - 4ac$.

Example 2: Classify the steady two-dimensional irrotational isentropic flow of gas.

Let us first collate the governing equations in Cartesian frame of reference. Since we are interested in classifying the steady state equation for compressible flow, the equation of continuity (Equation. (2.6)) simplifies to

$$(\rho u)_x + (\rho v)_y = 0. \tag{3.11}$$

As the flow considered is irrotational, therefore the vorticity is identically zero, i.e.

$$\omega_z = v_x - u_y = 0. \tag{3.12}$$

For the steady inviscid flow the momentum equations (Equations (2.19) and (2.20)) simplifies to the following equations for two-dimensional flows in the absence of body force:

$$uu_x + vu_y = -\frac{1}{\rho} p_x \tag{3.13}$$

and

$$uv_x + vv_y = -\frac{1}{\rho} p_y. \tag{3.14}$$

Since we are considering the isentropic flow, the pressure and density can be related by

$$p = K\rho^\gamma \quad \text{or} \quad dp = c^2 d\rho, \tag{3.15}$$

where γ is the ratio of the specific heat and $c\left[= \left(\frac{dp}{d\rho}\right)_s^{\frac{1}{2}}\right]$ is the speed of sound.

From Equations (3.13) and (3.14), by algebraic manipulation one obtains

$$(\rho u^2 u_x + 2uv\rho u_y + \rho v^2 v_y) = up_x + vp_y. \tag{3.16}$$

Also, from the isentropic relation,

$$p_x = c^2 \rho_x \quad \text{and} \quad p_y = c^2 \rho_y. \tag{3.17}$$

Furthermore, from the continuity equation,

$$u\rho_x + v\rho_y = -\rho(u_x + v_y). \tag{3.18}$$

Using Equations (3.12), (3.17) and (3.18) in (3.16) we get the following equation,

$$(u^2 - c^2)u_x + uvu_y + uvv_x + (v^2 - c^2)v_y = 0. \tag{3.19}$$

Now we are in a position to classify the equation by comparing Equations (3.19) with (3.1), and (3.12) with (3.2) to identify the coefficients a_1, a_2, b_1, b_2, etc. in Equation (3.7). Substitution and simplification yields

$$D = c^2(u^2 + v^2 - c^2). \tag{3.20}$$

If we define the local Mach number (M) by

$$M = \sqrt{\frac{u^2 + v^2}{c^2}}, \tag{3.21}$$

then Equation (3.20) can be rewritten as

$$D = c^4(M^2 - 1). \tag{3.22}$$

Thus the PDE representing the flow can be classified as hyperbolic, parabolic or elliptic depending on whether M > 1 (locally supersonic); M = 1 (locally sonic), or M < 1 (locally subsonic), respectively.

One major challenge to the CFD community till early '70s was to solve transonic flow problems that represented mixed flows, i.e., flows in which part of the flow was subsonic and part of the flow was supersonic. This was resolved brilliantly by Murman and Cole (1971).

3.1 Nature of Well-Posed Problems

It is important to examine the aspects of well-posedness before attempting to obtain the numerical solution of PDEs. In the governing equation, the initial and boundary conditions are well posed if the following three conditions are met:

(*a*) The solution exists.
(*b*) The obtained solution is *the solution*, i.e. the solution is unique.
(*c*) The solution depends continuously on the applied initial and boundary conditions.

Although there are no general proofs available, the existence of solution to 3D Navier-Stokes equation is assumed most often. It is only the simplification and additional modelling that raises the question of existence of solution.

The cause of non-uniqueness of the solution mostly arises due to poor matching of the governing differential equation type with the auxiliary conditions. For potential flow equations and for boundary layer equations the appropriate initial and boundary conditions are well established. For Navier-Stokes equation the wall boundary conditions are well known but the same cannot be said about the far-field boundary conditions. It is to be emphasized however that some physical situations warrant multiple solutions. Such situations come about due to flow instability and nonlinearity being important.

The third criterion implies that a small change in auxiliary condition should bring about a small change in the response of the system. However, it is well known that chaotic systems are characterized by large departure of response when small changes are brought about in initial conditions. For a description on this aspect, the readers' attention is drawn to the monograph by Holmes et al. (1998) which tries to analyze fluid dynamical system by dynamical system and chaos dynamics tools. Thus, this aspect should be considered as an open question at this point. The same must be true for fluid dynamic system that is undergoing transition.

3.2 The Non-Dimensional Form of Equations

In all CFD practices one always solves the conservation equations in non-dimensional form. The reason is obvious—the need to perform fewer simulations that represent an infinite number of combinations of physically relevant parameters. Also, the physics of the problem depends strongly on non-dimensional parameters, hence their direct appearance in governing equations facilitates physical understanding. For example, one can start with the Navier-Stokes equation as given by Equation (2.22). Now if one non-dimensionalises all lengths by a length scale L, and all velocity components by a velocity scale U, then the time scale is fixed as $\frac{L}{U}$ automatically. The pressure in Equation (2.22) is similarly non-dimensionalized by ρU^2. The process followed is then to represent governing equations in terms of the following non-dimensional quantities.

$$x_{\text{ND}} = \frac{x}{L};$$

$$u_{\text{ND}} = \frac{u}{U};$$

$$t_{\text{ND}} = \frac{Ut}{L};$$

$$p_{\text{ND}} = \frac{p}{\rho U^2}.$$

One can substitute these relations in Equation (2.22) and omit the subscripts for ease of representation. On doing this, we get the following non-dimensional form of Navier-Stokes equation for incompressible flow.

$$\frac{\partial \vec{V}}{\partial t} + (\vec{V}.\nabla)\vec{V} = -\nabla p + \frac{1}{Re}\nabla^2 \vec{V}. \tag{3.23}$$

In Equation (3.23), $Re = \frac{UL}{\nu}$, the Reynolds number is based on the characteristic length scale L, and the velocity scale U. It is to be noted that non-dimensionalization should follow physical interpretation of the flow events. For example, if one wants to solve Navier-Stokes equation for the lubrication problem, then Equation (3.23) is not the appropriate one. This is because various non-dimensionlization used in Equation (3.23) implies the relative importance of inertial terms and pressure gradient terms in comparison to viscous forces. Exercise 4 indicates the correct formulation for such problems.

Exercise

1. Classify the two-dimensional steady boundary layer equations for incompressible flow over a flat plate. Find the characteristic direction and discuss the physical implication of your results.
2. Classify the following equation of transonic aerodynamics for potential flows.

$$(\phi_x^2 - c^2)\phi_{xx} + 2\phi_x\phi_y\phi_{xy} + (\phi_y^2 - c^2)\phi_{yy} = 0, \tag{3.24}$$

where ϕ is the velocity potential and c is the speed of sound.

3. Consider the non-similar boundary layer equation of two-dimensional incompressible flow as given by the Falkner-Skan equation (see Cebeci and Bradshaw (1977))

$$f''' + \frac{m+1}{2} ff'' + m[1 - (f')^2] = x\left(f' \frac{\partial f'}{\partial x} - f'' \frac{\partial f}{\partial x}\right), \qquad (3.25)$$

where f is a non-dimensional stream function and the dimensionless pressure gradient parameter m is given by $m = \frac{x}{u_e}\frac{du_e}{dx}$ and treat it as a constant. A prime indicates differentiation with respect to the wall normal coordinate η. Classify this equation and obtain the characteristic direction.

4. Nondimensionalize Navier-Stokes equation by taking reference pressure as (i) dynamic pressure (ρU^2) and (ii) $\mu U/L$, where the symbols have their usual meaning. If both the equations are used to solve the Stokes flow problem ($Re \to 0$), which results would be more accurate?

References

[1] CEBECI, T., & BRADSHAW, P., *Momentum Transfer in Boundary Layers*, Hemisphere Publications, (1977).

[2] HOLMES, P., LUMELY, J.L. & BERKOOZ, G., *Turbulence, Coherent Structures, Dynamical Systems and Symmetry*, Cambridge University Press, (1998).

[3] MURMAN, E.M., & COLE, J.D., Calculation of plane steady transonic flows, *AIAA J.*, **9**, 114–121, (1971).

[4] RICHARDSON, L.F., *Weather Prediction by Numerical Process*, Cambridge University Press, reprinted Dover, 1965, (1922).

4 Additional Issues of CFD: Space–Time Resolution of Flows

All practical flows associated with nature or man-made engineering devices are essentially high Reynolds number flows (as defined in Equation (3.23)). It is well known that laminar flows become unstable as the Reynolds number increases. So, at high Reynolds numbers the flow is either transitional or turbulent. Unlike in laminar flows where the flow can be strictly classified as steady and unsteady, this is not possible for transitional or turbulent flows. For these flows, the exhibited space and time scales show a very broad spectrum. Resolution of this broadband spectrum shows up as a problem for computing these flows. The CFD tools employed to solve these problems must be *spectrally accurate* to resolve all these scales. But first, let us talk about the excited spatial and temporal scales. To aid our discussions, let us introduce some fundamental concepts of wave motions.

Waves are usually associated with hyperbolic PDEs. What have waves to do with CFD then ? It is interesting to note that for many CFD problems that are given by parabolic or elliptic PDEs, the numerical algorithm treats the problem as one akin to hyperbolic system. It will be explained later on, that the numerical characteristics of the equivalent hyperbolic system must include the single characteristics of a parabolic system, for the method to be stable and accurate. In the same way, solving an elliptic PDE by an iterative method is equivalent to solving a hyperbolic system. The following section can thus be viewed as an extension of the ideas proposed in Vichnevetsky & Bowles (1982) that strictly dealt with hyperbolic PDEs.

4.1 Element of Wave Mechanics

A wave may be viewed as a response of a system to imposed disturbances. The response could be in terms of physical deflection, pressure, velocity, vorticity or entropy perturbation propagating in the field. The response shows up as a variation of the property in space and time. Any arbitrary function of space and time can be written in terms of its Fourier-Laplace integral given by

$$f(x,t) = \int_{-\infty}^{\infty}\int_{-\infty}^{\infty} F(\omega,k)e^{i(kx-\omega t)}d\omega dk \tag{4.1}$$

for the response of such a system, whose property varies with time and only a single space dimension x. Van der Pol & Bremmer (1960) is an excellent treatise that also discusses issues relevant to the present discussion. In Equation (4.1), k is the wave number and ω is the circular frequency (not to be confused with vorticity, for which we have used the same notation). There is no need to go into greater details about the above expression and the limit of integration (as is usually given by Bromwich contour), rather than noting that any arbitrary functions of x and t can be interpreted as a combination of an infinitely large number of waves with different wave numbers and circular frequencies.

Consider a simple wave equation that is typical of a single wave given by

$$\frac{\partial f}{\partial t} + c\frac{\partial f}{\partial x} = 0, \tag{4.2}$$

which represent the governing equation in one dimension for the response in the physical ($x{-}t$) plane. In Equation (4.2), c is the phase speed of the disturbance propagation. If we use the representation given by Equation (4.1) in Equation (4.2), we get the following:

$$\int_{-\infty}^{\infty}\int_{-\infty}^{\infty} i(kc - \omega)Fe^{i(kx-\omega t)}d\omega dk = 0. \tag{4.3}$$

Since the integral is zero, it implies that the integrand must be zero and this is written down as

$$\omega = kc. \tag{4.4}$$

The expression given by Equation (4.4) is called the *dispersion relation* and is a property of the system in the spectral plane for which the physical plane relation is given by Equation (4.2). Thus every relation in the physical plane has a corresponding *dispersion relation* in the spectral plane relating the space–time behaviour. The physical implication of the dispersion relation is straightforward. It implies that the spatial and temporal behaviour of the system are not independent of each other, but are governed by relations like Equation (4.4) and tell us that each frequency component will travel in space with the corresponding phase speed. The system given by Equation (4.2) is simple because the phases of all wave numbers travel with the same phase speed c, hence such systems are called *non-dispersive*. In contrast, a *dispersive system* is one for which a different frequency travels with different speed and hence will be found dispersed with respect to each other as time progresses.

Now let us focus on the more general system. For such systems the governing differential equation determines the *dispersion relation*. Once the *dispersion relation* is known, can we say the same about the response of the system? The answer is 'yes' and let us illustrate this by considering two neighbouring wave numbers k_1 and k_2 of the response field. For a well-posed system (as discussed in Chapter 3), the solution depends continuously on the forcing and auxiliary conditions and then the corresponding circular frequencies ω_1 and ω_2 will also be two close-by neighbouring points in the spectral plane. The corresponding response components are given by

$$f_1 = a\cos(k_1x - \omega_1 t), \tag{4.5a}$$

$$f_2 = a\cos(k_2x - \omega_2 t). \tag{4.5b}$$

We have only considered the real part of the response and both the components have the same amplitude because we have considered these two harmonics to be separated infinitesimally apart. The total contribution coming from these two harmonics is then,

$$f = f_1 + f_2 = \left[2a\cos\frac{(k_1 - k_2)x - (\omega_1 - \omega_2)t}{2}\right]\cos\frac{(k_1 + k_2)x - (\omega_1 + \omega_2)t}{2}. \quad (4.6)$$

Thus the total effect of the two harmonics shows up as a combination of two changes. One, on a slow scale given by the first quantity inside the square bracket and the other, almost the same on the original scale given by the second periodic function in Equation (4.6). This slow changing quantity can be termed as the amplitude of the resultant wave. Since energy is proportional to the square of amplitude, so the speed at which the energy propagates is given by the speed at which the amplitude changes in space and time and is therefore given by

$$V_g = \frac{(\omega_1 - \omega_2)}{(k_1 - k_2)}. \quad (4.7)$$

The second quantity shown outside the square bracket in Equation (4.6) shows up as the phase variation of response on the original scale. Now if the response of the system is considered for an infinite number of harmonics which are dense in the spectral plane, then the above relation given in Equation (4.7) can be written as

$$V_g = \frac{d\omega}{dk}. \quad (4.8)$$

The physical implication of Equation (4.8) is obvious and the quantity represented by it is called the group velocity. Thus group velocity is the speed at which the energy of a group of waves, centered around the middle wave number, travels. Notice that for the dispersion relation given for the Equation (4.2),

$$V_g = \frac{d\omega}{dk} = c,$$

i.e., the group velocity is same as the phase speed for all wave numbers. This is the central property of all non-dispersive systems.

Any system that does not obey this, is called a dispersive system. In general, most fluid dynamical events are dispersive, e.g. the turbulent flows prevalent at high Reynolds numbers.

The idea that energy in a physical system propagates with group velocity was originally suggested by Rayleigh (1899, 1900). A recent and a good description of the same for waves in fluids are to be found in Lighthill (1978a).

4.2 Spatial Scales in Turbulent Flows

Scales of turbulent flows are related to eddy sizes in the evolving flow field. Let the largest length scale be associated with the dimension of the fluid dynamical system and be denoted by l.

In general, it is customary to represent the kinetic energy density (i.e. per unit volume) for turbulent flows in the spectral plane. One can show using Parseval's theorem (see for example Lighthill (1978b)) that the total energy remains the same whether represented in the real space

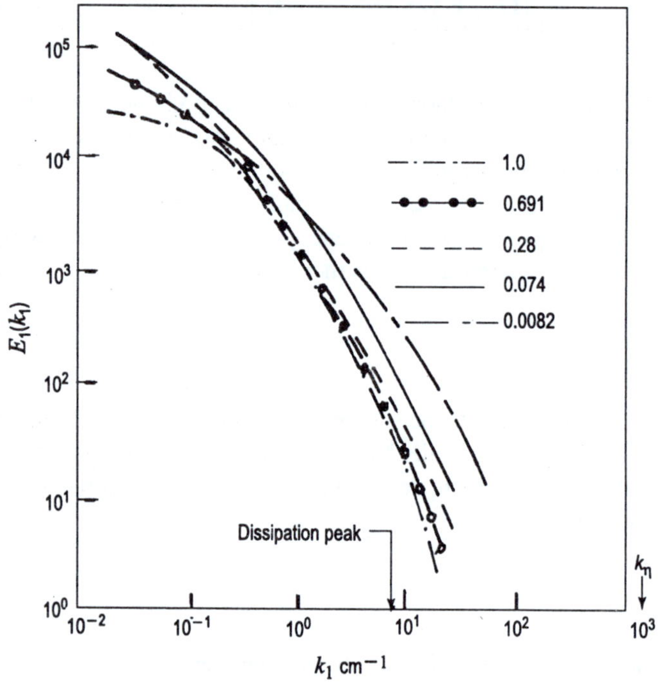

Fig. 4.1 *(a) Energy spectrum measured by Laufer (1954) for a pipe flow at high Reynolds number (500,000). The diameter of the pipe is 25.4 cm and the results shown here is for the longitudinal spectra. Note the dissipation peak and the wave number corresponding to Kolmogorov length scale defined by Equation (4.9).*

or in the spectral space. However, the spectral space is often chosen because it allows easier interpretation and a theory of turbulence in spectral space, when one looks at homogeneous turbulence, is available. In Fig. 4.1 the energy spectra has been shown for pipe flow (Laufer (1954)) and in Fig. 4.2 for turbulent boundary layer (Sadoughi and Veeravalli (1994)). For the pipe flow, the longitudinal spectra is shown, while for turbulent boundary layer, all the three components are shown. As shown for the pipe flow, the dissipation peak is at a higher wave number *as compared to the peak of energy spectrum.* In general the energy spectrum depends on the wave number (k), dissipation (ε) and molecular viscosity (ν). If we define l as the integral length scale (which corresponds to the largest eddies), and u as the velocity scale (which defines kinetic energy per unit mass), then we can obtain a Reynolds number given by

$$Re = \frac{ul}{v}.$$

It has been shown by Kolmogorov (see Tennekes & Lumley (1971)) that the smallest excited scale (known as Kolmogorov scale) is given by,

$$\eta = \left(\frac{\nu^3}{\varepsilon}\right)^{\frac{1}{4}}.$$

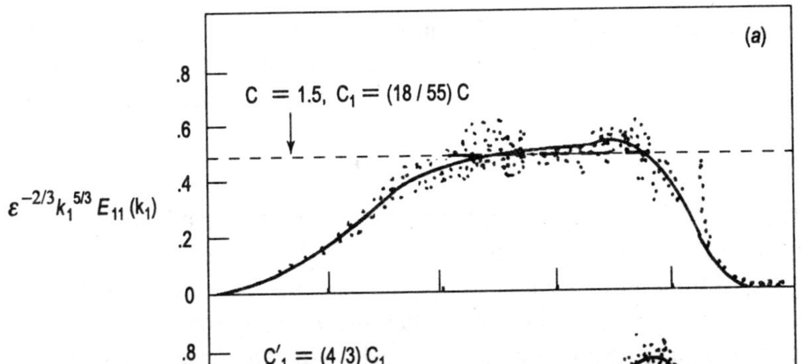

Fig. 4.2 *(b) Energy spectra along logitudinal and transverse direction measured by Saddoughi & Veeravalli (1994) for a flat plate boundary layer. The Reynolds number based on momentum thickness is 370,000. The measurements were made at $y^+ = 16{,}000$.* [Reprinted with permission from Saddoughi S.G and Veeravally S.V, Local isotropy in turbulent boundary layers at high Reynolds number, *Journal of Fluid Mechanics*, vol. 268, pp. 333–372(1994), © Cambridge University Press.]

The largest and the smallest length scales can be shown to be related by

$$\frac{l}{\eta} = (Re)^{\frac{3}{4}}. \tag{4.10}$$

In turbulent boundary layers there is a region very close to the wall where the velocity varies linearly with the distance from the wall. This region is called the *viscous sublayer* whose thickness ξ is related to l by,

$$\frac{l}{\xi} = Re. \tag{4.11}$$

For flow computations at high Reynolds numbers, it is mandatory that all of these scales are resolved. This is attempted in direct numerical simulation (DNS). If we represent the cut-off wave number by k_c (related to η), then the relation given by Equation (4.10) can also be written as

$$k_c l \approx Re^{\frac{3}{4}}. \tag{4.12}$$

Equation (4.12) is used for setting a limit for grid size in performing DNS. For example for DNS in 3 D, this shows the resolution requirement scales as $(Re^{3/4})^3$ or roughly about by Re^2.

In deriving the Kolmogorov's scaling theory, it is said that there exist length scales shorter than those that are directly excited (l), but larger than the Kolmogorov scale (η), where the energy spectrum is independent of viscous dissipation mechanism. On these intermediate scales – called the *inertial subrange* – the structure of the energy spectrum ($E(k)$) is determined solely by the nonlinear energy transfer (by vortex stretching) by a cascade process and the overall energy flux is shown to depend in the following way:

$$E(k,\varepsilon) = C_k k^{-\frac{5}{3}} \varepsilon^{\frac{2}{3}}. \tag{4.13}$$

The existence of inertial subrange suggests some form of universality, that exists beyond this length scale. This is exploited in large eddy simulation (LES) where the flow is computed by resolving all the way up to the inertial subrange and anything smaller than this is modeled via sub-grid scale stress (SGS) models.

4.3 Temporal Scales in Turbulent Flows

Temporal scales cannot be determined independent of the spatial scales, as we have shown already while discussing about the dispersion relation. However, if we consider the role of temporal scales as the independent variable then it is determined by the frequency of externally induced or self-induced unsteadiness. In flow induced vibration at high Reynolds numbers these two unsteadiness may be coupled. In general the unsteadiness can be *stochastic, highly organized or mixed (interacting)*. From the point of view of CFD it is the highest frequency that is the critical time scale. For general unsteady flows with organized unsteadiness, if the dominant frequency is f, then one can define a Strouhal number as,

$$St = \frac{fL}{U}.$$

For forced flows, such as that occur in rotor wing aerodynamics or in turbomachines, the dominant driven response is sinusoidal with higher harmonics becoming important as the blades/ aerofoils pass through the trailing vortices of the preceding blades/ aerofoils. These flows are characterized by a non dimensional frequency parameter defined as $\bar{k} = \frac{\omega c}{2\pi}$, where ω is the forcing circular frequency and c is the chord of the airfoil section. Let us enumerate some typical 'St' for unsteady flows:

(*a*) For an oscillating body in its own plane a shear layer forms whose thickness is of the order of $(O\sqrt{\nu/f})$, where ν is the kinematic viscosity.

(*b*) For a rigid circular cylinder: For $10^3 < Re < 5 \times 10^5$, $St \approx 0.2$ and for $5 \times 10^5 < Re < 10^7$, $St \approx 0.2-0.4$.

(*c*) For a sudden change in a boundary layer a Stokes layer is formed at the wall and its thickness is always less than the boundary layer thickness. The diffusive time scale is $O(\delta^2/\nu)$ and pertains to diffusion of momentum and vorticity from the body surface where it is usually created.

(*d*) For an impulsive start of a physical body, once again the Stokes layer is formed whose thickness is $O(\sqrt{\nu t})$ for small times.

A detailed description of these classes of flows are provided in Telionis (1981), and Schlichting (1979).

4.4 Computing Time-Averaged Flows and Unsteady Flows

Consider for the moment the incompressible flow at very high Reynolds number so that the flow can be considered as fully turbulent. The governing equations are rewritten here for convenience,

$$\nabla.\vec{V} = 0, \tag{4.14}$$

$$\frac{\partial \vec{V}}{\partial t} + (\vec{V}.\nabla)\vec{V} = -\nabla p/\rho + \nu\nabla^2\vec{V}. \tag{4.15}$$

We have not considered the presence of body force as its presence will not alter the subsequent analysis qualitatively. If the flow variables exhibit unsteadiness at very high frequencies – as experimental high Reynolds number flows exhibit – then we can double decompose the velocity field by

$$\vec{V}(\vec{X}, t) = U(\vec{X}) + v(\vec{X}, t). \tag{4.16}$$

On the right hand side of Equation (4.16) the variables are vector quantities and upper case variables are time-independent or are the time average of the total velocity field (in the sense defined below). Therefore this implies that the lower case velocity components are truly random fluctuating quantities, i.e. they have zero time averages. This type of splitting variables for turbulent flows was suggested by Reynolds (1895) and the process is thus called Reynolds averaging. Here the time averaging is defined in the following sense

$$\langle V(\vec{X}, t)\rangle = U(\vec{X}) = \lim_{T\to\infty} \frac{1}{T}\int_0^T \vec{V}(\vec{X}, t)dt. \tag{4.17}$$

We use the angular brackets to denote the time averaging operation. If we split the velocity field as given by Equation (4.16), and if the mean U satisfies the time-independent boundary conditions that $\vec{V}(\vec{X}, t)$ satisfies, then $v(\vec{X}, t)$ satisfies the homogeneous boundary conditions irrespective of whether the boundary conditions are Dirichlet or Neumann type. Similarly, if $\vec{V}$ is periodic in some direction then so is v.

Now suppose the time averages of time derivatives vanish and time averaging commutes with spatial derivative operations. Then we can time average Equations (4.14) and (4.15) to get

$$U.\nabla U + \langle v.\nabla v\rangle = -\nabla p/\rho + \upsilon \nabla^2 U \tag{4.18}$$

$$\text{and} \qquad \nabla.U = 0. \tag{4.19}$$

Note the presence of additional momentum flux terms on the left hand side of Equation (4.18) as a consequence of the presence of the fluctuation terms, $\langle v.\nabla v\rangle$. These are called the Reynolds stresses—a symmetric stress tensor like general stress system in fluids in motion. However, unlike viscous stress, Reynolds stresses do not depend on fluid property but are a property of the flow.

Time averaged Navier-Stokes equations, with the additional Reynolds stress term appearing due to turbulent fluctuations are called the *Reynolds-averaged Navier-Stokes (RANS) equation*. Many of the present day CFD procedures used in the industry are based on this equation. There are two drawbacks with this approach. Firstly, one doesn't know how to provide information about the Reynolds stresses since no additional principles are known about these *stress tensors*. In actual practice these are usually modelled or parameterized, a subject area of fertile activity known as *turbulence modeling*. The second problem is about defining the time averages of turbulent signals. In using the definition given by Equation (4.17), it is implicitly assumed that the time dependence of the flow field is via the fluctuating quantities $v(\vec{X}, t)$ and is essentially at very high frequencies. This procedure seems to be alright for flows which have truly time-independent mean flows, as is the case of unseparated flows. But the major problems come for truly unsteady flows, i.e., where the turbulent unsteadiness cannot be differentiated from the imposed unsteadiness in the flow. Thus this procedure will fail when the averaging time interval (T) is less than the period of mean unsteady motion. Therefore, when turbulence structures are unaffected by externally induced unsteadiness, there is a critical frequency below which steady flow turbulence models can be used for predicting unsteady (in the mean) turbulent flows.

It has been seen for attached turbulent boundary layers (see Tennekes and Lumley (1971) for the morphology of turbulent boundary layers), that there are coherent structures in the near-wall region of such flows. It is noted that the organized structures in turbulent flows carry about 20% of total turbulent kinetic energy (TKE) and hence their role in determining turbulent flow dynamics cannot be underestimated. These organized structures shows up as peaks and valleys in the near-wall region and in terms of wall units, they have lengths between 100 to 2000 units in the streamwise direction and have a spacing of about 50 units in the spanwise direction. This high energetic events occur at a height of about 20 to 50 units (i.e., in and around the buffer layer). These estimates are statistical estimates. Additionally, these near-wall events are interspersed by bursting of these structures. After burst new intermediate scale motions ensue in the buffer layer, which are also streamwise and/or hair-pin vortices. Thus the unsteadiness of turbulent boundary layer is characterized by the bursting frequency—even for an attached shear layer. For example in zero pressure gradient boundary layer this critical frequency is roughly between 20 to 100% of the turbulence burst frequency (F_b), where

$$F_b = \frac{U_\infty}{5\delta}$$

with δ as the shear layer thickness. For adverse pressure gradient flows this critical frequency is between 6 to 28% of F_b. Therefore, it is not surprising that for unsteady flows as in bluff body flows, the performance of RANS with turbulence model is less than satisfactory. Also, in the light of the existence of organized unsteady motion it is not clear whether a double decomposition as resorted to in Reynolds averaging is a correct procedure. For high Reynolds number flows, with or without separation, it is preferable to perform LES instead.

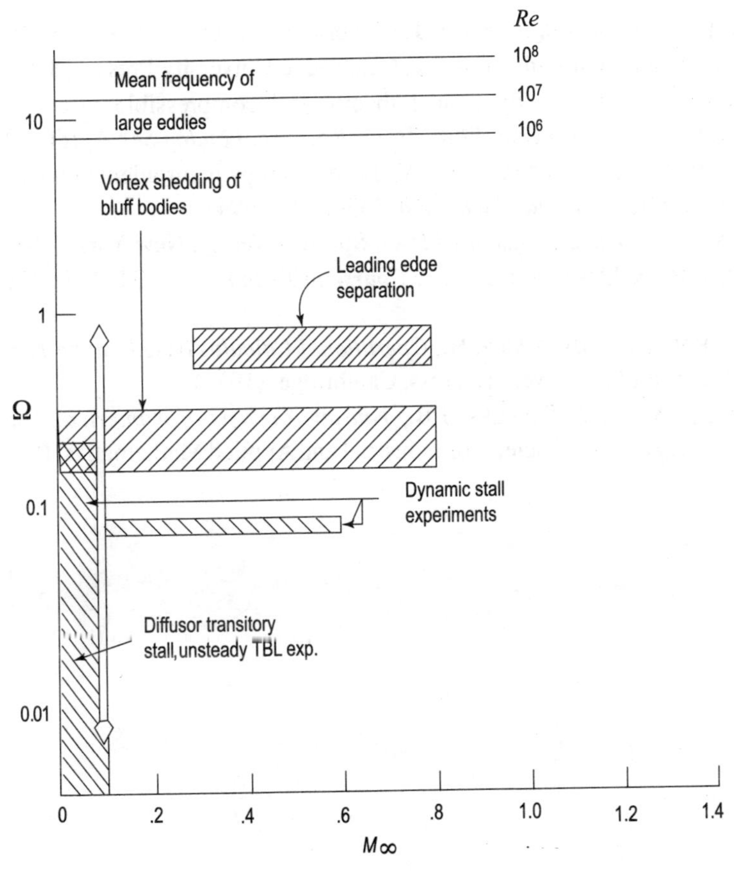

Fig. 4.3 *The ranges of temporal scales excited in some flows of engineering applications for different speed regimes and the corresponding frequency ranges (Ω). Also indicated in the figure is the band for mean frequencies of large eddies in high Reynolds number flows.*

In fig. 4.3 various temporal scales that are excited in many typical engineering flows are displayed, for different speed regimes. Also, for high Reynolds number flows, only the mean frequencies of large eddics are shown. It is quite clear that the ranges of non-dimensional frequencies span over three orders of magnitudes and any numerical method should be able to resolve these scales for high Reynolds number flows.

References

[1] LAUFER, J., The structure of turbulence in fully developed pipe flow, NACA Report. no **1174**, (1954).

[2] LIGHTHILL, M.J., *Waves in Fluids*, Cambridge University Press, (1978a).

[3] LIGHTHILL, M.J., *Fourier Analysis and Generalized Functions*, Cambridge University Press, (1978b).

[4] RAYLEIGH, LORD, *Scientific Papers* **1**, Cambridge University Press, (1889).

[5] RAYLEIGH, LORD, *Scientific Papers* **2**, Cambridge University Press, (1890).

[6] REYNOLDS, O., On the dynamical theory of incompressible viscous fluids and the determination of the criterion, *Phill. Trans. Roy. Soc. London Ser. A*, **186**, 123–164, (1895).

[7] SADDOUGHI, S.G., & VEERAVALLI, S.V., Local isotropy in turbulent boundary layers at high Reynolds number, *J. Fluid Mech.*, **268**, 333–372, (1994).

[8] TELIONIS, D.P., *Unsteady Viscous Flows,* Springer Verlag, New York, (1981).

[9] TENNEKES, H., & LUMLEY, J.L., *First Course in Turbulence*, MIT Press, Cambridge, USA, (1971).

[10] VAN DER POL, B., & BREMMER, H., *Operational Calculus based on the Two-Sided Laplace Integral*, Cambridge University Press, Cambridge, (1959).

[11] VICHNEVETSKY, R., & BOWLES, J.B., *Fourier Analysis of Numerical Approximations of Hyperbolic Equations*, Society for Industrial and Applied Mathematics, Philadelphia, USA, (1982).

5 Discretization of Partial Differential Equations

Unlike analytic solutions where we have information everywhere in the domain at all times, in CFD we can obtain answers only at *discrete points* in the domain, called the grid points. This discretization of the independent variable is the first step of CFD, and a typical section of a discrete grid of points are shown in the Cartesian two-dimensional plane below. (Fig. 5.1)

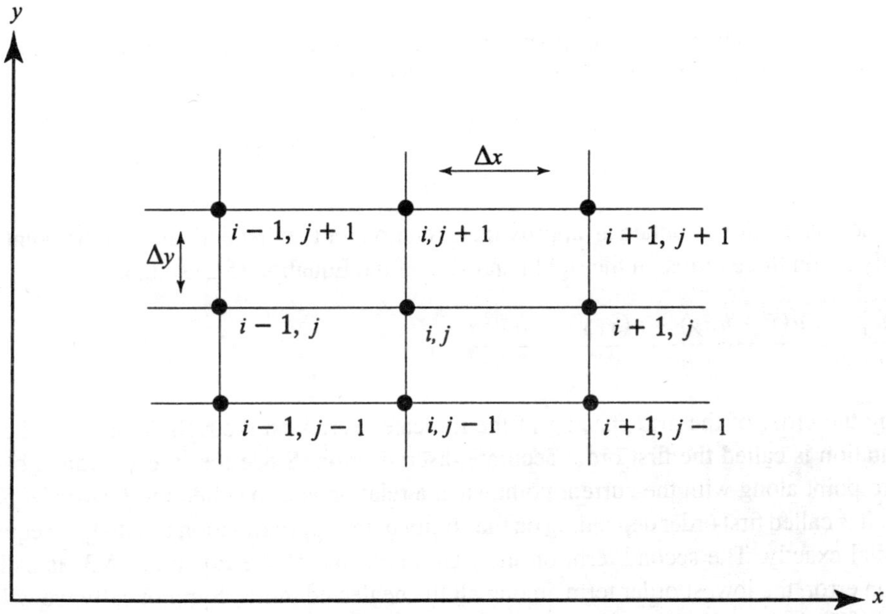

Fig. 5.1 *Discrete grid points.*

The approach in obtaining the numerical solution of differential equation rests on representing the operators by an equivalent algebraic expression, i.e., the differential equation is reduced to a set of algebraic relations that are eventually solved to obtain the unknown at the above grid points. This step of generating algebraic equation from differential equation is known as the discretization process. There is no unique way of doing so, hence there is no unique way of solving the governing differential equation. However, there are some overall guiding principles that allow CFD practitioners to choose some methods that provide the correct answer. The main guiding principle is the *consistency*, by which one means that the discretized differential equation should reproduce the original differential equation in the limit of vanishing distance in the grid network shown in Fig. 5.1.

Let us assume, for the sake of convenience, that the spacing of grid points in either direction is uniform, although it need not be the same in both the directions. Let the dependent variable u be represented by

$$u(x, y) = u(i\Delta x, j\Delta y) = u(ih, jk). \tag{5.1}$$

Symbolically, one represents $u(x, y) \equiv U_{ij}$. The discretization process followed can be explained either in terms of Taylor series expansion or by a polynomial representation. Let us first explain discretization in terms of Taylor's series.

5.1 Taylor Series Representation

If we know the representation of the unknown u and its derivatives at (x, y) then we can represent the function at a neighbouring point $(x + h, y)$ by the Taylor series,

$$u(x + h, y) = u(x, y) + h\frac{\partial u}{\partial x}\bigg|_{ij} + \frac{h^2}{2}\frac{\partial^2 u}{\partial x^2}\bigg|_{ij} + \frac{h^3}{3!}\frac{\partial^3 u}{\partial x^3} + O(h^4). \tag{5.2}$$

We can use this relation to obtain an approximate relation for the first derivative at the point (x, y), if we only retain three terms on the right hand side of the Equation (5.2) so that

$$\frac{\partial u}{\partial x}\bigg|_{ij} = \frac{u(x + h, y) - u(x, y)}{h} - \frac{h}{2}\frac{\partial^2 u}{\partial x^2}. \tag{5.3}$$

Noting the order of the grid spacing of the truncated series on the right hand side, the above representation is called the first order accurate discretization. Since the information is based on a forward point along with the current point, such a relation is also called the *forward difference formula*. It is called first order depending on the ability of the approximation to satisfy an equivalent polynomial exactly. The second term on the right hand side of the Equation (5.3) is called the truncation error, the lowest order term among all the neglected terms. Symbolically we will write Equation (5.3) as

$$\frac{\partial u}{\partial x}\bigg|_{ij} = \frac{U_{i+j} - U_{ij}}{h}. \tag{5.4}$$

Similarly we can represent that unknown at $(x - h, y)$ by another Taylor series given by

$$u(x - h, y) = u(x, y) - h\,\frac{\partial u}{\partial x}\bigg|_{ij} + \frac{h^2}{2}\,\frac{\partial^2 u}{\partial x^2}\bigg|_{ij} - \frac{h^3}{3!}\,\frac{\partial^3 u}{\partial x^3}\bigg|_{ij} + O(h^4). \tag{5.5}$$

One can also use Equation (5.5) to represent the first derivative by an equivalent *backward difference formula* given by

$$\left.\frac{\partial u}{\partial x}\right|_{ij} = \frac{U_{ij} - U_{i-1j}}{h}. \tag{5.6}$$

This is also a first order accurate expression for the derivative. We can also use Equations (5.2) and (5.5) to obtain the alternate expression given by

$$\left.\frac{\partial u}{\partial x}\right|_{ij} = \frac{u_{i+1j} - u_{i-1j}}{2h} - \frac{h^2}{6}\frac{\partial^3 u}{\partial x^3} + O(h^4). \tag{5.7}$$

This relation is called the *central difference* approximation for the derivative and is also said to be second order accurate based on the exponent of the grid spacing of the leading truncation term. The points that are involved in expressing the derivative are called the *stencil* of the formula. It is apparent that the higher the order of expression the larger will be the size of the stencil.

Using the expression given in Equations (5.2) and (5.5), we can also construct the following discretization for the second derivative as

$$\frac{\partial^2 u}{\partial x^2} = \frac{u_{i+1j} - 2u_{ij} + u_{i-1j}}{h^2} - \frac{h^2}{12}\frac{\partial^4 u}{\partial x^4}. \tag{5.8}$$

Note that by earlier convention of classifying the order of the discretization, it is second order. However, the above equation satisfies a third order polynomial exactly and hence should be called a third order accurate expression. We would use the latter convention.

In many CFD applications we will come across the Laplacian operator and in Cartesian frame it can be expressed as

$$\begin{aligned}\nabla^2 u &= \frac{\partial^2 u}{\partial x^2} + \frac{\partial^2 u}{\partial y^2} \\ &= \frac{u_{i+1j} - 2u_{ij} + u_{i-1j}}{h^2} + \frac{u_{ij+1} - 2u_{ij} + u_{ij-1}}{k^2} \\ &\quad - \frac{h^2}{12}\frac{\partial^4 u}{\partial x^4} - \frac{k^2}{12}\frac{\partial^4 u}{\partial y^4} + O[h^4 + k^4]\end{aligned} \tag{5.9}$$

5.2 Polynomial Representation

Next, let us represent the discretization procedure by polynomial expression. The coefficients of the polynomial are evaluated in terms of the grid point unknowns. For example consider a second order polynomial

$$f(x) = ax^2 + bx + c. \tag{5.10}$$

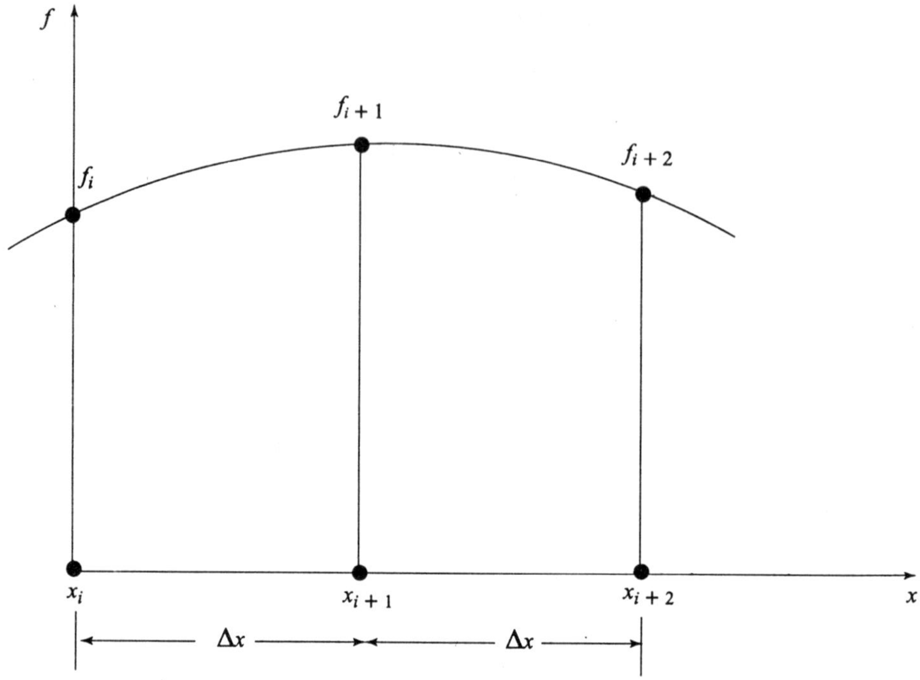

Fig. 5.2 *A second order polynomial approximating the actual function.*

In Fig. 5.2 we show the sketch of the same polynomials with the unknowns marked at three grid points, x_i, x_{i+1}, and x_{i+2}, with the origin fixed at x_i.

Thus,

$$\begin{aligned} f_i &= f(x_i) = c, \\ f_{i+1} &= ah^2 + bh + c, \\ f_{i+2} &= 4ah^2 + 2bh + c. \end{aligned} \tag{5.11}$$

These can be solved to obtain

$$a = \frac{f_{i+2} - 2f_{i+1} + f_i}{2h^2} \quad \text{and} \quad b = \frac{-f_{i+2} + 4f_{i+1} - 3f_i}{2h}. \tag{5.12}$$

Therefore,

$$\left.\frac{\partial f}{\partial x}\right|_i = b \quad \text{and} \quad \left.\frac{\partial^2 f}{\partial x^2}\right|_i = 2a. \tag{5.13}$$

Note that the procedure followed here is equivalent to obtaining one-sided difference formulae. What is the order of accuracy of these representations?

Discretization on a non-uniform grid

Central difference for y_i'

The Taylor series for the function $y = f(x)$ at $x_i + h_r$ and $x_i - h_l$ (see Fig. 5.3) expanded about x_i are:

$$y(x_i + h_r) = y_i + y_i' h_r + \frac{y_i'' h_r^2}{2!} + \frac{y_i''' h_r^3}{3!} + \ldots \tag{5.14}$$

$$y(x_i - h_l) = y_i - y_i' h_l + \frac{y_i'' h_l^2}{2!} - \frac{y_i''' h_l^3}{3!} + \ldots \tag{5.15}$$

where primes indicate differentiation with respect to x.

In terms of $\frac{h_r}{h_l} = \beta$, Equations (5.14) and (5.15) can be rewritten as

$$y(x_i + h_r) = y_{i+1} = y_i + y_i'(\beta h_l) + \frac{y_i''(\beta h_l)^2}{2!} + \frac{y_i'''(\beta h_l)^3}{3!} + \ldots \tag{5.16}$$

$$y(x_i - h_l) = y_{i-1} = y_i - y_i' h_l + \frac{y_i'' h_l^2}{2!} - \frac{y_i''' h_l^3}{3!} + \ldots \tag{5.17}$$

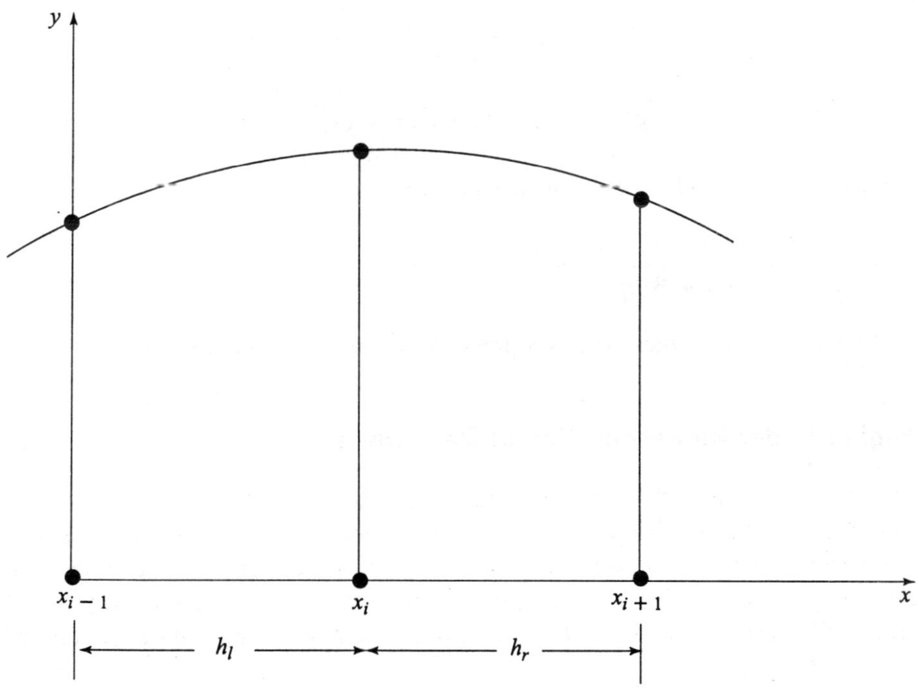

Fig. 5.3 *Unequally spaced grid points for a continuous and differentiable function* $y = f(x)$.

Subtracting $\beta^2\times$ Equation (5.17) from Equation (5.16) gives

$$y_{i+1} - \beta^2 y_{i-1} = y_i - \beta^2 y_i + y_i'(\beta h_l) + y_i'\beta^2 h_l + O(h_l^3).$$

Therefore,

$$y_{i+1} - \beta^2 y_{i-1} = y_i(1-\beta^2) + y_i'\beta h_l(1+\beta) + O(h_l^3).$$

Upon simplification one gets

$$y_i' = \frac{y_{i+1} - y_i(1-\beta^2) - \beta^2 y_{i-1}}{\beta(1+\beta)h_l} + O(h_l^2). \tag{5.18}$$

Central difference for y_i''

The addition of $\beta^3\times$ Equation (5.17) to Equation (5.16) gives

$$y_{i+1} + \beta^3 y_{i-1} = y_i(1+\beta^3) + y_i'\beta_l(1-\beta^2) + y_i''\frac{h_l^2}{2}\beta^2(1+\beta) + O(h_l^4). \tag{5.19}$$

Substituting y_i' from Equation (5.18) in Equation (5.19) and simplyfying gives

$$y_{i+1} + \beta^3 y_{i-1} = y_i(1+\beta^3) + \frac{y_{i+1} - y_i(1-\beta^2) - \beta^2 y_{i-1}}{\beta(1+\beta)h_l}\beta h_l \times (1-\beta^2) + y_i''\frac{h_l^2}{2}\beta^2(1+\beta) + O(h_1^4).$$

Further simplification provides the second derivative expression as

$$y_i'' = \frac{2\beta y_{i+1} - 2\beta(1+\beta)y_i + 2\beta^2 y_{i-1}}{\beta(1+\beta)h_l^2} + O(h_l^2). \tag{5.20}$$

One can check for the correctness of this expression by considering the case for $\beta = 1$.

5.3 Higher Order Representation of Derivatives

Using Taylor series or polynomial expressions in deriving expression for derivatives are easy when the order of the discretization is low. To arrive at higher order expressions even for the low order derivatives become cumbersome; we will instead introduce finite difference operators to simplify the representations. A comprehensive description of the notation and method can be obtained from (Hildebrand (1956) and Ames (1977)). Let us briefly discuss about some of the operators and their usage.

Let $y = y(x)$ be a functional representation of a variable of a function of its argument defined at discrete intervals. We will describe this in finite difference (FD) notation as

$$y_n = y(x_n),$$

representing the unknown y at a discrete point x_n.

Let us define the following operators:

$$\text{Forward difference operator: } \Delta y_n = y_{n+1} - y_n \tag{5.21a}$$

$$\text{Backward difference operator: } \nabla y_n = y_n - y_{n-1} \tag{5.21b}$$

$$\text{Central difference operator: } \delta y_n = y_{n+\frac{1}{2}} - y_{n-\frac{1}{2}} \tag{5.21c}$$

$$\text{Averaging operator: } \mu y_n = \frac{1}{2}(y_{n+\frac{1}{2}} + y_{n-\frac{1}{2}}) \tag{5.21d}$$

$$\text{Shift operator: } E y_n = y_{n+1} \tag{5.21e}$$

$$\text{Differential operator: } Dy = \frac{dy}{dx} \tag{5.21f}$$

Using the above expressions we can also relate various operators. For example,

$$\Delta y_n = y_{n+1} - y_n = E y_n - y_n = (E-1) y_n.$$

Hence the following identities are obtained.

$$\Delta = E - 1 \quad \text{and} \quad \nabla = 1 - E^{-1}. \tag{5.22}$$

One can similarly establish the following identities:

$$\mu = \frac{E^{\frac{1}{2}} + E^{-\frac{1}{2}}}{2} \quad \text{and} \quad \mu^2 = \frac{E + E^{-1} + 2}{4}. \tag{5.23a}$$

$$\delta = E^{\frac{1}{2}} - E^{-\frac{1}{2}} \quad \text{and} \quad \delta^2 = (E + E^{-1} - 2). \tag{5.23b}$$

$$\text{Therefore} \quad \mu = \left(1 + \frac{\delta^2}{4}\right)^{\frac{1}{2}}. \tag{5.23c}$$

Next, we obtain an expression of the differential operator in terms of the forward difference operator.

$$\text{As} \quad f(x+h) = f(x) + hf'(x) + \frac{h^2}{2!} f''(x) + \frac{h^3}{3!} f'''(x),$$

therefore,

$$Ef(x) = \left[1 + \frac{h}{1!}D + \frac{h^2}{2!}D^2 + \frac{h^3}{3!}D^3 + \ldots\right] f(x) = e^{hD} f(x),$$

which implies

$$D = \frac{1}{h} \ln E. \tag{5.24}$$

Using Equation (5.22) in this relation, one gets the desired relationship as

$$D = \frac{\ln(1+\Delta)}{h} = -\frac{\ln(1-\nabla)}{h}. \tag{5.25}$$

This is an exact relation between D and Δ or between D and ∇. Expanding the logarithmic term in a series and retaining terms up to some order we can obtain the derivative expansion up to that order. For example, we can write,

$$D = \frac{1}{h}\left[\Delta - \frac{\Delta^2}{2} + \frac{\Delta^3}{3} - \ldots\right], \tag{5.26}$$

$$\text{or} \quad = \frac{1}{h}\left[\nabla + \frac{\nabla^2}{2} + \frac{\nabla^3}{3} + \ldots\right]. \tag{5.27}$$

These expressions can also be used to obtain the higher derivatives. For example,

$$\begin{aligned}\frac{d^k y}{dx^k} &= D^k y \\ &= \frac{1}{h^k}\left[\Delta^k - \frac{k}{2}\Delta^{k+1} + \frac{k(3k+5)}{24}\Delta^{k+2} - \frac{k(k+2)(k+3)}{48}\Delta^{k+3} + \ldots\right]y.\end{aligned} \tag{5.28}$$

One can also show that $\delta = 2\sinh\frac{hD}{2}$ and hence

$$D = \frac{2}{h}\left[\frac{\delta}{2} - \frac{1}{6}\left(\frac{\delta}{2}\right)^3 + \frac{3}{40}\left(\frac{\delta}{2}\right)^5 - \frac{15}{336}\left(\frac{\delta}{2}\right)^7 + \ldots\right]. \tag{5.29}$$

Higher order upwind differences

Before closing this section we will derive some additional expressions for first derivative and higher order even derivatives which have symmetric stencils. These expressions will be used subsequently for generating higher order upwind schemes. First let us obtain the expression of the first derivative which is sixth order accurate by writing the following stencil for the purpose:

$$f_n' = a(f_{n+3} - f_{n-3}) + b(f_{n+2} - f_{n-2}) + c(f_{n+1} - f_{n-1}) \tag{5.30}$$

with a, b and c to be obtained for the expression with desired accuracy. We can collate the coefficients of the derivatives of various orders on the right hand side and equate them with those on the left hand side, noting that the stencil being symmetric, there will be only odd derivative terms on the right hand side.

We get the following coefficients for:

$$f_n' : 2h(3a + 2b + c) = 1 \tag{5.31a}$$

$$f_n''' : \frac{h^3}{3}(27a + 8b + c) \tag{5.31b}$$

$$f_n^{v} : \frac{h^5}{60}(243a + 32b + c) \tag{5.31c}$$

For a sixth order accurate representation these last two coefficients should be zero and thus constitute three equations for the three unknowns. On solving these equations, one can show that

$$a = \frac{1}{60h}; \quad b = -\frac{3}{20h} \quad \text{and} \quad c = \frac{3}{4h}.$$

Thus the sixth order representation of the first derivative (by the symmetric stencil) is,

$$\left(\frac{df}{dx}\right)_{\mathrm{CD}_6} = \frac{f_{n+3} - 9f_{n+2} + 45f_{n+1} - 45f_{n-1} + 9f_{n-2} - f_{n-3}}{60h}. \tag{5.32}$$

Next, we obtain the symmetric stencil for even derivatives and in the following we represent the sixth derivative as an example given by

$$\left(\frac{d^6 f}{dx^6}\right) = a(f_{n+3} + f_{n-3}) + b(f_{n+2} + f_{n-2}) + c(f_{n+1} + f_{n+1}) + df_n. \tag{5.33}$$

Once again collating the coefficients of derivatives of all three orders up to the sixth, we get (noting that the odd derivatives will be absent in this case),

$$2a + 2b + 2c + d = 0 \tag{5.34a}$$

$$9a + 4b + c = 0 \tag{5.34b}$$

$$81a + 16b + c = 0 \tag{5.34c}$$

$$729a + 64b + c = e \tag{5.34d}$$

where $e = \dfrac{360}{h^6}$.

Solving for a, b, c and d in terms of e we get the expression for the derivative as

$$\left(\frac{d^6 f}{dx^6}\right) = \frac{f_{n+3} - 6f_{n+2} + 15f_{n+1} - 20f_n + 15f_{n-1} - 6f_{n-2} + f_{n-3}}{h^6}. \tag{5.35}$$

A fifth order accurate scheme for first derivative is then obtained from

$$\left(\frac{df}{dx}\right)_{\mathrm{UD}_5} = \left(\frac{df}{dx}\right)_{\mathrm{CD}_6} + \theta\frac{h^5}{6!}\left(\frac{d^6 f}{d^6 x}\right)_{\mathrm{CD}},$$

where θ is the parameter.

The explanation for this choice will be given after discussion on numerical stability.

5.4 Numerical Errors

The most important errors that commonly occur in numerical solution are due to (a) rounding-off and (b) truncation.

(*a*) **Round-off error**
Round-off error is introduced because of the computer's inability to handle large number of significant digits. Typically, in single-precision, the number of significant figures retained ranges from 7 to 16, although it may vary from one computer to another. The round-off error arises due to the fact that a finite number of significant digits or decimal places are retained and all real numbers are rounded off by the computer. The last retained digit is rounded off if the first discarded digit is equal to or greater than 5. Otherwise, it is unchanged. For example, if five significant digits are to be kept in place, 5.37527 is rounded off to 5.3753, and 5.37524 to 5.3752. A more detailed discussion on round-off error is to be found in Hamming (1973) and Sterbenz (1974).

(*b*) **Truncation error**
Truncation error occurs due to the replacement of an exact mathematical expression by a numerical approximation. This error has already been discussed earlier in this chapter with respect to finite difference approximations. Basically, it is the difference between an exact expression and the corresponding truncated form (for example, truncated Taylor series) used in the numerical solution.

Exercise

1. Obtain the first derivative for fourth order accurate representation and another expression for the fourth derivative, using symmetric stencils.
2. Consider non-uniform spacing among the points such that

$$x_{i+1} - x_i = h \quad \text{and} \quad x_{i+2} - x_{i+1} = ah.$$

   Obtain the expression for the first and second derivative using polynomial expression. Using Taylor series expansion, obtain the order of truncation error.
3. Repeat the above procedure for obtaining the equivalent central difference formulae.
4. Following the steps in Exercise 1, construct a third order upwind expression for the first derivative.

References

[1] AMES, W. F., *Numerical Methods for Partial Differential Equations*. 2nd edition, Academic Press, New York, (1977).

[2] HAMMING, R. W., *Numerical Method for Science and Engineering*, McGraw Hill, New York, (1973).

[3] HILDEBRAND, F.B., *Introduction to Numerical Analysis*, McGraw-Hill, New York, (1956).

[4] STERBENZ, P.H., *Floating Point Computations*, Prentice Hall, New Jersy, (1974).

6 Solution Methods for Parabolic PDEs and their Analysis

In Chapter 3 we had classified quasi-linear partial differential equations into parabolic, elliptic and hyperbolic equations. In the present chapter we will discuss methods for solving parabolic PDEs. Parabolic PDEs are important for various flow (development of shear layer) and heat transfer (conduction) problems. Historically some of the early progress made in developing concepts and numerical methods came while treating parabolic PDEs. The presentation here tries to follow some of these developments.

To analyse the discretization schemes pertinent to a given differential equation obtained by applying techniques discussed in Chapter 5 for PDEs, we must answer the following question first.

(*a*) Are the solutions accurate?
(*b*) Is the method/algorithm a true representation of the physical system?
(*c*) Does the solution process proceed without divergence?

The answer to the first question is related to the *convergence* of the numerical solution to the actual solution in the limit of grid being refined sufficiently. The second question was considered in the previous chapter and we term it as the consistency property. It is the third question that we would like to answer in detail in this chapter. This is related to the *numerical stability* of the obtained solution. The equivalence theorem due to Lax provides a relationship among these properties of numerical methods, which is stated later in this chapter.

We are now in a position to analyse different discretization schemes. As an example let us look at the following problem:

$$u_t = u_{xx} \tag{6.1}$$

defined in the domain $0 \leq x \leq 1$ and $0 < t < T$ and subject to the following auxiliary conditions,
Initial condition:

$$u(x, 0) = f(x) \tag{6.2}$$

Boundary conditions:

$$u(0, t) = g(t) \tag{6.3a}$$

$$u(1, t) = h_1(t) \tag{6.3b}$$

Here we have considered only the Dirichlet type of boundary conditions in Equations (6.3a) and (6.3b) for the ease of analysis. Other types of boundary conditions would be considered later. It is easy to show that the above equation is parabolic and this equation is usually obtained for one-dimensional diffusion problem, either for heat or momentum transfer.

Even before we start discretizing the above equation, it is worthwhile to investigate what the true solution does when $t \to \infty$? To answer this question let us construct an energy functional,

$$E(t) = \frac{1}{2}\int_0^1 u^2(x, t)dx. \tag{6.4}$$

Such a function mimics kinetic energy, if u represents a velocity. However if this function represents temperature, this will not be an energy functional. Even in that case $E(t)$ is a non-negative function of time.

Obviously, for a real solution of the problem, $E(t) \geq 0$. In addition to this, the solution must also satisfy some other physical attributes. For example, the time rate of change of the energy function may be another quantity of interest, which is given by

$$\frac{dE(t)}{dt} = \int_0^1 u\frac{du}{dt}dx = \int_0^1 uu_{xx}dx$$

since

$$uu_{xx} = \frac{\partial}{\partial x}(uu_x) - u_x^2.$$

Therefore

$$\frac{dE}{dt} = \int_0^1 \left[\frac{\partial}{\partial x}(uu_x) - u_x^2\right]dx.$$

Integrating by parts and using the boundary conditions given by Equations (6.3a) and (6.3b) in the above, one gets,

$$\frac{dE}{dt} = h_1(t)u_x(1, t) - g(t)u_x(0, t) - \int_0^1 u_x^2 dx. \tag{6.5}$$

If $h_1(t) = g(t) = 0$ then Equation (6.5) shows that the time rate change of the energy function is a decaying function of time, i.e., as time progresses the energy of the system decreases, thereby establishing the system to be physically stable. Note also that the physical stability of the system is determined by the functions $h_1(t)$ and $g(t)$ taking such values so that the right hand side of Equation (6.5) is non-positive. From now on we will focus our attention to systems that are only physically stable, if it is not stated otherwise. For such a physically stable system it is quite natural to expect that the numerical method does not lead to unstable solutions. Thus for parabolic system, like the one we are discussing here, one of the major requirements is *stability* of the numerical method. We have already stated in Chapter 5 that all discretization schemes also must be *consistent*, i.e. the discrete system reverts to the continuum in the limit of Δx, $\Delta t \to 0$. We will come back to this aspect while discussing Dufort-Frankel method in this chapter.

6.1 Algorithm for Numerical Solution of Equation (6.1)

The exact form of the difference equation is obtained by many alternative methods. Here we begin our discussions by considering Forward-in-Time and Central-in-Space (FTCS) schemes as given by

$$\frac{u_{lj+1} - u_{lj}}{k} - \frac{k}{2}\frac{\partial^2 u}{\partial t^2} + O(k^2) = \frac{u_{l+1j} - 2u_{lj} + u_{l-1j}}{h^2} - \frac{h^2}{12}\frac{\partial^4 u}{\partial x^4} + O(h^4). \tag{6.6}$$

In Equation (6.6) first subscript (l) is indicating the space variable and the second subscript (j) the time variable.

We have used the following notations: $k = \Delta t$ and $h = \Delta x$ with the understanding that a uniform grid network is considered as the one shown in Fig. 6.1.

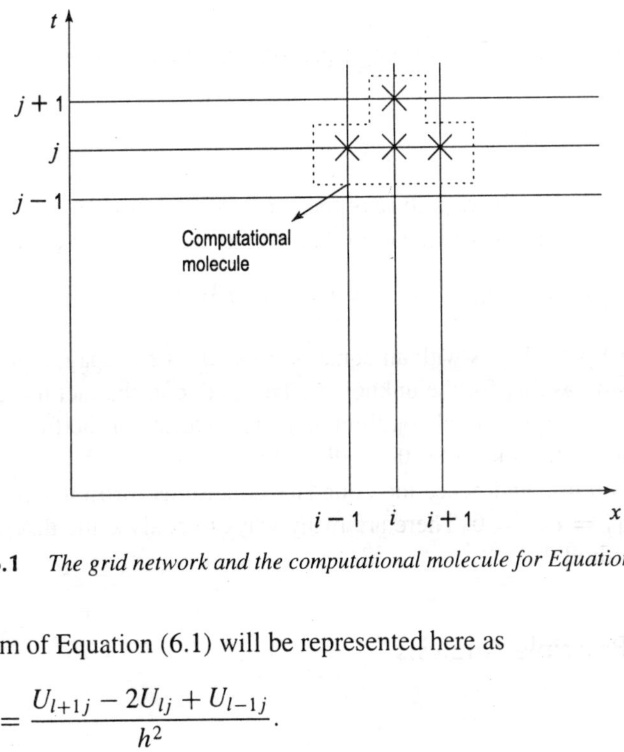

Fig. 6.1 *The grid network and the computational molecule for Equation (6.6).*

The discrete form of Equation (6.1) will be represented here as

$$\frac{U_{lj+1} - U_{lj}}{k} = \frac{U_{l+1j} - 2U_{lj} + U_{l-1j}}{h^2}. \tag{6.7}$$

The convention followed here on will be that upper case variables indicate the numerical approximation of the variables.

Here, $h = 1/(N-1)$ and $k = T/(M-1)$, with N and M as the total number of points taken in the x and t directions. The above discretized Equation (6.7) is used to advance the solution in time for the FTCS scheme by

$$U_{lj+1} = rU_{l-1j} + (1-2r)U_{lj} + rU_{l+1j}, \tag{6.8}$$

where $r = k/h^2$. This is called the *difference equation* and is the discrete analogue of the continuum Equation (6.1); this is central to all numerical computing. This particular way of advancing the solution in time is called the *explicit method* because obtaining the unknown at a given node at the next time step is by an explicit relation where every quantity on the right hand side is known at the previous time step ($t = j\Delta t$).

Thus Equation (6.8) is employed to obtain solution for the new time level $(j + 1)$ in terms of the quantities at the jth time level for all the interior node points (from $l = 2$ to $l = N - 1$). Note that at $j = 1$, the solution is known from the initial condition given by Equation (6.2). Hence Equation (6.8) is solved for $j = 2$ to $j = M$. Also $u_{1,j} = g_j$ and $u_{N,j} = h_{1j}$ are the discrete version of the boundary conditions given by Equations (6.3a) and (6.3b), to be used when using Equation (6.8) at $l = 2$ and $l = N - 1$ respectively.

Error analysis

To analyse the discrete form of the governing equation, one studies the behaviour of error, defined by

$$e_{lj} = u_{lj} - U_{lj}. \tag{6.9}$$

where the error is nothing but the departure of the computed solution (U) from the exact solution (u). Subtracting Equation (6.7) from Equation (6.6) we get the evolution equation for the error as,

$$e_{lj+1} = re_{l+1j} + (1 - 2r)e_{lj} + re_{l-1j} + O(k^2 + kh^2). \tag{6.10}$$

Thus, Equation (6.10) provides us with an equation to study the propagation of error and which has identical expression as that for the unknowns. This is due to the fact the governing equation is linear. However, because the initial condition is given in terms of the function itself, one does not make any error at $t = 0$. Thus $e_{l1} = 0$ for all l's. In the same way the boundary conditions are given as Dirichlet condition and hence the boundary conditions for Equation (6.10) will also be homogeneous, i.e. $e_{1j} = e_{Nj} = 0$. There are many ways to analyse the discrete scheme. Let us begin by a simple method.

6.2 Maximum Principle Analysis

If we define

$$\max_{l=2,N-1} |e_{lj}| = \|e_j\|,$$

then from Equation (6.10) we can write the following:

$$\|e_{j+1}\| \le r\|e_j\| + (1 - 2r)\|e_j\| + r\|e_j\| + 0(k^2 + kh^2). \tag{6.11}$$

In arriving at the above expression, we have taken all the coefficients in Equation (6.10) to be positive. That is the case only when $r \le \frac{1}{2}$, and then Equation (6.11) reduces to

$$\|e_{j+1}\| \le \|e_j\| + 0(k^2 + kh^2). \tag{6.12}$$

Since $\|e_1\| = \max_l \|e_{l1}\| \equiv 0$, so Equation (6.12) gives for $j = 1$,

$$\|e_2\| \le 0(k^2 + kh^2).$$

A repeated application of Equation (6.12) yields the following relation:

$$\|e_j\| \le (j-1)k * 0(k + h^2),$$

and since the current time is $t = (j-1)k$, the above reveals that the error is a linearly increasing function of time and bounded according to

$$\|e_j\| \le T\{0(\Delta t + \Delta x^2)\}. \tag{6.13}$$

The error goes to zero as $k, h \to 0$. Thus the finite difference solution *converges* to actual solution for $r \le 1/2$.
When we discuss numerical stability, we should distinguish between the following two definitions (Ames (1977)).

Concept due to John (1953): Fix $T < \infty$ and take finer and finer mesh, i.e. k and h are reduced further and further. If there exists any error which can become unbounded for $j = M$, then the algorithm is *unstable* in the sense of John. Previous converged solution is stable in this sense, only if $r \le 1/2$.

Concept of Crandall (1953): If we allow the computation to go on for very large t, i.e. $j \to \infty$, then $U_{ij} \to \infty$ while u_{ij} remains bounded, then this implies *stepwise instability*. Usually this definition of instability is used. In this sense the algorithm given by Equation (6.13) is not stable due to accumulation of truncation errors, if the truncation error is considered to be the same for all time. However, this is never the case.

6.3 Fourier Stability Method (due to Von Neumann)

We illustrate this by once again considering Equation (6.1) along with the auxiliary conditions given by Equations (6.2) and (6.3).

As the problem under consideration is linear, separation of variables applies.

If we separate the error into a time dependent and space dependent part as $e_{lj} = s_l t_j$ and substitute it in Equation (6.10), we get,

$$\frac{t_{j+1}}{t_j} = r\frac{s_{l+1}}{s_l} + (1-2r) + r\frac{s_{l-1}}{s_l} = \alpha \text{ (constant)}. \tag{6.14}$$

Using this splitting, it is easy to show that the error can be written as

$$e_{lj} = e^{il\theta h} e^{\alpha jk} \tag{6.15}$$

with $i = \sqrt{-1}$. Substituting Equation (6.15) in Equation (6.10) one gets,

$$e^{\alpha k} = re^{i\theta h} + (1 - 2r) + re^{-i\theta h}. \tag{6.16}$$

The expression on the left hand side of Equation (6.16) is called the amplification factor. From Equation (6.15) it is apparent that the error increases by a factor $e^{\alpha k}$ after advancing the solution by one time step (k) using the algorithm given by Equation (6.10). Hence this nomenclature of amplification factor is used to define $e^{\alpha k}$ for the following conditions:

- If this is less then one then the error decays and the method is *stable*.
- If this is exactly equal to one, then the method is *neutrally stable*.
- If, on the contrary, the amplification factor is greater then one the error grows with time and the method is *unstable*.

Furthermore, Equation (6.16) can be written as

$$e^{\alpha k} = 1 - 2r(1 - \cos\theta h) = 1 - 4r\sin^2\frac{\theta h}{2}.$$

Since $r > 0$ therefore the amplification factor is always less then one.
However, the following sub-conditions need to be considered:

(a) If $0 \le e^{\alpha k} < 1$, then the error will decay monotonically and then we have a stable algorithm.
(b) If $-1 < e^{\alpha k} < 0$, then we get oscillatory decay of error.
(c) Finally, if $e^{\alpha k} < -1$, then the error will be oscillatory as well as unstable.

To achieve the first two conditions, we must have $e^{\alpha k} \ge -1$, i.e.,

$$r \le \frac{1}{2}\text{cosec}^2\frac{\theta h}{2}. \tag{6.17}$$

Obviously if we choose $r \le 1/2$, then the above inequality will always be satisfied. This was also the condition obtained using maximum principle in the previous section. To achieve the first condition alone, i.e. to achieve monotonic decay of error, we must have

$$\left.\begin{aligned} e^{\alpha k} &> 0 \\ \text{i.e. } r &\le \frac{1}{4}\sin^2\frac{\theta h}{2}. \end{aligned}\right\} \tag{6.18}$$

Thus for monotonic decay of error, we must have $r \le 1/4$ and that will satisfy the above inequality always.

We conclude the discussion of stability analysis of difference schemes for the time being. However, in a later chapter we will introduce a more powerful analysis method that is based on spectral analysis, a concept which has already been introduced in Chapter 4.

6.4 Treating Neumann Boundary Conditions

Here we discuss the method of treating boundary conditions that are given in terms of derivative of the unknown. One simple way of treating such a boundary condition would be to represent the derivative by the one-sided difference formula that we derived in Equations (5.4) or (5.6) for the first derivative for low order representation. We could have also used either Equation (5.26) or (5.27) to represent the same derivative by one-sided higher order representation by appropriately truncating the series representations. However, in the following we discuss another method where we would like to use second order central differencing to represent the derivative.

Let us look at the differential equation given by Equation (6.1) and the initial condition given by Equation (6.2). However, let us now consider the following boundary conditions:

$$\frac{\partial u}{\partial x}(0, t) = g(t), \tag{6.19a}$$

$$u(1, t) = h_1(t). \tag{6.19b}$$

Consider the grid network of Fig. 6.2

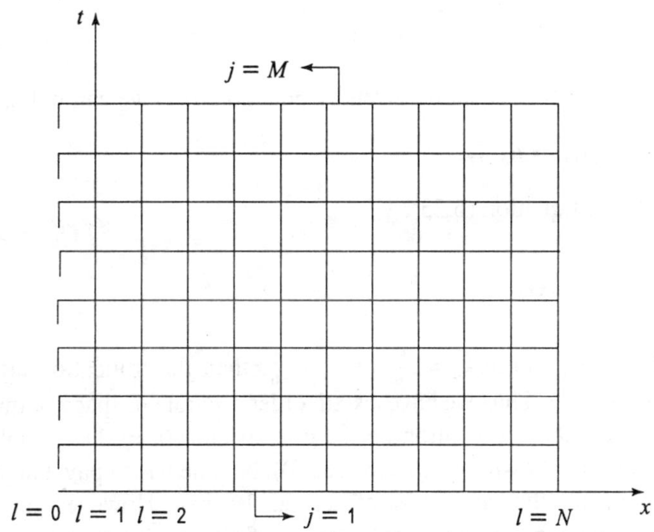

Fig. 6.2 *The ghost boundary introduced at $l = 0$ to apply Neumann boundary condition at $x = 0$.*

Construct another vertical grid line ($l = 0$) next to the $x = 0$ line as shown by the dotted line. This way of extending the boundary is called the method of using a *ghost boundary*. Now one can apply the boundary conditions given by Equation (6.19a) at $x = 0$ by employing

$$\left.\frac{\partial u}{\partial x}\right|_{1,j} = \frac{U_{2j} - U_{0j}}{2h} + O(h^2). \tag{6.20}$$

The above equation can be used to represent the unknown of the *ghost boundary* as

$$U_{0j} = U_{2j} - 2hg_{1j}. \tag{6.21}$$

Now one can write the difference Equation (6.8) for the unknown column of grid points at $l = 1$ by using

$$U_{1j+1} = rU_{2j} + (1 - 2r)U_{1j} + r\{U_{2j} - 2hg_{1j}\}. \tag{6.22}$$

Other Methods for Solving Parabolic PDEs

6.5 Milne's Method (1953)

This is a higher order method which utilizes the choice of time and space steps in the truncation error to achieve higher order accuracy. Let us illustrate this by using the idea for Equation (6.1) by noting that the leading order truncation error of the finite difference representation given by Equation (6.6) is

$$\text{Truncation error} = \frac{k}{2}\left[ku_{tt} - \frac{h^2}{6}u_{xxxx}\right]. \tag{6.23}$$

From the governing differential equation given by Equation (6.1) we can write,

$$u_{tt} = (u_{xx})_t = (u_t)_{xx} = u_{xxxx}. \tag{6.24}$$

Using Equation (6.24) in Equation (6.23),

$$\text{T.E.} = \frac{k}{2}\left[k - \frac{h^2}{6}\right]u_{xxxx}. \tag{6.25}$$

This suggests that if we chose $k = \frac{h^2}{6}$ or $r = \frac{1}{6}$, then the truncation error changes from $O(k^2 + kh^2)$ to $O(k^3 + kh^4)$. Thus we have a sixth order accurate in space or third order accurate in time scheme. And the above condition on the grid spacing ($r = 1/6$) implies the error will decrease monotonically (as shown by Equation (6.18)), but one has to pay a higher price in terms of increased computing for the same level of acceptable error. However, one can use a much coarser grid, thereby saving in computational time significantly. This technique to achieve higher accuracy is often resorted to in CFD application for particular choice of grid parameters.

6.6 Richardson's Method (1910)

This is an interesting method which was reported in Richardson (1910) that used desk calculators manually operated. However, this turns out to be a numerically unstable method. This is achieved here by simply replacing the first order accurate time advancement strategy by a second order accurate central scheme.

This is illustrated for Equation (6.1). In this method the time discretization is performed via second order central differencing. Thus instead of Equation (6.6) we get the following difference equation,

$$\frac{U_{lj+1} - U_{lj-1}}{2k} = \frac{U_{l+1j} - 2U_{lj} + U_{l-1j}}{h^2}. \tag{6.26}$$

Since the alternate time steps are interrelated, this method is thus known as *overlapping step method.* Obviously, the time accuracy of the representation has increased. But this need not be the only consideration! For example one can write the corresponding error evolution equation as

$$e_{lj+1} - e_{lj-1} = 2r \lfloor e_{l+1j} - 2e_{lj} + e_{l-1j} \rfloor. \tag{6.27}$$

If one performs Von Neumann's stability analysis for this case, as discussed in Section 6.3, then one obtains the following equation for the amplification factor

$$e^{\alpha k} - e^{-\alpha k} = 4r(\cos \lambda h - 1). \tag{6.28}$$

This is a quadratic equation for the amplification factor given by

$$e^{2\alpha k} + 8r \sin^2 \frac{\lambda h}{2} e^{\alpha k} - 1 = 0. \tag{6.29}$$

The product of the roots is given by the third term on the left hand side. Since the product of the two amplification factors is equal to -1, it implies that if one root is less than one (stable) then the other root is greater than one (unstable) and vice versa. Hence the overall method is unstable and unsuitable for computations! This also brings in the complexity that the alternate time steps follow independent evolution. In the context of hyperbolic PDEs, this method is known as *leap frog method.*

6.7 Dufort-Frankel Method (1953)

In this explicit method, in the difference equation given by Equation (6.7), the following average is used for enhanced time accuracy.

$$U_{lj} = \frac{1}{2}(U_{lj+1} + U_{lj-1}).$$

Substituting this in the difference equation yields the following:

$$\frac{U_{lj+1} - U_{lj-1}}{2k} = \frac{U_{l+1j} - U_{lj+1} - U_{lj-1} + U_{l-1j}}{h^2}. \tag{6.30}$$

Writing down the Taylor series expansion of the expressions about the node (l, j) and simplifying, one gets

$$u_t + \frac{k^2}{6}u_{ttt} + \ldots\ldots = \frac{h^2 u_{xx} + k^2 u_{tt} + \frac{h^4}{12}u_{xxxx} - \frac{k^4}{12}u_{tttt} + \ldots}{h^2}. \tag{6.31}$$

Thus the truncation error of difference Equation (6.31) is

$$= -\frac{k^2}{6}u_{ttt} + \frac{h^2}{12}u_{xxxx} + \frac{k^2}{h^2}u_{tt} - \frac{k^4}{12h^2}u_{tttt} + \ldots \tag{6.32}$$

This expression brings forth a very interesting possibility, namely that the truncation error will not vanish even in the limit of $k, h \to 0$. This is because $\frac{k}{h}$ or $\frac{k^2}{h}$ may tend to a finite limit (let us say β and γ respectively) and then instead of solving the problem given by Equation (6.1), we will be solving the following continuum equation:

$$u_t - u_{xx} + \beta^2 u_{tt} + \gamma^2 u_{tttt} = 0 \tag{6.33}$$

which is not even a parabolic PDE. This is the problem of *consistency*, which we have mentioned as a basic requirements of computations at the beginning of the chapter.

In Appendix B we have some of the popular explicit methods listed for the solution of 1D heat equation. One can make a detailed analyses of them for studying consistency, convergence and stability.

Let us now discuss a new class of methods that is different from these explicit methods. These are the implicit methods and in CFD practices they are often the preferred methods of solution.

Implicit methods

It was noticed while solving Equation (6.1) subject to the auxiliary conditions given by Equations (6.2) and (6.3a-b) by explicit methods, that for stability and convergence of the numerical solution we have a restriction on the grid size imposed through the requirement $r \leq \frac{1}{2}$. Why is this so? Apart from the analyses performed before, there is another elegant analysis that is due to Crandall (1956), which explains the restriction.

For the explicit algorithm given by Equation (6.7), the initial condition is given along EF as shown in Fig. 6.3

The boundary conditions are given along ER and FQ. However, knowing the initial condition alone, solution at point P can be obtained, i.e. the solution at point P does not depend on the boundary conditions explicitly. Thus the explicit scheme acts like a hyperbolic system with two real characteristic EP and FP for the point P. We note the original problem is parabolic with the single characteristic (t = constant). Thus the continuum solution at P should not be known till we obtain the solution along QR completely. Note that for a fixed r, as $\Delta t \to 0$, the finite difference characteristic EP and FP approaches the physical characteristics.

The foundation of implicit method lies in the attempt to bring finite difference characteristics closer to physical characteristics. Also in doing so the restriction on grid size is relaxed considerably. In this method, Equation (6.1) is discretized in the following manner:

$$u_t|_{ij} = \lambda u_{xx}|_{ij+1} + (1-\lambda)u_{xx}|_{ij} \qquad \text{with } 0 \leq \lambda \leq 1 \tag{6.34}$$

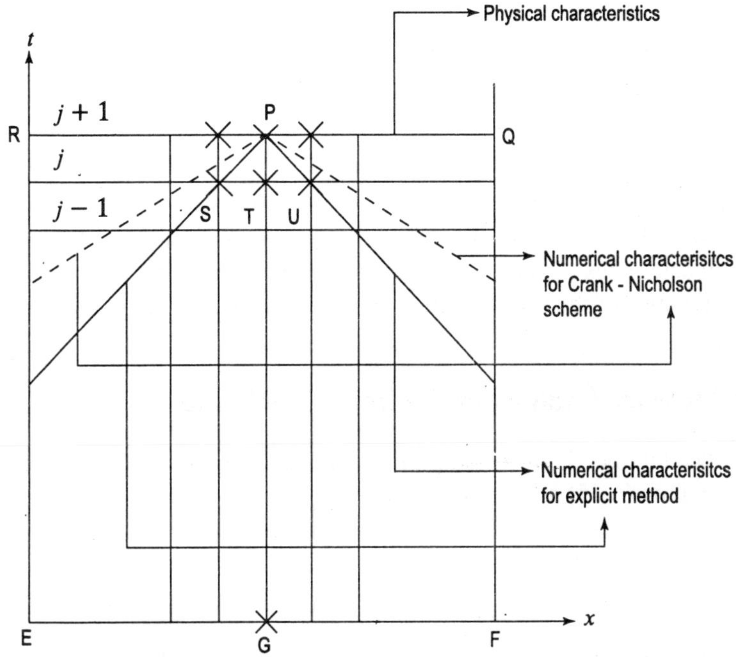

Fig. 6.3 *Various computational molecules and their characteristics.*

so that the difference equation is given by

$$\begin{aligned} &U_{lj+1} - U_{lj} \\ &= r[\lambda(U_{l-1j+1} - 2U_{lj+1} + U_{l+1j+1}) + (1-\lambda)(U_{l-1j} - 2U_{lj} + U_{l+1j})]. \end{aligned} \tag{6.35}$$

The implicit computational molecule is indicated by the points marked by six crosses in Fig. 6.3. Since we are trying to obtain the information at the advanced time level $(j + 1)$ in terms of the old time level values, then we can simplify the above, keeping all the unknowns on the left hand side as

$$\begin{aligned} &-r\lambda U_{l-1j+1} + (1 + 2r\lambda)U_{lj+1} - r\lambda U_{l+1j+1} \\ &\quad = r(1-\lambda)U_{l-1j} + [1 - 2r(1-\lambda)]U_{lj} + r(1-\lambda)U_{l+1j}. \end{aligned} \tag{6.36}$$

This equation cannot be solved *explicitly* point by point. Computationally, for fixed $(j + 1)$, the above equation is an *implicit* relation among the unknowns and constitutes a set of coupled algebraic equation given by

$$[A]\{U\} = \{\text{R.H.S}\}, \tag{6.37}$$

where R.H.S implies the right hand side of the linear algebraic equation given above.

Note the following features of the above equation:

(a) The matrix A is *sparse* and in this case it is a *tridiagonal* matrix.

(b) If we put $\lambda = 0$, we recover the explicit scheme.
(c) For $\lambda = 1/2$, we get what is known as Crank-Nicholson scheme (1947).
(d) For $\lambda = 1$, we get the *fully implicit scheme* (due to O' Brien et al (1951)).
(e) For arbitrary λ, we get Crandall's method (1956).

It is possible to directly solve the Equation (6.37) exactly. An algorithm – Thomas algorithm – is given at the end of this chapter which solves the above equation. Note that the numerical characteristics approaches the physical characteristics as λ increases. For fully implicit method the numerical characteristics are identical to the physical characteristics. In the figure above, the numerical characteristics of the Crank-Nicholson method is indicated by the dashed line.

6.8 Fourier Stability Analysis of the Implicit Scheme

Following the procedure adopted for the explicit schemes, we can perform the Fourier stability analysis using Equation (6.15) and show that

$$e^{\alpha k} = \frac{1 - 4r(1-\lambda)\sin^2\frac{\theta h}{2}}{1 + 4r\lambda\sin^2\frac{\theta h}{2}}. \tag{6.38}$$

This is obtained by noting that the error evolution in this case is also governed by an equation similar to Equation (6.37). As $0 \leq \sin(\theta\ h) \leq 1$, the amplification factor is limited from above by one. Furthermore, to ensure numerical stability, we must have $e^{\alpha k} > -1$. When this is required in Equation (6.38) we get the following inequality:

$$\frac{1}{2} > r(1-2\lambda)\sin^2\frac{\theta h}{2}. \tag{6.39}$$

Immediately one can see from the above that for $\lambda = 1/2$ (i.e. for the Crank-Nicholson method) this condition is unconditionally satisfied, i.e. it does not depend on the value of r (the grid parameter) chosen. Thus, it is safe to choose any grid spacing from numerical stability consideration but still one should remember that increasing the grid spacing will increase the truncation error and hence the accuracy of the solution will suffer. Also notice that a choice of $r = \frac{1}{2(1-2\lambda)}$ provides the critical stability criteria. Any value of λ will also provide unconditional stability. This is one of the reasons that *implicit methods are preferred* over explicit methods. The other important aspects of implicit methods are given as exercises at the end of the chapter.

Exercise 5 should convince you that although the stability restrictions are relaxed for the numerical method, the accuracy of the solution actually degrades. This is one of the reason that for high accuracy solutions explicit methods are preferred.

We will close this chapter stating an important theorem due to Lax (as given in Richtmeyer (1967)), which gives a relationship among the three cardinal requirements of the numerical solutions posed at the beginning of the chapter: "Given a properly posed initial boundary value problem and a finite difference approximation to it that satisfies the *consistency* condition, *stability* is the necessary and sufficient condition for *convergence*."

The results obtained in this chapter should be also important from the point of view of obtaining the unsteady solution of the full Navier-Stokes equation. The viscous dissipation terms in conjunction with the unsteady term has similar property as shown here by the heat diffusion equation. Some of the guidelines stated in this chapter are thus relevant in choosing grid parameters from the point of view of obtaining solutions with numerical stability that are convergent, and the overall chosen method must be consistent.

Thomas Algorithm for Tridiagonal Matrix (Thomas (1949))

The equations that are solved for the numerical solution are given by

$$a_i u_{i-1} + b_i u_i + c_i u_{i+1} = d_i$$
$$\text{for } 1 \leq i \leq N \quad \text{with} \quad a_1 = c_N = 0.$$

The algorithm is as follows:
First, compute, for $i = 2$ to N

$$\beta_i = b_i - \frac{a_i c_{i-1}}{\beta_{i-1}} \qquad \text{with} \quad \beta_1 = b_1$$

and for $i = 2$ to N

$$\gamma_i = \frac{d_i - a_i \gamma_{i-1}}{\beta_i} \qquad \text{with} \quad \gamma_1 = \frac{d_1}{b_1}.$$

The values of the dependent variable are then computed from

$$u_N = \gamma_N \qquad \text{and} \qquad u_i = \gamma_i - \frac{c_i u_{i+1}}{\beta_i}. \qquad \text{for } i = N-1 \text{ to } 1.$$

In solving $AX = b$, with A as a tridiagonal matrix, we note that the Thomas algorithm solves the linear algebraic equation by replacing A as a product of a upper and a lower bidiagonal matrix.

Exercise

1. Construct the A matrix when the Neumann boundary condition given by Equation (6.20a) is used. Can you use the Thomas algorithm directly in this case?
2. What happens to Thomas algorithm when you have periodic boundary conditions $u(0,t) = u(1,t)$ for the solution of Equation (6.1)?
3. Obtain the exact expressions for the amplification factors for Richardson's method.
4. Reason out that the time steps chosen always have to be smaller than the space step for both explicit and implicit method that uses $\lambda < 1/2$.
5. For a fixed value of r and h, show that the rate of change of the amplification factor with respect to λ is always positive. What is the physical implication of this observation? Obtain the similar rates with respect to h and r.
6. Classify the equation (6.34). How would you interpret the numerical results of this?

References

[1] AMES, W.F., *Numerical Methods for Partial Differential Equations*, 2nd Edition, Academic Press, New York, (1977).

[2] CRANDALL, S. H., *J. Math. Phy.*, **32**, 80, (1953).

[3] CRANDALL, S. H., A practical method for numerical evaluation of solutions of partial differential equations of the heat conduction type, *Engineering Analysis*, (382–395), McGraw-Hill, New York, (1956).

[4] CRANK, J. & NICHOLSON, P., *Proc. Camb. Phil. Soc.* **32**, 50, (1947).

[5] DUFORT, E.C., & FRANKEL, S.P., Stability conditions in the numerical treatment of parabolic differemtial equations, *Math. Table NRC*, Washington, **7**, B5, (1953).

[6] JOHN, F., *Comm. Pure Appl. Math.*, **5**, 155, (1953).

[7] MILNE, W.E., *Numerical Solution of Differential Equations*, Wiley, New York, (1953).

[8] O' BRIEN, G.G., HYMAN, M.A., KAPLAN, S., A study of the numerical solution of partial differential equations, *J. Math. Phys.* **29**, 223, (1951).

[9] RICHARDSON, L.F., The approximate arithmatical solution by finite differences of physical problems involving differential equations, with an application to the stresses in a masonry dam, *Ph. Trans. R. Soc.,* **A210**, 307, (1910).

[10] RICHTMEYER, R. D., *Difference Methods for Initial Value Problems*, 2nd ed., Wiley (Interscience), New York, (1949).

[11] THOMAS, L. H., Elliptic problems in linear difference equations over a nature, Watson Scientific Computation Lab. Rept., Columbia University, New York, (1949).

7 Solution Method for Elliptic PDEs

For elliptic partial differential equations, we have noted in Chapter 3 that the characteristics are complex and hence there are no real lines in the physical plane along which information propagates. This therefore implies that every point influences all other points. One of the most simple forms of this class of problem is given by the Laplace's equation. For example, it is known that this equation needs to be solved for incompressible irrotational flows and is usually done in panel methods using boundary element formulation. However, in this chapter we will show various solution methods for elliptic equation by finite difference methods, by taking Laplace or Poisson equation as examples. Poisson equation is another example for an elliptic equation that is solved routinely for grid generation, solving the stream function equation in (ψ–ω) formulation for 2D flows. In primitive variable formulation of Navier-Stokes equation, often the pressure Poisson equation is required to be solved, as was explained in Chapter 2. If loads are required, then for derived variable formulations pressure Poission equation is to be solved.

Let us discuss the solution procedure for the following Poisson equation expressed in the Cartesian coordinates by

$$\frac{\partial^2 u}{\partial x^2} + \frac{\partial^2 u}{\partial y^2} = -f(u, x, y), \tag{7.1}$$

with the uniform grid spacing, $h = \Delta x$ and $k = \Delta y$. For the ease of analyses, let us consider all the boundary conditions to be of Dirichlet type. The difference equation for the Equation (7.1) is given by

$$\frac{U_{m+1n} - 2U_{mn} + U_{m-1n}}{h^2} + \frac{U_{mn+1} - 2U_{mn} + U_{mn-1}}{k^2} = -F_{mn}, \tag{7.2}$$

where $x = m\Delta x$ and $y = n\Delta y$.

The computational domain and the associated grids are shown in Fig. 7.1.

Note that the computational molecule used in Equation (7.2) is a five point molecule identified by a hatched line with $m = 1$ and $m = M$ and $n = 1$ and $n = N$ constituting the boundary lines. Thus the above difference equation needs to be written down for $m = 2$ to $M - 1$ and $n = 2$ to $N - 1$. We will store these variables in the lexicographic manner, i.e., ordering the unknowns from left to right and bottom to top for the nodes shown in Fig. 7.1. The above difference equation when assembled for all the points will constitute the linear algebraic equation

$$[A]\{U\} = \{b\}, \tag{7.3}$$

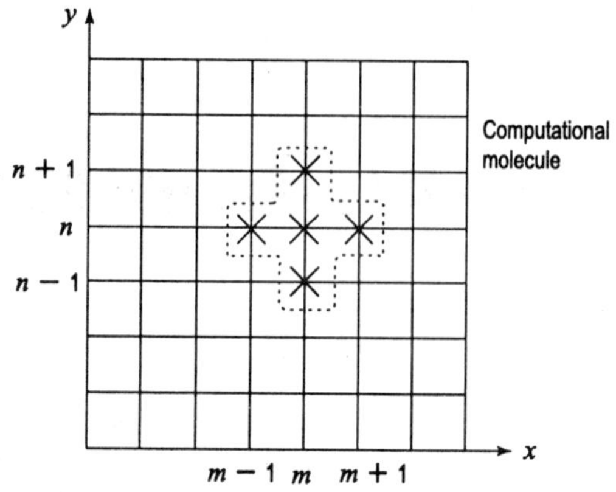

Fig. 7.1 *Computational grid network used for solving Equation (7.2).*

where the assembled $[A]$ matrix has the following *pentadiagonal structure*:

$$\begin{bmatrix} -\left(\frac{2}{h^2}+\frac{2}{k^2}\right) & \frac{1}{h^2} & & \frac{1}{k^2} & & 0 \\ \frac{1}{h^2} & -\left(\frac{2}{h^2}+\frac{2}{k^2}\right) & \frac{1}{h^2} & 0 & \frac{1}{k^2} & \\ & \frac{1}{h^2} & -\left(\frac{2}{h^2}+\frac{2}{k^2}\right) & \frac{1}{h^2} & & \frac{1}{k^2} \\ \frac{1}{k^2} & 0 & \frac{1}{h^2} & -\left(\frac{2}{h^2}+\frac{2}{k^2}\right) & \frac{1}{h^2} & \\ 0 & \frac{1}{k^2} & & \frac{1}{h^2} & -\left(\frac{2}{h^2}+\frac{2}{k^2}\right) & \end{bmatrix}$$

Solution of Equation (7.3) by direct inversion is prohibitive for the usual choices of M and N. For an $(L \times L)$ matrix, direct inversion requires operations of the order of L^3. For the linear algebraic Equation (7.3), $L = (M-2) \times (N-2)$ and hence a direct solution will involve massive number of operations. Furthermore, A matrix has good *banded* structure and is also highly *sparse*. These two properties suggest that instead of direct solution, one should resort to iterative methods. At intermediate steps the linear algebraic equation will be approximately satisfied. The idea behind iterative methods is that, for a convergent strategy, after a large number of iterations the solution U^∞ will satisfy Equation (7.3) exactly; i.e.

$$\{b\} - [A]\{U^\infty\} = 0, \tag{7.4}$$

while at an lth level of iteration the right hand side of Equation (7.4) will not necessarily be zero. This non-zero part is called the *residue* (R^l), which is dependent on the iteration level, i.e.,

$$\{b\} - [A]\{U^l\} = \{R^l\}. \tag{7.5}$$

Now let us also define the error at any level of iteration by

$$\{c^l\} = \{U^\infty\} - \{U^l\}. \tag{7.6}$$

Then subtracting Equation (7.4) from Equation (7.5) we get the linear algebraic equation for the error propagation through the iteration as

$$[A]\{c^l\} = \{R^l\}. \tag{7.7}$$

After a very large number of iterations ($l \to \infty$), the iterated solution will presumably converge and then both the solution residue and error vector will vanish, i.e.,

$$R^\infty, c^\infty \to 0.$$

Therefore in Equation (7.7) both the right and left hand side are identically zero for very large l. Thus if we are looking at error evolution with iteration, it does not matter whether we use the $[A]$ matrix or some other convenient matrix in Equation (7.7). All classical iterative methods are based on this simple observation and we discuss a few of them.

For some time, the classical methods went out of favour with the advent of newer and faster methods, some of which are also discussed in this chapter. However, with the popularization of parallel computing, some of the classical methods are experiencing resurgence due to better parallelizability. It has been observed that the classical methods become slow in converging as the grid is refined. In the classical methods one choose $A = \bar{N}$, where A is not easily invertible but $\bar{N}$ is an alternate matrix that can be easily inverted. In fact, in actual computations, one does not really invert a matrix, but the algorithm that is used for iteration is equivalent to inversion.

In general, we can write down the $[A]$ matrix as a sum of lower triangular matrix L, upper triangular matrix $\tilde{U}$ and the diagonal matrix D, i.e.

$$A = L + D + \tilde{U}.$$

The chosen $\tilde{N}$ matrix is related to L, D and $\tilde{U}$ matrices. This relationship is explored next.

7.1 Jacobi (1884) or Richardson (1965) Iteration

For Equation (7.3) we make the following replacement for $[A]$ matrix.

$$[\tilde{N}] = -\left(\frac{2}{h^2} + \frac{2}{k^2}\right)[I] = -\frac{2}{h^2}(1 + \beta^2)[I]. \tag{7.8}$$

Thus in the Jacobi iteration the $[A]$ matrix is replaced by the diagonal matrix D. In Equation (7.8) β is the cell aspect ratio that is equal to h/k. In actual computation the following iteration is performed with Equation (7.2) rewritten for the unknown interior nodes at each $(l+1)$th iteration (indicated by a superscript in the following) as

$$\left(\frac{2}{h^2} + \frac{2}{k^2}\right) U_{mn}^{l+1} = \frac{U_{m+1n}^l + U_{m-1n}^l}{h^2} + \frac{U_{mn+1}^l + U_{mn-1}^l}{k^2} + F_{mn}. \tag{7.9}$$

7.2 Gauss-Seidel (1874) Iteration

In the iteration sequence given by Equation (7.9), if we store our unknowns in a lexicographic sequence, then when we are working at the node (m, n) we already have new values of the unknown at the $(m-1, n)$ and $(m, n-1)$ node. In Gauss-Seidel iteration this information is used. This is used in the operational sequence as

$$\left(\frac{2}{h^2}+\frac{2}{k^2}\right)U_{mn}^{l+1}=\frac{U_{m+1n}^{l}+\bar{U}_{m-1n}^{l+1}}{h^2}+\frac{U_{mn+1}^{l}+\bar{U}_{mn-1}^{l+1}}{k^2}+F_{mn}. \tag{7.10}$$

The above sequence is equivalent to choosing $\bar{N}=L+D$ (but not exactly, as explained below). Notice also that the unknowns at the current level on the right hand side are known and hence have been identified by an additional over-bar. Thus the $[\bar{N}]$ matrix for the Gauss-Sieldel method approximate $[A]$ better than the Jacobi method and hence the Gauss-Sieldel method will perform better than the Jacobi method wherever both of them work.

In these two iteration sequences the unknowns are obtained point-by-point and hence these are known as *point iteration methods*. The following two iteration sequences introduced below, are those for which the unknowns are obtained line-by-line and hence are known as *line iteration methods*. Notice that the equivalent difference equations for all the points along a line are assembled together. The resultant equations constitute a tridiagonal matrix equation and can be solved by the Thomas algorithm, given in Chapter 6.

7.3 Line Jacobi Method

Here the iteration sequence is achieved when we decide to iterate along the horizontal or vertical nodes line-by-line. Here we explain the procedure when we solve for all the horizontal nodes together by using the following algorithm

$$\frac{U_{m+1n}^{l+1}-2U_{mn}^{l+1}+U_{m-1n}^{l+1}}{h^2}+\frac{U_{mn+1}^{l}-2U_{mn}^{l+1}+U_{mn-1}^{l}}{k^2}=-F_{mn}. \tag{7.11}$$

One can similarly write down the corresponding line Gauss-Seidel iteration sequence. This is left as an exercise.

7.4 Basic Methodology of Iterations and their Analysis

In Equation (7.7) if we replace the $[A]$ matrix by $[\bar{N}]$ matrix, and if we define $c^k=U^{k+1}-U^k$, then the iteration sequence is equivalent to solving the following matrix equation

$$[\bar{N}]\{U^{l+1}-U^l\}=R^l. \tag{7.12}$$

Therefore,

$$\begin{aligned}\{U^{l+1}\} &= \{U^l\} + [\bar{N}]^{-1}\{R^l\} \\ &= \{U^l\} + [\bar{N}]^{-1}\{b - AU^l\} \\ &= (I - \bar{N}^{-1}A)\{U^l\} + \{\bar{N}^{-1}b\}.\end{aligned}$$

Defining

$$B = \bar{N}^{-1}A \quad \text{and} \quad C = \bar{N}^{-1}b,$$

the above iteration sequence can be rewritten as

$$\{U^{l+1}\} = (I - B)\{U^l\} + C. \tag{7.13}$$

Compare this with the general linear iteration given by

$$\{U^{l+1}\} = G_l\{U^l\} + r_l, \tag{7.14}$$

where G_l is the amplification or gain matrix that signifies growth or decay of error with each iteration. If we define

$$\{e^l\} = \{U^l\} - [A]^{-1}\{b\}, \tag{7.15}$$

then comparing Equations (7.14) and (7.13) we get,

$$\{U^{l+1}\} = G_l\{U^l\} + C.$$

Using Equation (7.15) in this we get

$$\{e^{l+1}\} + A^{-1}b = G_l(\{e^l\} + A^{-1}b) + C$$

or

$$\{e^{l+1}\} = G_l\{e^l\} + G_lA^{-1}b + C - A^{-1}b.$$

Using the definition of G_l and C in the above and simplifying, we get

$$\{e^{l+1}\} = G_l\{e^l\} + (I - \bar{N}^{-1}A)A^{-1}b - A^{-1}b + \bar{N}^{-1}b,$$

$$\text{i.e.,} \quad \{e^{l+1}\} = G_l\{e^l\}. \tag{7.16}$$

This is the recursive relation between the errors at successive iteration and this can be simplified to

$$\{e^l\} = G_{l-1}G_{l-2}G_{l-3}\ldots\ldots\ldots G_2G_1\{e^0\}.$$

We define any iteration as a *stationary linear iteration* (SLI) if all the $G_l's$ are independent of iteration and thus the error at any level of iteration for SLI is given by

$$\{e^l\} = [G]^l\{e^0\}. \tag{7.17}$$

Here on the right hand side, the amplification matrix G is raised to the power 'l'. Thus the performance of the iteration scheme will totally depend on the G-matrix, which is nothing but $(I - \bar{N}^{-1}A)$. The performance parameter of the matrix is nothing but the spectral radius (i.e. the maximum eigenvalue) of the G matrix. Note that the iteration sequence will be better when the $\tilde{N}$ matrix is closer to the A matrix and therefore G is a very small perturbation matrix. This is the reason that the Gauss-Seidel method is better than the Jacobi method. If λ_i's are the eigenvalues of the G matrix, then the initial error can be related to the error at the lth level of iteration by

$$\{e^l\} = \sum \alpha_i \lambda_i^l \varepsilon_i.$$

ε_i's are the eigenvectors corresponding to the eigenvalues λ_i, and in terms of the maximum eigenvalue or the spectral radius (λ_M) this can be expressed as

$$= \lambda_M^l \left[\alpha_M \varepsilon_M + \sum_{i \neq M} \alpha_i \left(\frac{\lambda_i}{\lambda_M}\right)^l \varepsilon_i\right]. \tag{7.18}$$

Let us state a theorem, without proof, which will help us in selecting a strategy for solving elliptic PDEs, leading to stationary linear iteration.

7.5 Convergence Theorem for Stationary Linear Iteration

"The stationary linear iteration $u^l = Gu^{l-1} + Mv$ converges, that is say $G^l X \to 0$ for any arbitrary X if and only if each eigenvalue λ_i of G is less than one in absolute value".

If G is symmetric, it has linear independent eigen modes even though not all the eigenvalues are necessarily distinct. For the Jacobi iteration, the error component can be written as

$$e_{li} = U_i^{(l)} - U_i = \frac{1}{\alpha_{ii}}[-\alpha_{i1}e_{l-11} - \alpha_{i2}e_{l-12} - \ldots\ldots - \alpha_{iN}e_{l-1N}], \tag{7.19}$$

where U_i is the ith component of the exact solution. If we set

$$\theta_i = \sum_{j=1, j \neq 1}^{N} |g_{ij}| = \sum_{j=1}^{N}\left|\frac{\alpha_{ij}}{\alpha_{ii}}\right|,$$

then for convergence, we need θ_i to be as small as possible compared to one. This is the reason *diagonal dominance* is so important for numerical methods that leads to stationary linear iterations. *Diagonal dominance* requires that the diagonal term is greater than the sum of the absolute value of all off-diagonal terms of a matrix, row-by-row or column-by-column. We will quote some more important theorems from linear algebra that are useful in selecting an appropriate numerical method. The following is due to Stein and Rosenberg (see Ames (1977)). If $A = [\alpha_{ij}]_{N \times N}$ satisfies the following condition (*a*) $\alpha_{ii} > 0, \alpha_{ij} \leq 0,$

(*b*) $\alpha_{ii} \geq \sum_{j=1}^{N} |\alpha_{ij}|,$

(*c*) A is irreducible,

and G_J and G_S are the amplification matrices associated with the Jacobi and Gauss-Seidel iterations, respectively, then one and only one of the following mutually exclusive conditions hold:

(*a*) $\lambda(G_J) = \lambda(G_S) = 0.$
(*b*) $0 < \lambda(G_S) < \lambda(G_J) < 1.$
(*c*) $1 = \lambda(G_J) = \lambda(G_S).$
(*d*) $1 < \lambda(G_J) < \lambda(G_S).$

This theorem clearly states that Gauss-Seidel iteration converges(diverges) faster if both the methods are convergent (divergent). There is also a "good news" if $[A]$ is a symmetric positive definite matrix. Then the Gauss-Seidel iteration always converges.

A symmetric matrix $[A]$ is positive definite if for any vector $X \neq 0, X^T[A]X > 0$ and semidefinite if the scalar product is equal to zero.

7.6 Relaxation Methods

This is an acceleration technique for solving the problem by modifying the iteration procedure given in Equation (7.12). Instead of directly using the total residue on the right hand side of Equation (7.12), one uses the following equation:

$$[N][\{U^{l+1}\} - \{U^l\}] = \alpha\{R^l\} = \alpha(b - [A]^{-1}\{U\}^l). \tag{7.20}$$

The choice of α is dictated by the following theorem due to Ostrowski-Varga (1962): If $A = [\alpha_{ij}]$ is symmetric and $\alpha_{ii} > 0$ for all $i = 1, 2, \ldots N$, then $\lambda(G_\alpha) < 1$ *iff* A is positive definite and $0 < \alpha < 2$, G_α is defined by Equation (7.22) below.

If $\alpha < 1$, we have what is known as *under-relaxation scheme* and if $1 < \alpha < 2$ we have the corresponding *over-relaxation scheme*. One can rewrite Equation (7.20) using Equation (7.5) as

$$\{U^{l+1}\} = (I - \alpha N^{-1}A)\{U^l\} + \alpha N^{-1}b. \tag{7.21}$$

Comparing this with Equations (7.13) and (7.14) we see that

$$G_\alpha = I - \alpha N^{-1}A \quad \text{and} \quad r_\alpha = \alpha N^{-1}b. \tag{7.22}$$

By a proper choice of the relaxation parameter α, one can accelerate the convergence of the iteration process.

In an actual computational environment the relaxation method (also known as *successive over relaxation* (SOR) method) is used in the following fashion:

$$\{U^{l+1}\} = \{U^l\} + \alpha\lfloor\{\bar{U}^{l+1}\} - \{U^l\}\rfloor. \tag{7.23}$$

As before, the quantity with the over-bar indicates the computed quantity evaluated at the present level of iteration—a predicted quantity. This has the usual predictor–corrector structure. Let us illustrate the application of SOR method as applied to the basic Gauss-Seidel iteration.

Using Equation (7.14), the Gauss-Seidel iteration can be written as,

$$\bar{U}_i^{(k)} = \sum_{j=1}^{i-1} g_{ij} U_j^{(k)} + \sum_{j=i+1}^{N} g_{ij} U_j^{(k-1)} + r_i. \tag{7.24}$$

When SOR is used along with it,

$$U_i^{(k)} = (1-\alpha) U_i^{(k-1)} + \alpha \left[\sum_{j=1}^{i-1} g_{ij} U_j^{(k)} + \sum_{j=i+1}^{N} g_{ij} U_j^{(k-1)} + r_i \right]. \tag{7.25}$$

The matrix notational form and the corresponding gain matrix has been given as an exercise (Exercise 7.3) at the end of the chapter.

7.7 Rate of Convergence

In all iterative methods of solving linear algebraic equation, the most important parameter of concern is the *rate of convergence* because this decides the total computational effort in solving elliptic PDE by an iterative method. The estimate for the rate of convergence in turn, depends upon two other parameters:

(*a*) The computational work required per iteration and
(*b*) number of iteration necessary for convergence to the correct solution.

Usually the work required per iteration is dictated by the number of multiplications and divisions, and for a $(L \times L)$ matrix this is of the order of L^2 if the matrix is full. However, for finite difference (FD) or finite volume (FV) methods the matrix is not only sparse, but is banded too. The algorithm is constructed in such a way that these banded matrices are either tridiagonal or pentadiagonal. For such matrices, the operation count is between $3L$ and $5L$.

For the linear algebraic equation given by Equation (7.3), the initial error e_0 propagates after a sufficiently large number (k) of stationary linear iterations to e_k, such that they are related by their norm as,

$$\|e_k\| = \|G^k e_0\| \leq \|G^k\| \; \|e_0\|. \tag{7.26}$$

Thus the way error propagates, i.e. how the convergence is reached, is determined by the spectral radius of the gain matrix, G as explained by Equation (7.18). It can also be shown that for large values of k, $\frac{\|e_{K+1}\|}{\|e_k\|}$ averages to $\lambda(G)$. Therefore, on an average the error decreases by the factor $\lambda(G)$ at each step of the iteration. Now, if the spectral radius $\lambda(G)$ is less than one, then $-\log_{10} \lambda$ would be the number of decimal digits of accuracy gained at each step. Based on this observation, Young (Ames (1977)) has suggested the quantitative measure of *rate of convergence* $R(G)$ by,

$$R(G) = -\log_{10} \lambda(G). \tag{7.27}$$

Some spectral estimates for the Laplacian operato

Next we will quote some results (without proof) related to the solution of the Laplace's equation (Ames (1997))

$$\nabla^2 u = 0 \quad \text{in the domain} \quad 0 \leq x, y \leq \pi.$$

which gives us estimates about the efficiency of each of the classical methods. If we take equispaced grids of size $h = \frac{\pi}{N}$, then we can show that the eigenvalues of the corresponding linear algebraic equation $AU = b$ are given by

$$\mu_{pq} = \frac{4}{h^2}\left[\sin^2 \frac{ph}{2} + \sin^2 \frac{qh}{2}\right].$$

Having obtained these eigenvalues, it is easy to provide the quantitative estimates of the following methods.

Jacobi method: Here $G_J = -D^1(L + \tilde{U}) = -D^{-1}(A - D) = I - D^{-1}A$, since $D = \frac{4}{h^2}I$.

$$\text{So} \quad G_J = I - \frac{h^2}{4}A$$

and

$$\lambda(G_J)_{pq} = 1 - \frac{h^2}{4} + \frac{4}{h^2}\left[\sin^2 \frac{ph}{2} + \sin^2 \frac{qh}{2}\right] \approx \frac{1}{2}(\cos(ph) + \cos(qh)).$$

Therefore the spectral radius is given by,

$$\lambda(G_J) = \max |\lambda_{pq}| \approx \cos(h) \approx 1 - \frac{h^2}{2},$$

and the rate of convergence is given by

$$R(G_J) = -\log \lambda = \frac{h^2}{2}.$$

Since for the Gauss-Seidel method, $\lambda(G_S) \approx \lambda^2(G_J)$, one can easily show that

$$R(G_S) = h^2.$$

Thus for this problem, the point Gauss-Seidel method is twice as fast in convergence as the point-Jacobi method. It can also be shown that the successive over-relaxation method as applied to Gauss-Seidel method has the following maximum rate of convergence:

$$R(G_{\text{SSOR}}) = 2h.$$

We would like to point out the implications of the quoted results. Firstly, the rate of convergence degrades severely as the mesh is refined. Thus these methods are hardly preferred for practical large scale problem solving activities using sequential computing. The maximum convergence is obtained for the Gauss-Seidel successive over-relaxation (GSSOR) method. But *apriori* one does not know the value of the optimum relaxation factor, which not only depends on the problem but also depends on the boundary conditions, discretization, etc. However, there is one method due to Lyusternik (1947) that tries to relate the relaxation parameter α with the spectral radius λ.

7.8 Method of Acceleration due to Lyusternik

We have noted above that after large number of iterations, $e_{k+1} \approx \lambda e_k$. Now if we are solving the matrix equation $Au = v$ iteratively, then $u = A^{-1}v = \bar{u}^{(k)} + e_k$. Similarly for the $(k+1)$th iteration we can write

$$A^{-1}v = \bar{u}^{k+1} + e_{k+1} = \bar{u}^{(k+1)} + \lambda e_k + \varepsilon_k. \tag{7.28}$$

Notice that ε_k is distinct from e_k and is the small correction term, and so this can be neglected. We can substitute for $e_k = A^{-1}v - u^{(k)}$ in Equation (7.28) so that

$$u^{(k+1)} = A^{-1}v = \frac{\bar{u}^{k+1}}{1-\lambda} - \frac{\lambda}{1-\lambda}u^{(k)} = \alpha\bar{u}^{-(k+1)} + (1-\alpha)u^{(k)}.$$

So if we can find out the spectral radius, then it is easy to find out the optimum relaxation parameter. In the next section where we introduce the alternate direction implicit (ADI) method, we will provide a practical and fast method of calculating the spectral radius of any given matrix.

7.9 Alternate Direction Implicit (ADI) Method

We have noted in the previous section that the rate of convergence of classical iteration schemes, severely degrades as the mesh is progressively refined. Even the best method—the successive over relaxation Gauss Seidel method (GSSOR)—has a rate of convergence that is directly proportional to the mesh size. In contrast the ADI method first proposed by Peaceman and Rachford (1955) has very high rate of convergence. We will discuss this in the context of solving the elliptic PDE in self-adjoint form given by

$$G(x,u)u - \frac{\partial}{\partial x}\left[A(x,y)\frac{\partial u}{\partial x}\right] - \frac{\partial}{\partial y}\left[C(x,y)\frac{\partial u}{\partial y}\right] = S(x,y), \tag{7.29}$$

where $G \leq 0$ and $A, C > 0$. A comprehensive description of the method can be obtained from Birkhoff et al. (1962). Here a brief description is given. We discretize the derivatives terms with $\Delta x = h$ and $\Delta y = k$ as

$$-hk\frac{\partial}{\partial x}\left[A\frac{\partial u}{\partial x}\right] = [H]\{U\} = -a_{mn}u_{m+1n} + 2b_{mn}u_{mn} - c_{mn}u_{m-1n}, \tag{7.30a}$$

$$-hk\frac{\partial}{\partial y}\left[C\frac{\partial u}{\partial y}\right] = [V]\{U\} = -\alpha_{mn}u_{mn+1} + 2\beta_{mn}u_{mn} - \gamma_{mn}u_{mn-1}, \tag{7.30b}$$

where, to retain the diagonal dominance of the $[H]$ and $[V]$ matrices, the following second order accurate discretization is performed:

$$a_{mn} = \frac{k}{h}A_{m+1/2n} \qquad c_{mn} = \frac{k}{h}A_{m-1/2n} \qquad \text{and} \quad 2b_{mn} = a_{mn} + c_{mn}$$

$$\alpha_{mn} = \frac{h}{k}C_{mn+1/2} \qquad \gamma_{mn} = \frac{h}{k}C_{mn-1/2} \qquad \text{and} \quad 2\beta_{mn} = \alpha_{mn} + \gamma_{mn}$$

This discretization makes $[H]$ and $[V]$ symmetric positive definite matrices. For $[H]$ and $[V]$ matrices, the strict diagonal dominance is ensured from the first and last rows of the matrices when the boundary conditions are the Dirichlet type. For example,

$$[H][U]_{i1} = \begin{bmatrix} 2b_{11} & -a_{11} & & & & \\ -c_{21} & 2b_{21} & -a_{21} & & & \\ & c_{31} & 2b_{31} & a_{31} & & \\ & & \cdots & \cdots & \cdots & \\ & & & c_{n-11} & 2b_{n-11} & -a_{n-11} \\ & & & & c_{n1} & 2b_{n1} \end{bmatrix} \begin{Bmatrix} U_{11} \\ U_{21} \\ U_{31} \\ \vdots \\ \vdots \\ \vdots \\ U_{N-11} \end{Bmatrix}.$$

The FD analogue of Equation (7.29) is

$$(H + V + \bar{D})U = K, \tag{7.31}$$

where $\bar{D}$ is the diagonal matrix arising from the term hkG_{mn}. Thus this term adds to the diagonal dominance and we will not consider it any further because its presence can only improve the convergence of the iteration process. The matrix K arises from the source term on right hand side in Equation (7.29). We thus consider the linear algebraic equation

$$(H + V)U = K. \tag{7.32}$$

This equation, in the ADI method, is solved iteratively in alternate directions. Equation (7.32) is clearly equivalent (for any arbitrary D and E matrices) to each of the two linear algebraic equations given below.

$$(H + D)U = K - (V - D)U, \tag{7.33a}$$
$$(V + E)U = K - (H - E)U, \tag{7.33b}$$

provided $(H + D)$ and $(V + E)$ are non-singular. Peaceman and Rachford (1955) first used Equations (7.33a) and (7.33b) for the particular choice of $D = E = \rho I$. They in fact proposed a sequence of positive numbers ρ_k used to obtain the sequence of vectors U^k, $U^{k+1/2}$, and U^{k+1} from the following equations:

$$(H + \rho_k I)U^{k+1/2} = K - (V - \rho_k I)U^k, \tag{7.34a}$$
$$(V + \rho_k I)U^{k+1} = K - (H - \rho_k I)U^{k+1/2}. \tag{7.34b}$$

Although the Equations (7.34a) and (7.34b) are not the same as Equation (7.32) in the intermediate stages of iteration, but at convergence we get back the original difference equation. Thus the Peaceman & Rutherford method is defined by the two following half steps:

$$U^{k+1/2} = (H + \rho_k I)^{-1}[K - (V - \rho_k I)U^k], \qquad (7.35a)$$

$$U^{k+1} = (V + \rho_k I)^{-1}[K - (H - \rho_k I)U^{k+1/2}]. \qquad (7.35b)$$

Although in the above mathematical notations we see the presence of inverses of matrices on right hand side, for actual computations this is not required as the concerned matrices are tridiagonal and whose analytical inverses can be obtained by few operations using Thomas algorithm. The above sequence implies that one first solves Equation (7.35a) in the vertical direction, followed by solving the Equation (7.35b) in the horizontal direction. This justifies the naming of the method as ADI.

7.10 The Analysis of the ADI Method

At convergence ($k \to \infty$), Equations (7.34a) and (7.34b) can be rewritten as

$$(H + \rho_k I)U^{\infty} = K - (V - \rho_k I)U^{\infty}, \qquad (7.36a)$$

$$(V + \rho_k I)U^{\infty} = K - (H - \rho_k I)U^{\infty}. \qquad (7.36b)$$

Subtracting Equation (7.34a) from Equation (7.36a) and Equation (7.34b) from Equation (7.36b) and denoting $e^{k+1/2} = U^{\infty} - U^{k+1/2}$ and so on, we get

$$(H + \rho_k I)e^{k+1/2} = -(V - \rho_k I)e^k, \qquad (7.37a)$$

and

$$(V + \rho_k I)e^{k+1} = -(H - \rho_k I)e^{k+1/2}. \qquad (7.37b)$$

Therefore

$$e^{k+1} = -(V + \rho_k I)^{-1}(H - \rho_k I)e^{k+1/2}$$
$$= (V + \rho_k I)^{-1}(H - \rho_k I)(H + \rho_k I)^{-1}(V - \rho_k I)e^k.$$

Or we can write this as

$$e^{k+1} = T^{(k)}e^{(k)},$$

where the *error reduction matrix*, $T^{(k)}$, decides the error propagation for an iteration step via

$$\|e^{(k+1)}\| \le \|T^{(k)}\| \; \|e^{(k)}\|. \qquad (7.38)$$

The first factor on the right hand side is nothing but the spectral radius of the *error reduction matrix*. We know furthermore, that the spectral radius does not change by a similar transformation, i.e., A and CAC^{-1} have the same spectral radius for any matrix C. Thus $T(\rho_k)$ and $\tilde{T}(\rho_k)$ have the same spectral radius, where

$$\begin{aligned}\tilde{T}(\rho_k) &= (V + \rho_k I)T(\rho_k)(V + \rho_k I)^{-1}\\ &= (V + \rho_k I)(\{(V + \rho_k I)^{-1}(H - \rho_k I)(H + \rho_k I)^{-1}(V - \rho_k I)\}(V + \rho_k I)^{-1}\\ &= [(H - \rho_k I)(H + \rho_k I)^{-1}][(V - \rho_k I)(V + \rho_k I)^{-1}].\end{aligned} \tag{7.39}$$

Both products in the square brackets in Equation (7.39) are symmetric matrices and hence have real eigenvalues. As H and V are commutative and symmetric matrices, they also have the same eigenvectors v. If H and V have eigenvalues μ and ν, then the eigenvalues of $\tilde{T}$ are of the form

$$\left(\frac{\mu - \rho}{\mu + \rho}\right)\left(\frac{\nu - \rho}{\nu + \rho}\right).$$

So after l iterations the error vectors are related by,

$$\lambda(\mu, \nu) = \prod_{i=1}^{l} \frac{(\mu - \rho_i)(\nu - \rho_i)}{(\mu + \rho_i)(\nu + \rho_i)}. \tag{7.40}$$

Thus,

$$\left\|\prod_{i=1}^{l} T(\rho_i)\right\| = \Lambda\left(\prod T(\rho_i)\right) \leq \max_{u,v} \prod_{i=1}^{l} \left|\frac{(\mu - \rho_i)(\nu - \rho_i)}{(\mu + \rho_i)(\nu + \rho_i)}\right|, \tag{7.41}$$

where μ and ν ranges over all eigenvalues of H and V respectively. In practice there are usually that many eigenvalues of H and V as are the number of points in H and V directions. The idea behind the choice of variable acceleration parameters ρ_k, is to selectively annihilate the error components corresponding to the eigenvalue spectrum of H and V matrices. In Fig. 7.2 we have shown the plot of $\frac{\mu-\rho_k}{\mu+\rho_k}$ versus μ for particular choices of the acceleration parameters. Note that the choice of $\rho_k = \mu_3$ makes $|T(\rho)| \equiv 0$ for the error component which is proportional to μ_3. Thus it is tempting to choose the sequence of acceleration parameters corresponding to each and every eigenvalues of H and V matrices. It is, however, not practical to consider them individually even when they are known. This is due to the fact that some of these eigenvalues will have multiplicities, while the eigen modes will still be inearly independent. This is also related to the aliasing problem when solving an equation of the type given by Equation (7.29). This last aspect is discussed in Chapter 10. In passing, we note that after each iteration the error spectrum will distribute error component due to aliasing. It is however, practical to estimate upper and lower bounds for the eigenvalues of H and V. Thus, having estimated a, b, α, β such that

$$a \leq \mu \leq b \quad \text{and} \quad \alpha \leq \nu \leq \beta,$$

one wishes to minimize

$$\Psi_l(a, b, \alpha, \beta, \vec{\rho}) = \max_{a \leq \mu \leq b, \alpha \leq \mu \leq \beta} \Pi_{i=1}^{l} \left|\frac{(\mu - \rho_i)(\nu - \rho_i)}{(\mu - \rho_i)(\nu - \rho_i)}\right|,$$

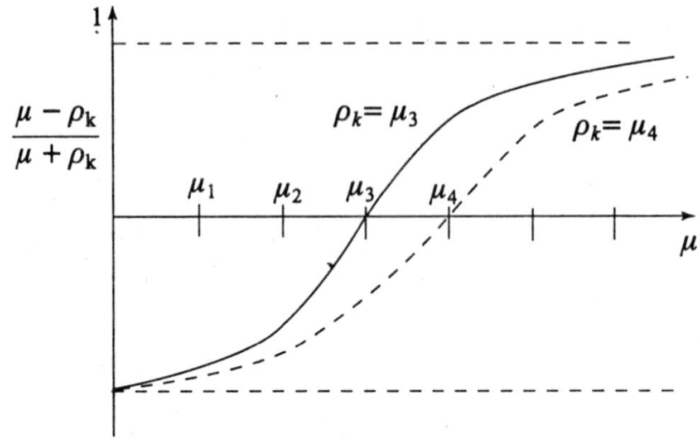

Fig. 7.2 *The effect of choice of different accleration paramenters for ADI scheme.*

where

$$\vec{\rho} = |\rho_1 \rho_2 \ldots\ldots\ldots\ldots\ldots \rho_N|^T.$$

Frequently it is convenient to use the inequality

$$\Psi_l(a, b, \alpha, \beta, \vec{\rho}) \leq [\Phi_l((\bar{a}), (\bar{b}), \vec{\rho})]^2$$

where

$$\bar{a} = \min\,(a, \alpha) \quad \text{and} \quad \bar{b} = \max\,(b, \beta) \quad \text{and}$$

$$\Phi_l(\bar{a}, \bar{b}, \vec{\rho}) = \max_{\bar{a} \leq \gamma \leq b} \prod_{i=1}^{l} \left| \frac{\gamma - \rho_i}{\gamma + \rho_i} \right|.$$

7.11 Choice of Acceleration Parameters

We choose a sequence of acceleration parameters $\rho_1, \rho_2, \ldots\ldots\ldots \rho_N$ for the optimum performance of the ADI scheme. Peaceman and Rachford suggest the following choice of parameters:

$$\rho_i^{(P)} = \bar{b}\left(\frac{\bar{a}}{\bar{b}}\right)^{\frac{(2i-1)}{2N}} \qquad \text{for } i = 1, 2, \ldots\ldots N. \tag{7.42}$$

N is chosen so that the rate of convergence is maximized. It has been shown in Varga et al. (1962) that N has to be found out from the following equation (which leads to maximization of rate of convergence) based on the suggestion of Peaceman and Rachford.

$$(\sqrt{2} - 1)^{2N} \leq c, \quad \text{where} \quad c = \frac{\bar{a}}{\bar{b}}.$$

Wachspress (1966) suggests choosing N from the inequality

$$(\sqrt{2}-1)^{2(N-1)} \leq c.$$

It is noted furthermore that the Wachspress parameters provide higher rate of convergence.

7.12 How to Estimate Maximum and Minimum Eigenvalues

This is done by using the theorems due to Gerschgorin (see Wilkinson (1971)). If we have a matrix $[A]$ with elements a_{ij}, then the following two theorems help in locating the maximum and minimum eigenvalues of the matrix.

Theorem 1: Every eigenvalue of the matrix A lies in at least one of the circular discs with centers a_{ii} and radii $\sum_{j \neq i} |a_{ij}|$ for the ith row.

Theorem 2: If s of the circular discs of Theorem 1 form a connected domain which is isolated from the other discs, then there are precisely s eigenvalues of A within this connected domain.

7.13 Further Explanatory Notes on ADI and Variant Methods

For the ADI method the error value at the end of each half step is given by Equations (7.37a) and (7.37b). Notice that the process of iteration of an elliptic equation can be viewed as time marching. For example, in solving the Laplace's equation by an iterative method, we are in effect solving the following equation

$$u_t = u_{xx} + u_{yy}. \tag{7.43}$$

If we write the error vector as

$$e_{ij}^{(n)} = E^n \sin(\pi i p h) \sin(\pi j q k), \tag{7.44}$$

then the finite difference form of (7.43), with the operator notation can be written as

$$\partial_x^2 u_{ij}^{(n+1/2)} + \partial_y^2 u_{ij}^{(n)} = \left[\frac{u_{ij}^{(n+1/2)} - u_{ij}^{(n)}}{\Delta t/2}\right].$$

Plugging Equation (7.44) in the above yields

$$\frac{E^{(n+1/2)}}{E^{(n)}} = \frac{1 - 2r_2 \sin^2(\pi q k/2)}{1 + 2r_1 \sin^2(\pi p h/2)} \approx \frac{1-a^2}{1+b^2}, \tag{7.45}$$

where $r_1 = \frac{\Delta t}{h^2}$ and $r_2 = \frac{\Delta t}{k^2}$.

Similarly,

$$\frac{E^{(n+1)}}{E^{n+1/2}} = \frac{1-b^2}{1+a^2}. \tag{7.46}$$

In each of the half step of ADI, as shown by Equations (7.45) and (7.46), it is not guaranteed that the ratio on the right hand side is less than one. However, for a complete sweep

$$\frac{E^{(n+1)}}{E^{(n)}} = \left(\frac{1-a^2}{1+a^2}\right)\left(\frac{1-b^2}{1+b^2}\right). \tag{7.47}$$

Now each of the quantities within brackets in Equation (7.47) are less than one irrespective of the fact if a^2 or b^2 is greater or less than one. Thus in ADI, *out of two unstable algorithm we get one stable algorithm.*

Such a thing is possible only for two-dimensional problems because of inherent symmetry of the problem. If one tries to generalize this for a 3D problem the symmetry is lost and no such possibilities exist directly. Also, notice that in ADI we have resorted to what is known as additive splitting. For 3D problems however we can resort to multiplicative splitting of the operator and this is the basis of the *fractional step method* pioneered by Yanenko (1971) and the Russian school.

7.14 Method of Fractional Steps

Consider the following problem.

$$u_t = L(u) \tag{7.48}$$

where $L(u)$ is any spatial operator.

In the ADI method one employs the additive splitting of the spatial operator, i.e.

$$L = L_x + L_y.$$

In the fractional steps method, one uses, instead, the following splitting.

$$L = L_1 L_2 L_3. \tag{7.49}$$

This type of splitting has been tried also in Beam & Warming (1978) scheme and in McCormack (1981) scheme for solving hyperbolic systems. In the method suggested by Yanenko and his colleagues, the following time evolution equation, is used.

$$u_t = L_x(u) + L_y(u) + L_z(u) \tag{7.50}$$

For example, if $L_x(u) = D_x u_{ijk}$, then let us define the following notations:

$$A_x = I - \frac{\Delta t}{2} D_x, \qquad A_y = I - \frac{\Delta t}{2} D_y, \qquad A_z = I - D_z \frac{\Delta t}{2}.$$

$$B_x = I + \frac{\Delta t}{2} D_x, \qquad B_y = I + \frac{\Delta t}{2} D_y, \qquad B_z = I + \frac{\Delta t}{2} D_z.$$

In the fractional step method the following sequence of computations are performed.

$$A_x u^{n+1/3} = B_x u^n; \quad A_y u^{n+2/3} = B_y u^{n+1/3}; \quad A_z u^{n+1} = B_z u^{n+2/3}. \tag{7.51}$$

If $D = D_x + D_y + D_z$, then the splitting performed by Equation (7.51) is equivalent to solving the equation

$$\frac{u^{n+1} - u^n}{\Delta t} = D\left[\frac{u^{n+1} + u^n}{2}\right] - \frac{\Delta t^2}{4}(D_x D_y + D_x D_z + D_y D_z)\left[\frac{u^{n+1} - u^n}{\Delta t}\right]$$
$$+ \frac{\Delta t^2}{4} D_x D_y D_z (u^{n+1} + u^n)/2. \tag{7.52}$$

Thus the above splitting gives rise to additional dispersion effect via the last term on the right hand side of Equation (7.52).

7.15 General Introduction

Multi-Grid Methods (Hackbusch (1985))

In many natural and man-made systems one can observe the multi-level processes. For example in smooth running of a country we have administration at different levels. Global decisions are taken at the central or federal levels while issues pertaining to the very local levels can be tackled by the authorities at the local level itself. We recognize that the problems and solutions at any level depend on input from other levels. But it is also well appreciated that the local issues are solved more efficiently at the local level.

What is the relevance of previous discussions to the solution of PDEs? Iterative procedures are very efficient in solving large and complicated problems in CFD. The solutions are tackled independently at different levels. Then they are forced to communicate to each other so that some constraints are satisfied (the auxiliary conditions of Section 2.9). The inter-level communication is also enforced due to the presence of unwanted aliasing error. After the communications stage, the solution proceeds again at the different levels based on the above mentioned constraints. This is where the *iterative* nature of the problem solving shows up.

Most of the present day application of multi-level algorithms are seen in efficient algorithms used in solving large algebraic systems arising in discretizing PDE's—especially those describing steady state physical systems. Here, the full hierarchical nature described by the discrete separated harmonics in the system, is not at all generally explicit in the problem itself or in its most general governing equations. However, a spectral analysis of the underlying iterative process will clearly reveal the same, including the presence of aliasing error.

Steady state problems and solutions of their discretized equations are examined from two points of view.

1. First, while studying the reasons for slow convergence of various relaxation solvers it becomes clear that relaxation is a "local" process that cannot efficiently treat "global" or smooth solution components.
2. Secondly—which is a complimentary view point—the errors that are slow to converge by the relaxation method on some fine grid can be approximated more efficiently on a coarser grid, where the mesh size is comparable to their scale and hence the convergence rate of this scale is not slow.

We hasten to add that working on a coarser grid entails incurring larger truncation error, thereby degrading overall accuracy of the obtained solutions. A competitive method, like ADI also works on similar principles of error control but always works on the fine grid. But our subsequent analysis will reveal that this is also the reason for relatively slower functioning of the ADI method as compared to the multi-grid method. Because ADI only works on the finest grid, it makes the associated truncation errors lower compared to multi-grid method. What can one say about the role of truncation errors in the stabilization of error propagation by these two methods?

A natural hierarchy of levels emerges, based on viewing the solution from each of the linear boundary value problem as a linear combination of components with different scales. Each component is most effectively controlled by grids with mesh size comparable to its scale and efficient multi-level control can thus be realized in the form of multi-grid processing.

We explain these with respect to the following model problem from Hackbusch (1985).

$$u''(x) = -f(x) \quad \text{in} \quad \Omega = \lfloor 0, 1 \rfloor \tag{7.53}$$

with the boundary condition

$$u(x) = g \quad \text{on} \quad x = \partial\Omega.$$

To solve this problem we form a sequence of grids as shown below.

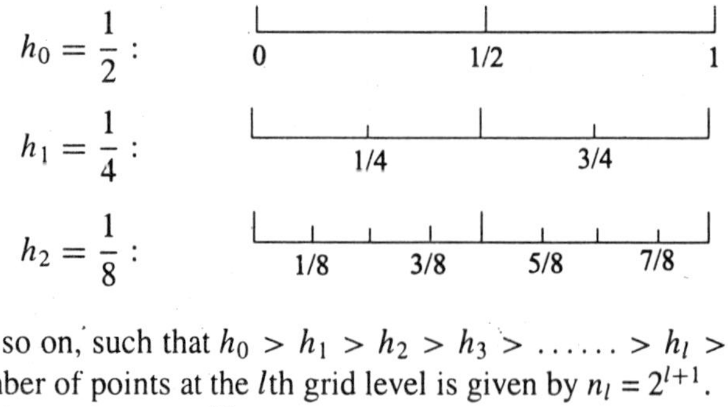

and so on, such that $h_0 > h_1 > h_2 > h_3 > \ldots\ldots > h_l > \ldots$, where $h_l = 1/2^{l+1}$ and the total number of points at the lth grid level is given by $n_l = 2^{l+1}$.

The grid at level 'l' is given by: $\quad \Omega = \{\nu h_l : \nu = 0, 1, 2, 3 \ldots (2^{l+1} - 1)\}$.

The difference equation for Equation (7.53) is:

$$\frac{-u(x_{\nu-1}) + 2u(x_\nu) - u(x_{\nu+1})}{h_l^2} = f_l(\nu h_l).$$

If we adopt the notations

$$\left.\begin{aligned} u_l &= u_l(\nu h_l) \\ f_l &= f_l(\nu h_l) \end{aligned}\right\} \quad \text{for} \quad \nu = 1 \quad \text{to} \quad n_l\left(= \frac{1}{h_l} - 1\right),$$

then the above difference equation can also be written in the matrix notations

$$\underbrace{\frac{1}{h_l^2}\begin{bmatrix} 2 & -1 & \dots & \dots & \dots \\ -1 & 2 & -1 & \dots & \dots \\ \dots & -1 & 2 & -1 & \dots \\ \dots & \dots & -1 & 2 & -1 \end{bmatrix}}_{[L_l]} \underbrace{\begin{Bmatrix} u_{1l} \\ u_{2l} \\ \vdots \\ u_{nl} \end{Bmatrix}}_{\{u_l\}} = \underbrace{\begin{Bmatrix} f_{1l} \\ f_{2l} \\ \vdots \\ f_{nl} \end{Bmatrix}}_{\{f_l\}},$$

where $[L_l] = -\frac{1}{h_l^2}[C]$. Since L_l is tridiagonal, the above equation can be directly solved. However, we have chosen this simple problem for illustrative purposes only. To understand the problem solving strategy by the iterative method let us evaluate the eigen values of the $[C]$ matrix.

Let

$$g_m = \sin\frac{m\pi\nu}{(n_l+1)} \qquad m, \nu = 1, 2, 3 \dots n_l. \tag{7.54}$$

Then $(C - \lambda_\nu I) = 0$ gives

$$g_{m-1} - (2 + \lambda_\nu)g_m + g_{m+1} = 0. \tag{7.55}$$

Using Equations (7.54) in Equation (7.55) we obtain

$$\lambda_\nu = -4\sin^2\frac{\pi\nu}{2(n_l+1)}.$$

Since $\quad h_l = \dfrac{1}{(n_l+1)},$

so $\quad \lambda_\nu = -4\sin^2\dfrac{\pi\nu h_l}{2}$

as $\quad L_l = -\dfrac{C}{h_l^2}.$

Therefore the eigenvalues of the L_l are

$$\lambda_\nu(L_l) = \frac{4}{h_l^2}\sin^2\frac{\nu\pi h_l}{2} \tag{7.56}$$

and the corresponding eigenvectors are: $e_l^m = \sin(\pi m\nu h_l)$ for $m, \nu = 1, 2 \ldots n_l$.

7.16 Smoothing Effects of Classical Iterations

Let us consider point-Jacobi and point-Jacobi with under-relaxation schemes for

$$L_l u_l = f_l.$$

Let us represent: $L_l = D_l - B_l$, where D_l is the diagonal matrix associated with the L_l matrix and $-B_l$ is the rest of the L_l matrix. Then $D_l u_l = B_l u_l + f_l$. In the point-Jacobi method, the iteration scheme is

$$\begin{aligned} u_l^{j+1} &= D_l^{-1}(B_l u_l^j + f_l) \\ &= D_l^{-1}[-(L_l - D_l)u_l^j + f_l] \\ &= u_l^j - D_l^{-1}L_l u_l^j + D_l^{-1} f_l \\ &= u_l^j - D_l^{-1}[L_l u_l^j - f_l]. \end{aligned} \tag{7.57}$$

Let us define the following terms:

Correction:	$u_l^{j+1} - u_l^j$
Defect:	$L_l u_l^j - f_l$
Residue:	$f_l - L_l u_l^j$

Notice that *residue* and *defect* are the same but with opposite signs. *Under-relaxation* scheme with Jacobi method is given by

$$u_l^{j+1} = u_l^j - \Theta D_l^{-1}[L_l u_l^j - f_l] \quad \text{with} \quad \Theta \in (0, 1). \tag{7.58}$$

For our example, $D_l^{-1} = \frac{h_l^2}{2} I$, and if we denote $\omega = \left(\frac{\Theta}{2}\right)$, then from Equation (7.58)

$$\begin{aligned} u_l^{j+1} &= u_l^j - \omega h_l^2(L_l u_l^j - f_l) \\ &= u_l^j - \omega h_l^2 L_l u_l^j + \omega h_l^2 f_l \\ &= (I - \omega h_l^2 L_l)U_l^j + \omega h_l^2 f_l \\ &= M_l u_l^j + N_l f_l, \end{aligned}$$

where M_l is called the iteration matrix. This matrix is also called the *amplification matrix* G of the SLI defined in Section 7.4. The eigenvalues of L_l and M_l matrix are related. It is already shown earlier in Equation (7.17) that

$$u_l^j - u_l = [M_l]^j (u_l^0 - u_l),$$

where the superscript on amplification factor is the exponent; u_l^0 is the initial guess and u_l is the exact solution. Eigenvalues of M_l matrix are obtained using Equation (7.56) as

$$\lambda_\nu(\omega) = 1 - \omega h_l^2 \left(\frac{4}{h_l^2} \sin^2 \frac{\pi \nu h_l}{2}\right) = 1 - 4\omega \sin^2 \frac{\pi \nu}{2(N+1)}.$$

These are plotted in Fig. 7.3 for two values of ω.

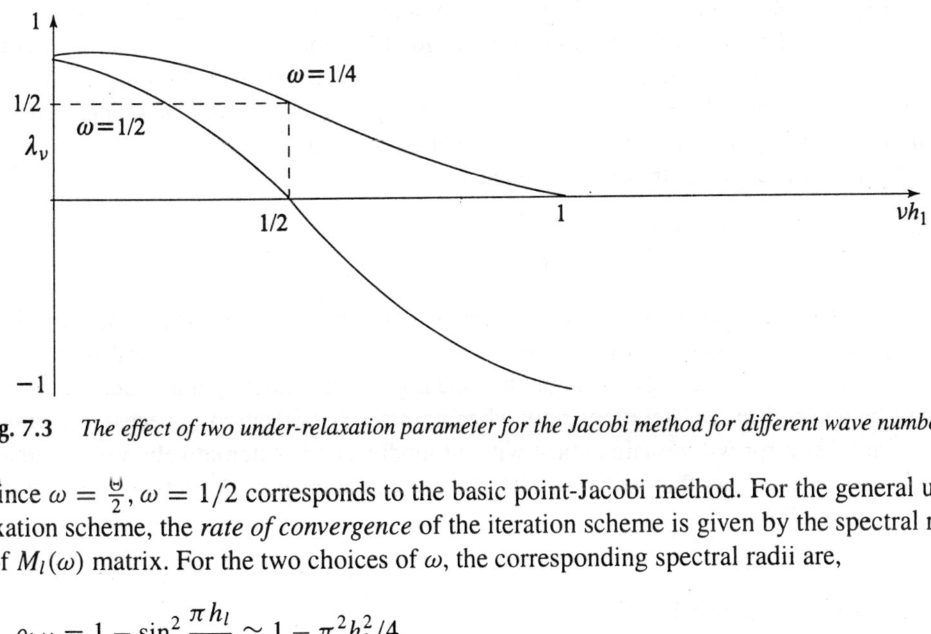

Fig. 7.3 *The effect of two under-relaxation parameter for the Jacobi method for different wave numbers.*

Since $\omega = \frac{\omega}{2}$, $\omega = 1/2$ corresponds to the basic point-Jacobi method. For the general under-relaxation scheme, the *rate of convergence* of the iteration scheme is given by the spectral radius ρ_ω of $M_l(\omega)$ matrix. For the two choices of ω, the corresponding spectral radii are,

$$\rho_{1/4} = 1 - \sin^2 \frac{\pi h_l}{2} \simeq 1 - \pi^2 h_l^2/4,$$

$$\rho_{1/2} = 1 - 2\sin^2 \frac{\pi h_l}{2} \simeq 1 - \pi^2 h_l^2/2.$$

Thus $\omega = 1/4$ needs twice as many iterations as for $\omega = 1/2$ case. From the previous figure, one can see that $\lambda_\nu \leq 1/2$ for $\nu h_l \geq 1/2$, i.e. the error components with wave numbers, $\nu \geq \frac{1}{2h_l}$, reduces by more than a factor of 1/2 per iteration, i.e., the smaller wave number damp rather slowly. Thus the classical relaxation processes are *local* and cannot efficiently treat *global* solution components. In the ADI method, this was attempted by choosing the accelerating parameter judiciously. However, in the ADI method this was done in the same finest grid for all the error components. In multi-grid methods, this has been alleviated by solving the problem in different grid sequences which are natural to suppressing the corresponding wave number components of error. Furthermore, notice that the *global* (small wave numbers) components show small residuals, but the individual errors are large. This is explained in Fig. 7.4.

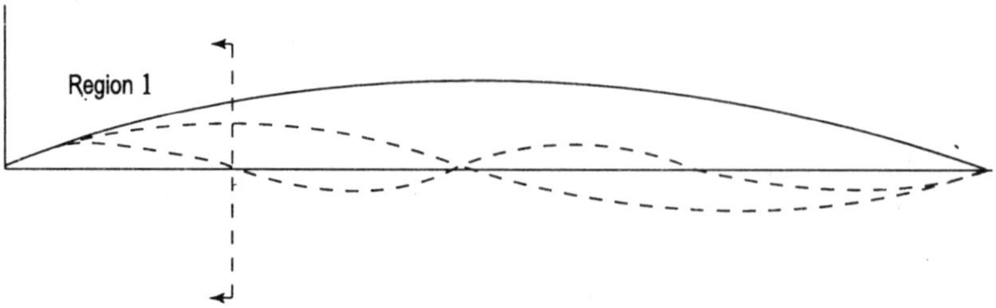

Fig. 7.4 *The effect of smoothing operation on different length scales by iterative methods.*

For example, in Region I all the three sketched components of error reinforce each other giving rise to large error. The preceding discussion holds good for linear boundary value problems. In case of nonlinear problems, the error spectrum will be refilled after each iteration. Even for linear problems in transformed plane, aliasing causes the error spectrum to be *refilled* after each iteration if it requires product evaluation. It is shown in Section 9.4 that the Laplacian operator (a linear operator) can be writen in the transformed plane as

$$\nabla^2 = \frac{1}{h_1 h_2}\left[\frac{\partial}{\partial \xi}\left(\frac{h_2}{h_1}\frac{\partial}{\partial \xi}\right) + \frac{\partial}{\partial \eta}\left(\frac{h_1}{h_2}\frac{\partial}{\partial \eta}\right)\right]. \tag{7.59}$$

Since h_1 and h_2 are functions of ξ and η, in general, one notices that this linear operator requires product evaluation. This product evaluation introduces aliasing error and spectral broadening at every iteration level. It is therefore apparent that the high wave number components of the initial error will be damped in few iterations only when no product evaluation is necessary, while the low wave number error will remain almost without modification. Alternatively, we state that the error $(u_l^\nu - u_l)$ is smoother than $(u^0 - u_l)$. Thus the iteration given by Equation (7.58) serves as a smoothing operation.

7.17 Two-Grid Method

We have so far seen that the damped Jacobi iteration with $\omega = 1/4$, i.e. $\Theta = 1/2$, gives an iteration sequence which is efficient in removing high wave number error components, and that convergence is particularly poor with respect to low wave numbers. Therefore one should combine this iteration with a second one which will effectively reduce the low wave number error components. Such a complimentary iteration procedure is obtained by working with a coarser grid with step size,

$$h_{l-1} = 2h_l.$$

We follow the calculation sequences for the two-grid method in the following fashion:

(a) Let us start with u^{old}.

(*b*) A small number of iterations with h_l-grid will result in $\bar{u}_l$. If $u_l = L_l^{-1} f_l$ is the exact solution, then $v_l = (\bar{u}_l - u_l)$ is smoother than $(u^{\text{old}} - u_l)$, where v_l is the exact correction.

(*c*) The defect is $d_l = L_l \bar{u}_l - f_l$. Now,

$$L_l v_l = L_l \bar{u}_l - L_l u_l = L\bar{u}_l - f_l.$$

Therefore,

$$L_l v_l = d_l.$$

Thus we need to solve this equation with v_l for a lot smoother function than u^{old}.

(*d*) Approximate the above problem $L_l v_l = d_l$ by a coarse grid equation

$$L_{l-1} v_{l-1} = d_{l-1}$$

where now d_{l-1} has to be chosen reasonably from the data available on d_l. A linear mapping r, called *restriction*, defines d_{l-1} by

$$d_{l-1} = r d_l.$$

The simplest possible choice for restriction is the trivial *injection* r_{inj} defined as

$$r_{\text{inj}}.d_l(x) = d_{l-1}(x) \quad \text{for all} \quad x \in \Omega_{l-1} \subset \Omega_l.$$

This is possible since coarse-grid points are also fine-grid points. Hackbusch (1985) noted that "although this is the easiest to perform, it has some disadvantages, specially when one uses a particular grid point ordering like red–black ordering. It is safer to take into account the components $d_l(x)$ at $x \in \Omega/\Omega_{l-1}$". He suggested using the *restriction*, r, given by,

$$[rd_l](x) = \frac{1}{4}[d_l(x - h_l) + 2d_l(x) + d_l(x + h_l)]$$

The corresponding matrix is

$$r = \frac{1}{4}\begin{bmatrix} 1 & 2 & 1 & \cdots & \cdots & \cdots & \cdots & \cdots & \cdots & \cdots & \cdots \\ \cdots & \cdots & 1 & 2 & 1 & \cdots & \cdots & \cdots & \cdots & \cdots & \cdots \\ \cdots & \cdots & \cdots & \cdots & 1 & 2 & 1 & \cdots & \cdots & \cdots & \cdots \\ \cdots & \cdots & \cdots & \cdots & \cdots & \cdots & \cdots & \cdots & \cdots & \cdots & \cdots \\ \cdots & \cdots & \cdots & \cdots & \cdots & \cdots & \cdots & \cdots & 1 & 2 & 1 \end{bmatrix}$$

It is a special weighted *restriction*. The implication of the above assertion from Hackbusch (1985) is further explored in Exercise 5, at the end of this chapter. It can be shown that the above operation is equivalent to adding second order dissipation. The red–black ordering gives rises to odd–even oscillations—an error type that is at highest wave numbers (at Nyquist limit) of that level. This will be smoothed out by the restriction operator.

(*e*) Having defined d_{l-1} from the *defect* d_l, we obtain

$$v_{l-1} = L_{l-1}^{-1} d_{l-1}$$

as the exact solution.

We expect v_{l-1} to be an approximation to the exact correction v_l. Since v_{l-1} is only defined on the coarse grid Ω_{l-1}, we have to interpolate this coarse grid function by

$$\tilde{v}_l = p v_{l-1},$$

where the *prolongation operator p* describes a coarse to fine interpolation. The simplest interpolation is the piecewise linear one:

$$[P_{l-1}^l v_{l-1}](x) = \begin{cases} v_{l-1}(x) & x \in \Omega_{l-1}, \\ [v_{l-1}(x - h_l) + v_{l-1}(x + h_l)]/2 & \text{otherwise.} \end{cases}$$

p can also be represented by the rectangular n_l X n_{l-1} matrix,

$$p_{l-1}^l = \frac{1}{2} \begin{bmatrix} 1 & \cdots & \cdots & \cdots & \cdots & \cdots \\ 2 & \cdots & \cdots & \cdots & \cdots & \cdots \\ 1 & 1 & \cdots & \cdots & \cdots & \cdots \\ \cdots & 2 & \cdots & \cdots & \cdots & \cdots \\ \cdots & 1 & 1 & \cdots & \cdots & \cdots \\ \cdots & \cdots & 2 & \cdots & \cdots & \cdots \\ \cdots & \cdots & \cdots & \cdots & \cdots & \cdots \\ \cdots & \cdots & \cdots & \cdots & \cdots & \cdots \\ \cdots & \cdots & 1 & \cdots & \cdots & 1 \\ \cdots & \cdots & \cdots & \cdots & \cdots & 2 \\ \cdots & \cdots & \cdots & \cdots & \cdots & 1 \end{bmatrix}.$$

The above structure of the matrix is due to the fact $v_{l-1}(0) = v_{l-1}(1) = 0$. One can also easily show that the prolongation operation implies implicit additional smoothing, in the same way the restriction operator smoothed the solution.

(*f*) Therefore one can construct the new solution at the fine grid level as,

$$u_l^{\text{new}} = \bar{u}_l - \tilde{v}_l.$$

The step from $\bar{u}_l$ to u_l^{new} is called the *coarse grid correction*. Hence one can summarise the whole process of iteration by the following steps.

$$\begin{aligned} u_l^{\text{new}} &= \bar{u}_l - \tilde{v}_l \\ &= \bar{u}_l - p v_{l-1} = \bar{u}_l - p[L_{l-1}^{-1} d_{l-1}] = \bar{u}_l - p[L_{l-1}^{-1}\{r d_l\}] = \bar{u}_l - p[L_{l-1}^{-1}\{r(L_l \bar{u}_l - f_l)\}]. \end{aligned}$$

See Fig. 7.5 for the error time-history with levels of iterations.

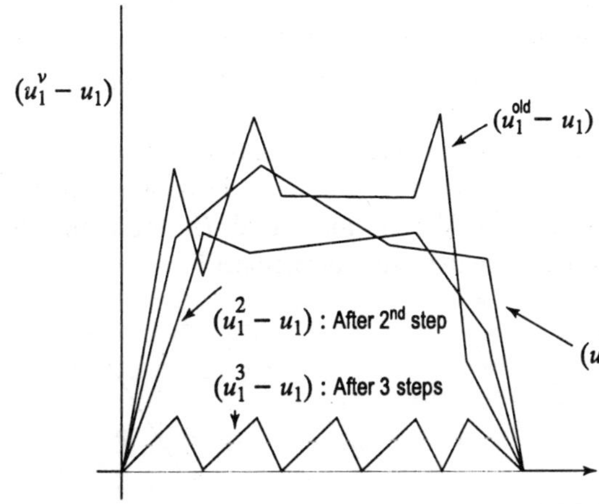

Fig. 7.5 *Error history as a function of smoothing iterations.*

Although coarse grid correction seems to be efficient, it cannot be used as an iteration by itself because (Hackbusch (1985)):

"The coarse grid correction is a non-convergent iteration."

It is a combination of smoothing iteration and coarse grid correction that is rapidly convergent, whereas both components by themselves converge slowly or not at all. The combination is called a *two-grid iteration* since only two levels are involved.

7.18 Multi-Grid Method

For the case of multidimensional boundary value problems, the two-grid iteration is impractical because the exact solution of $L_{l-1}v_{l-1} = d_{l-1}$ is required. The solution v_{l-1} is needed to compute the correction $\tilde{v}_l = pv_{l-1}$. Since $\tilde{v}_l$ is only an approximation to $v_l(= L_l^{-1}d_l)$, there is no need for the exact computation of first $\tilde{v}_l$ and then of v_{l-1}. It suffices to calculate an approximate solution $\tilde{v}_{l-1}$ of

$$L_{l-1}v_{l-1} = d_{l-1}.$$

For example, the approximation $\tilde{v}_{l-1}$ can be obtained by some iterative process:

$$v_{l-1}^0 = 0 \longmapsto v_{l-1}^1 \longmapsto \ldots \longmapsto v_{l-1}^\gamma = \tilde{v}_{l-1} \quad : \gamma = \text{number of iteration.}$$

Note that the coarse grid system is of the same form as the original problem

$$L_l u_l = f_l,$$

where one replaces l by $(l - 1)$ and f_l by d_l.

For $l > 2$, one can repeat the operation used in $(l, l-1)$ for the next grid levels $(l-1, l-2)$. This leads to three-grid method. This process can be repeated until all $(l+1)$ levels $\{l, l-1, l-2, \ldots\ldots 2, 1, 0\}$ are involved. Equations at level 0 must be solved exactly or approximated by some other iteration. Level 0 corresponds to the coarsest grid and therefore to the smallest system of equations.

The resulting $(l+1)$-level iteration is now called a *multi-grid iteration*. The sequence of operations during one step of multi-grid iteration (for $\gamma = 1$) is depicted in Fig. 7.6 for the example $l = 4$. This is also known as the V-cycle multi-grid. (In the figure: S-smoothing; R-restriction, E-exact solution at level 0; P-coarse grid correction.)

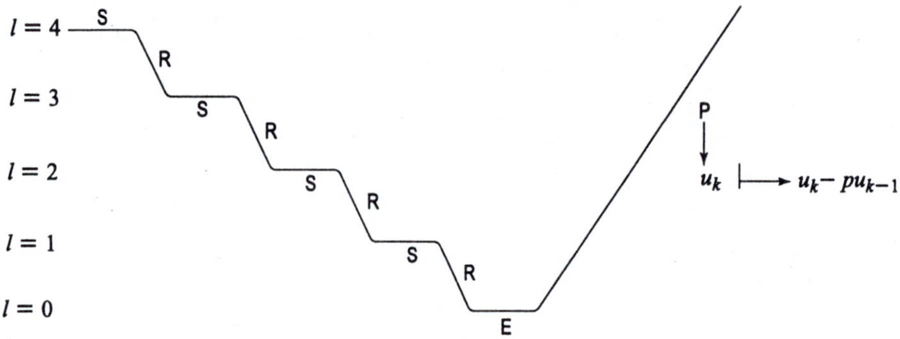

Fig. 7.6 *V-cycle multi-grid operation for $\gamma = 1$ and $l = 4$.*

Next, another illustration of the case when $\gamma = 2$ involving two iteration at level 3 is shown in Fig. 7.7.

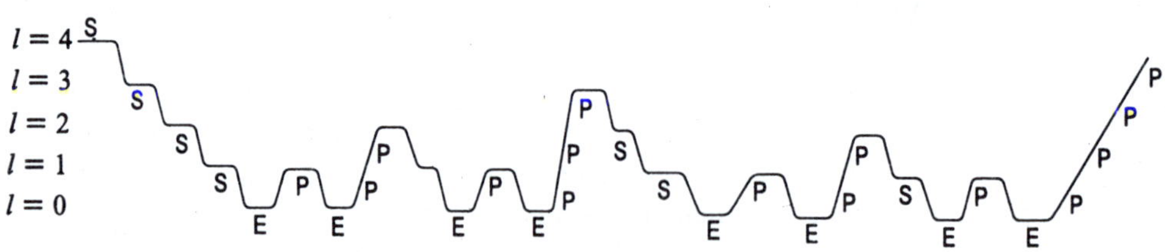

Fig. 7.7 *Multi-grid W-cycle with $\gamma = 2$ at level 3.*

The corresponding algorithm is called the W-cycle. Note that it has two iterations at level 3, four iterations at level 2 and eight iterations at level 1.

7.19 Other Classifications of Multi-Grid Methods

A multi-grid method is also classified according to whether one starts from the coarsest grid or the finest grid. The former is called the *full multi-grid* (FMG) method while the latter is called the *cycling algorithm*. In FMG one solves for the unknowns using multiple grid and hence is useful for both linear and nonlinear problems, while in *cycling algorithm* one solves for the corrections and hence apart from the initial guesses it is only the correction that is stored at each level of iteration. This method is suitable only for linear problems.

In this chapter only a few representative methods for solving elliptic PDE are discussed. In Chapter 13 we will discuss the conjugate gradient method of solving linear algebraic equation. In addition, there are many other methods that are currently in use. However, the basic principles are the same.

Exercise

1. Show that $\bar{N_J} = D$ and $\bar{N_{GS}} = (D + L)$ for Equations (7.8) and (7.10) corresponding to the Jacobi and Gauss-Seidel iteration.
2. Show that $G_S = I - (D + L)^{-1}A$ for the Gauss Seidel iteration.
3. Write down Equation (7.25) in the matrix notation in terms of $[A]$ and 'b' vector as,

$$\{u^{(k)}\} = (D + L\alpha)^{-1}[\{u^{(k-1)}[(1-\alpha)D - \alpha U] + \alpha b].$$

   Hence show the gain matrix as

$$G_\alpha = (D + \alpha L)^{-1}[(1-\alpha)D - \alpha U].$$

4. Draw the W-cycle when $\gamma = 2$ involving two iteration at level 2 for the multi-grid method.
5. Show that the application of restriction operator in multi-grid is equivalent to a simple addition of lowest order dissipation. What is the consequence of such an operator on the overall accuracy of the method?
6. What do you conclude about the prolongation operation used in the multi-grid method, in terms of accuracy of solutions.

References

[1] AMES, W.F., *Numerical Methods for Partial Differential Equation (2nd Edn.)*, Academic Press, New York, (1977).

[2] BEAM, R. M., WARMING, R.F., An implicit factored scheme for the compressible Navier - Stokes equation, *AIAA Journal*, **16**, 393–401, (1978).

[3] BIRKHOFF, G., VARGA, R.S., & YOUNG. D., In *Advances in Computer*, **3**, 189–273, (1962).

[4] HACKBUSCH, W., *Multigrid Methods and Applications*, Springer Verlag, Berlin, (1985).

[5] JACOBI, C. G. J., *Gessamelte Werke*, **3**, 467 Berlin, (1884).

[6] LYUSTERNIK, L.A., *Trudy. Mat. Inst. V.A. Steklova*, **20**, 49, (1947).

[7] MACCORMACK, R.W., A numerical method for solving the equations of compressible viscous flow. *AIAA*, paper no. **81–0110**, New York, (1981).

[8] PEACEMAN, D.W. & RACHFORD Jr., H.H., The numerical solution of parabolic and elliptic differential equations. *J. SIAM*, **3**, No.1, 28–41, (1955).

[9] RICHARDSON, L. F., *Weather Prediction by Numerical Process*, Dover Publication, (1965).

[10] SEIDEL, L., *Abh. Bayer Akad. Wiss.*, **II**, 81, (1874).

[11] VARGA, R.S., *Matrix Iterative Numerical Analysis*, Wiley, New York, (1962).

[12] WACHSPRESS, E.L., *Iterative Solution of Elliptic Systems*, Prentice Hall, Englewood Cliffs, New Jersy, (1966).

[13] WILKINSON, J.H., *Algebraic Eigenvalue Problem*, Oxford University Press, Oxford, (1971).

[14] YANENKO, N.N., *The Method of Fractional Steps: The Solution of Problems of Mathematical Physics in Several Variables* (Translation edited by M. Holt), Springer verlag, New York, (1971).

8 Solution of Hyperbolic PDEs

One of the classical method of solving hyperbolic PDEs is the method of characteristics. The beauty of the method of characteristic is, apart from being a graphical construction method (and hence easy to construct and visualise the solution) the solution obtained by this method is true for nonlinear differential equation without the need for quasi-linearization as required by computational methods. If the flow is truly hyperbolic, as in convergence-divergence nozzle flow downstream of sonic line and the flow is irrotational, then the solution is obtained very quickly. However, the method being a semi-graphical construction method, it does not allow formation of shock. Furthermore, if the flow is rotational then the solution method is iterative and is equivalent to solving the shock-free Euler equation, and is as computation intensive as any other modern CFD method. Hence, the CFD methods are more versatile because it can solve mixed-flow problem with shocks. Therefore we restrict our attention to CFD methods of solving hyperbolic equation.

Let us consider the simplest possible model equation representing a hyperbolic PDE:

$$\frac{\partial u}{\partial t} = -a\frac{\partial u}{\partial x}; \quad \text{for } a > 0. \tag{8.1}$$

The real characteristic of the equation is $x - at = \text{const}$. First let us discuss some representative classical methods of solving hyperbolic PDEs with the help of this model equation.

8.1 Explicit Methods

Forward-time-centered-space (FTCS) scheme: The scheme is given by

$$\frac{u_i^{n+1} - u_i^n}{\Delta t} = -a\frac{u_{i+1}^n - u_{i-1}^n}{2\Delta x}. \tag{8.2}$$

This being a linear equation, we can perform the spectral stability analysis. The spectral analysis method will be introduced in Chapter 10. But the essential introduction is already given in Section 4.1. In this analysis method, the dependent variable is represented by

$$u(x,t) = \int U(k,t)e^{ikx}dk$$

and the amplification factor or the gain is defined by

$$G(k) = \frac{U(k, t + \Delta t)}{U(k,t)}.$$

In the continuum sense ($\triangle t \to 0$) one expects

$$G \equiv 1.$$

Hence any good numerical should have the attribute of having a neutrally stable algorithm. The gain, $G(k)$, of the FTCS scheme is given by

$$G(k) = 1 - \sqrt{-1} N \sin k\triangle x. \tag{8.3}$$

In the expression $N = \frac{a\triangle t}{\triangle x}$ is the famous Courant-Friedrich-Lewy (CFL) number, Courant et al (1967). This signifies that this is an *unconditionally unstable method* since $|G| > 1$.

First order upwind difference method: Here the difference equation for Equation (8.1) is given by

$$\frac{u_i^{n+1} - u_i^n}{\triangle t} = -a \frac{u_i^n - u_{i-1}^n}{\triangle x}. \tag{8.4}$$

Note that the spatial derivative is replaced by a first order accurate one-sided difference formula. For Equation (8.1) the information propagates from left to right, hence we have used a backward difference formula.

$$G = 1 - N(1 - \cos k\triangle x) - iN \sin k\triangle x \tag{8.5a}$$

or $$|G|^2 = 1 + 4N(N-1)\sin^2 \frac{k\triangle x}{2}. \tag{8.5b}$$

Note that the choice of $N = 1$ produces the most appropriate result, i.e., $G \equiv 1$. Vanishing values of CFL number ($N \to 0$) also has the same feature.

Midpoint leap frog method (Courant et al (1967)): Here the time discretization is based on second order accurate central differencing and the corresponding difference equation is

$$\frac{u_i^{n+1} - u_i^{n-1}}{2\triangle t} = -a \frac{u_{i+1}^n - u_{i-1}^n}{2\triangle x}. \tag{8.6}$$

Since this is a three-time level scheme, the amplification factor G is given as the root of the quadratic equation

$$G^2 - 2iNG \sin k\triangle x - 1 = 0 \tag{8.7}$$

whose solution is given by

$$G_{1,2} = iN \sin k\triangle x \pm \sqrt{1 - N^2 \sin^2 k\triangle x}.$$

Once again, for $N \simeq 0$; $G_{1,2} \simeq 1$ and for $N \simeq 1$; $G_{1,2} = \pm[\cos k\triangle x \pm i \sin k\triangle x] = \pm e^{\pm ik\triangle x}$. Thus, this method also displays neutral stability like first order upwind scheme.

Lax-Wendroff method (1960): Here the finite difference equation (FDE) is obtained from the Taylor series expansion of the unknown given by

$$u_i^{n+1} = u_i^n + \frac{\partial u}{\partial t}\Delta t + \frac{\partial^2 u}{\partial t^2}\frac{\Delta t^2}{2} + 0(\Delta t^3).$$

Utilizing the governing differential Equation (8.1), we obtain

$$u_i^{n+1} = u_i^n - a\Delta t\frac{\partial u}{\partial x} + \frac{a^2\Delta t^2}{2}\frac{\partial^2 u}{\partial x^2} + \dots \tag{8.8}$$

Retaining the first three terms in the Taylor series expansion and using central differencing for the derivatives, we get the FDE as,

$$u_i^{n+1} = u_i^n - \frac{N}{2}[u_{i+1}^n - u_{i-1}^n] + \frac{N^2}{2}[u_{i+1}^n - 2u_i^n + u_{i-1}^n]. \tag{8.9}$$

The amplification factor is given by

$$|G|^2 = 1 + 4N^4 \sin^4\frac{k\Delta x}{2} - 4N^2 \sin^4\frac{k\Delta x}{2}. \tag{8.10}$$

8.2 Implicit Methods

Fully implicit FTCS method: Here the spatial derivative is evaluated at the current time step and then the difference equation is given by

$$\frac{N}{2}u_{i-1}^{n+1} - u_i^{n+1} - \frac{N}{2}u_{i+1}^{n+1} = -u_i^n. \tag{8.11}$$

We can write it as an implicit equation for the unknowns, taking three points at a time. When we write the FDE for all the unknown nodes and collate, we get a tri-diagonal matrix equation which can be solved exactly without the need for iteration, using Thomas algorithm.

Crank-Nicholson method (1947): Here the spatial derivative is evaluated at the mid-point $(n + 1/2)$-level and the FDE is given by

$$\frac{N}{4}u_{i+1}^{n+1} + u_i^{n+1} - \frac{N}{4}u_{i-1}^{n+1} = u_i^n - \frac{N}{4}(u_{i+1}^n - u_{i-1}^n). \tag{8.12}$$

This also constitutes a tridiagonal matrix equation and hence can be solved exactly without iteration. We can perform the spectral stability analysis, as performed on the heat equation.

Finally, we discuss a multi-step algorithm that has the predictor–corrector construct and was known for its relative superior performance at the time of its introduction. This method is due to MacCormack (1981).

MacCormack scheme: This two-stage scheme consists of the following steps:

(a) The predictor step: Here the spatial derivative is forward differenced and the time marching yields a intermediate *predicted value of the unknown* (u^*) and the difference equation is

$$\frac{u_i^* - u_i^n}{\Delta t} = -a\frac{u_{i+1}^n - u_i^n}{\Delta x}.$$

With the help of this predicted value, u^*, we can obtain the value of the unknown at the mid-point time level $(n + 1/2)$ as:

$$u_i^{n+1/2} = \frac{1}{2}(u_i^n + u_i^*).$$

(b) The corrector step: In this step the spatial derivative is backward differenced to yield the solution at the next time level.

$$\frac{u_i^{n+1} - u_i^{n+1/2}}{\Delta t/2} = -a\frac{u_i^* - u_{i-1}^*}{\Delta x}.$$

In terms of the CFL number, the above difference equations can be written as

$$\left.\begin{aligned} u_i^* &= u_i^n - N(u_{i+1}^n - u_i^n); \\ u_i^{n+1} &= \frac{1}{2}[(u_i^n + u_i^*) - N(u_i^* - u_{i-1}^*)] \end{aligned}\right\}. \tag{8.13}$$

It is to be noted that this method is identical to the Lax-Wendroff method if the governing differential equation is linear. It is only for nonlinear equation that the superior performance of the MacCormack scheme is noted.

General characteristics of various methods for linear problems

(a) First order accurate methods have built-in dissipation.

(b) In general, hyperbolic equations are characterized by wave nature and hence dispersion is a very important property. We will discuss about it in greater detail later. Higher order methods, without dissipation are dispersive.

(c) Note that the presence of dissipation has the property of reducing the intensity of dispersion error but can seriously affect dispersion property, as will be shown in Chapter 10.

(d) Implicit methods are more stable and hence allow taking larger time steps; but implicit solutions have larger error. This was also found to be the case with the parabolic equation solution method.

8.3 Nonlinear Problems

As an illustration consider the Burgers equation given by

$$\frac{\partial u}{\partial t} = -u\frac{\partial u}{\partial x} = -\frac{\partial E}{\partial x}. \tag{8.14}$$

where $E = \left(\frac{u^2}{2}\right)$.

We have already pointed out that MacCormack's method produces very good quality results for nonlinear problems. Hence we will discuss this method for Equation (8.14). Here the following steps are used:

Predictor step: $$u_i^* = u_i^n - \frac{\Delta t}{\Delta x}(E_{i+1}^n - E_i^n)$$

Corrector step: $$u_i^{n+1} = \frac{1}{2}\left[u_i^n + u_i^* - \frac{\Delta t}{\Delta x}(E_i^* - E_{i-1}^*)\right]$$

For nonlinear problems this method is not same as the Lax-Wendroff method. Also notice that the MacCormack's method by its very structure is amenable to conservative differencing and would be preferable.

We will discuss the recent methods of computing hyperbolic PDEs in Chapter 11. One of the major issue of solving hyperbolic PDEs is capturing discontinuous solutions, as one can see from the solution of Equation (8.14). Such discontinuites or hydrodynamic shocks are captured by special methods. The subject area is huge and is worthy of a seperate volume and is not attempted here.

Exercise

1. By inspection of Equation (8.10) can you draw any conclusions about the stability of this algorithm? What are the actual amplification factors?
2. Looking at the amplification factors of Exercise 1, what do you conclude for the case of $N = 1$ & $N \simeq 0$?
3. Plot the amplification factor for the Lax-Wendroff method given by Equation (8.10) for different values of N.
4. Obtain the amplification factors for the FDEs given by Equations (8.11) and (8.12).

References

[1] COURANT, R., FRIEDRICH, K. & LEWY, H., On the partial differential equations of mathematical physics, *IBM Journal*, 215–234 (1967).

[2] CRANK, J & NICHOLSON, P., A practical method for numerical evaluation of solutions of partial differential equations of the heat conduction type, *Proc. of Camb. Phil. Soc.*, **43(50)**, pp 50–67 (1947).

[3] LAX, P.D. & WENDROFF, B., Systems of conservations laws, *Comm. on Pure and App. Math*, **13**, 217–237 (1960).

[4] MACCORMACK, R.W., A numerical method for solving the equations of compressible viscous flow, *AIAA*, paper no. **81–0110**, New York (1981).

9 Curvillinear Coordinates and Grid Generation

In general, for solving the governing equation of motions in CFD we need to resolve all the excited length and time scales as was discussed in Chapter 4. However, even for flow past a flat plate, if one takes a uniform grid to accomodate the smallest energy-carrying length scales, one would be forced to take too many grid points. The solution of the Euler and Navier-Stokes equations for external flow problems requires the outer boundary to be far away where the asymptotic boundary conditions apply, and this puts even a larger requirment on the grid. Apart from the fact that such problems are poorly convergent, it may not also be within the range of the available storage of the largest supercomputer. This can be circumvented by using non-uniform grids. For example, we may decide to take finer grids inside the shear layer (or within the inner layer of a turbulent boundary layer), and take fewer points in the inviscid part of the flow. And the moment we decide to use non-uniform grid in the physical plane, we usually transform the governing equation to a transformed plane where the spacing is uniform. This also makes writing a code (that actually solves the problem) much easier because of the uniform grid in the transformed plane. Then one needs to write a single code for the transformed plane problem, and for the flow past different geometries we can use the same code provided we generate the grid mapping from the physical plane to the transformed plane. Thus the main solver is grid-independent, and all one needs to do is to generate the appropriate grid transformation either analytically or numerically.

There is another important reason for us to go from a simple Cartesian frame of reference to a complicated curvilinear coordinate system. This is due to the fact that accurate solution of viscous flow problems requires accurate satisfication of the boundary conditions. Thus we require a coordinate transformation such that the boundary line is one of the grid lines. This is the motivation for *body-conforming* grid generation that will be discussed in this chapter.

There is a third reason for which we require grid transformation. For some problems the boundary may move with time. Therefore, the accurate satisfaction of boundary condition would require generating body-conforming grid at each time instant, and this may be prohibitive computationally, or the accuracy of the solution may not be good enough. However, it is sometimes possible to arrive at a transformation in such a way that the boundary conditions in the transformed plane is time-independent.

The above mentioned issues provide sufficient motivation for performing a transformation by specifying a generalised coordinate system that maps a curvilinear system in the physical plane to a rectangular uniform grid spacing in the computational domain. Figures 9.1 and 9.2 show such transformations used for computation.

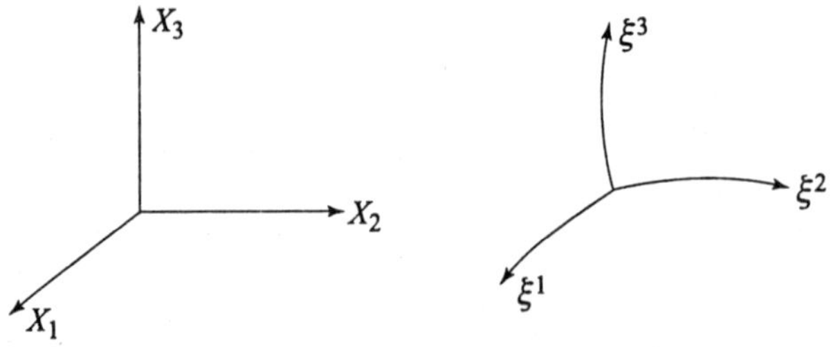

Fig. 9.1 *The Cartesian and a corresponding curvilinear coordinates system.*

The body-fitted coordinate system shown in Fig. 9.2 is such that the ξ- and ζ- lines are parallel to the surface of the body along the stream-wise and the circumferential direction while the third direction, η, is normal to the surface. Additionally, the following issues are found to be important for solving equations of motion in the transformed plane.

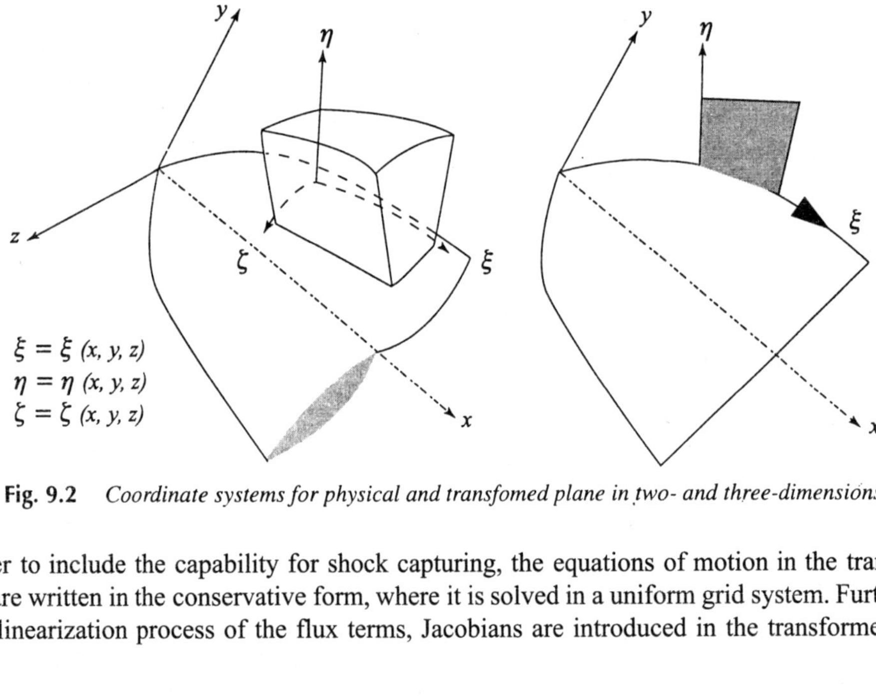

Fig. 9.2 *Coordinate systems for physical and transfomed plane in two- and three-dimensions.*

In order to include the capability for shock capturing, the equations of motion in the transformed plane are written in the conservative form, where it is solved in a uniform grid system. Furthermore, in the linearization process of the flux terms, Jacobians are introduced in the transformed plane.

9.1 Generalised Curvilinear System

Here we provide a brief review of the general curvilinear coordinate system. The details can be found in Aris (1962) or any standard book on mathematical physics. Consider the curvilinear coordinate system as shown in Fig. 9.1 given by,

$$\xi^i = \xi^i(x_1, x_2, x_3)$$

such that

$$d\xi^i = \left(\frac{\partial \xi^i}{\partial X_k}\right) dX_k.$$

The quantity in the parenthesis above, is called the Jacobian of the transformation or metrics. If $\mathbf{l}_j$'s are the unit vectors in the Cartesian coordinate system, then any vector $\mathbf{r}$ can be written as

$$\mathbf{r} = X_j \mathbf{l}_j$$

with X_j's as the Cartesian components. For the curvilinear system ξ^i, the base vector $\mathbf{e}_i$ is defined by

$$\mathbf{e}_i = \frac{\partial \mathbf{r}}{\partial \xi^i}.$$

Notice that these are not unit vectors. These can also be written as

$$\mathbf{e}_i = \frac{\partial}{\partial \xi^i}(X_j \mathbf{l}_j).$$

And thus,

$$\mathbf{e}_i.\mathbf{e}_j = \left(\frac{\partial X_k}{\partial \xi^i}\mathbf{l}_k\right).\left(\frac{\partial X_m}{\partial \xi^j}\mathbf{l}_m\right)$$

$$= \frac{\partial X_k}{\partial \xi^i}\frac{\partial X_m}{\partial \xi^j}\delta_{km} = \frac{\partial X_k}{\partial \xi^i}\frac{\partial X_k}{\partial \xi^j} = g_{ij}.$$

g_{ij}'s are the metrics and has six components as they are symmetric. δ_{ij} is the Kronecker delta, that is equal to one for $i = j$ and are zero otherwise.

9.2 Reciprocal or Dual Base Vectors

The dual base vectors $\mathbf{e}^i$ are defined by

$$\mathbf{e}_j.\mathbf{e}^i = \delta_{ij}$$

i.e., the dual is orthogonal to the two components of base vectors whose indices are different. Note that in general $\mathbf{e}^i$ is not orthogonal to $\mathbf{e}_j$. In the curvilinear system, the gradient of a vector (which is now a tensor) is defined as

$$\nabla() = \mathbf{e}^k \frac{\partial}{\partial \xi^k}()$$

Therefore,

$$\nabla(\xi^k) = \mathbf{e}^k \frac{\partial}{\partial \xi^k}(\xi^k) = \mathbf{e}^k,$$

and in the Cartesian frame,

$$\mathbf{e}^k = l_m \frac{\partial \xi^k}{\partial X_m}.$$

Hence

$$\mathbf{e}^i.\mathbf{e}^j = \frac{\partial \xi^i}{\partial X_k}\frac{\partial \xi^j}{\partial X_m}\delta_{km} = g^{ij}.$$

Again, g^{ij} is a symmetric tensor. The metrics of the base and reciprocal vectors are related by

$$g_{ij}g^{jk} = \left(\frac{\partial X_k}{\partial \xi^i}\frac{\partial X_k}{\partial \xi^j}\right)\left(\frac{\partial \xi^j}{\partial X_m}\frac{\partial \xi^k}{\partial X_m}\right)$$

$$= \frac{\partial X_k}{\partial \xi^i}\frac{\partial X_k}{\partial X_m}\frac{\partial \xi^k}{\partial X_m} = \frac{\partial X_k}{\partial \xi^i}\frac{\partial \xi^k}{\partial X_k} = \frac{\partial \xi^k}{\partial \xi^i} = \delta_{ik}.$$

Now, any vector $\mathbf{V}$ can be expressed by its components in either the base or the dual system. When expressed in the base system the components are called the *contravariant components.* In the dual system these are called the *covariant components.* Therefore

$$\mathbf{V} = V^i\mathbf{e}_i = V_i\mathbf{e}^i.$$

It is easy to see that

$$V^j = V_i g^{ij} \quad \text{and} \quad V_i = V^j g_{ij}.$$

Thus the base and dual vectors are related by similar relations as given above.

9.3 Geometric Interpretation of Metrics

In physical space, the arc length $(\triangle s)$ between two points is given by

$$\triangle s^2 = \sum_{k=1}^{3}\triangle X_k \triangle X_k,$$

where

$$\triangle X_k = \frac{\partial X_k}{\partial \xi^i}\triangle \xi^i.$$

Therefore,

$$\triangle s^2 = \sum_{k=1}^{3}\left(\frac{\partial X_k}{\partial \xi^k}\triangle \xi^i\right)\left(\frac{\partial X_k}{\partial \xi^j}\triangle \xi^j\right) = g_{ij}\triangle \xi^i \triangle \xi^j,$$

where

$$g_{ij} = \sum_{k=1}^{3}\frac{\partial X_k}{\partial \xi^i}\frac{\partial X_k}{\partial \xi^j}.$$

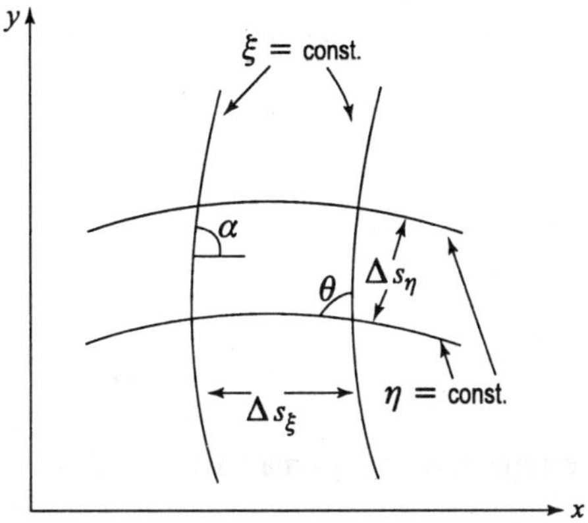

Fig. 9.3 *Geometrical intrepetation of metrices of transformation.*

In the physical plane if Δs_ξ and Δs_η are the distances between two neighbouring η = const. and ξ = const. lines respectively (as shown in Fig. 9.3 for a two-dimensional map), then

$$\Delta s_\eta = (x_\eta^2 + y_\eta^2)^{1/2}\Delta\eta = \sqrt{g_{22}}\Delta\eta$$

and

$$\Delta s_\xi = (x_\xi^2 + y_\xi^2)^{1/2}\Delta\xi = \sqrt{g_{11}}\Delta\xi.$$

In the transformed plane, the area of the cell is $(\Delta\xi\,\Delta\eta)$ and the corresponding area in the physical space is given by

$$\text{Area} = \|g\|^{1/2}\Delta\xi\,\Delta\eta.$$

The included angle between the ξ = const. line and the x-axis is given by

$$\cos\alpha = \frac{x_\xi}{\sqrt{g_{11}}}.$$

The ratio

$$\text{AR} = \frac{\Delta s_\eta}{\Delta s_\xi} = \sqrt{\frac{g_{22}}{g_{11}}}$$

is called the cell aspect ratio in the physical plane. The ease or difficulty of computations in CFD is determined by the magnitude of the cell aspect ratio in the physical plane. Higher is the value, more difficult is the convergence for iterative solver, or it will exhibit numerical instability for evolution equations. This is because the eigenvalue spectrum of the corresponding linear algebraic system is determined by this ratio.

The angle between the curvilinear coordinate lines is given by

$$\cos\theta = \frac{g_{12}}{\sqrt{g_{11}g_{22}}}$$

for the general curvilinear coordinate system in two dimensions. The Laplacian operator for a general curvilinear system is given by

$$\nabla^2 = \frac{1}{\sqrt{g}}\frac{\partial}{\partial\xi^i}\left[\sqrt{g}\; g^{ij}\frac{\partial}{\partial\xi^j}\right],$$

where g is defined as the determinant of the $[g]$ matrix.

9.4 Orthogonal Grid System

It is a grid system in which the base vector, $\mathbf{e}_i$'s form an orthogonal triad, i.e.,

$$\mathbf{e}_i.\mathbf{e}_j = K\delta_{ij}.$$

In this system we normalize the base vector $\mathbf{e}_i$'s such that,

$$\mathbf{e}_i = \frac{\partial\mathbf{r}}{\partial\xi^i} = \left|\frac{\partial\mathbf{r}}{\partial\xi^i}\right|\underline{\mathbf{e}}_i = h_i\underline{\mathbf{e}}_i.$$

h_i's are called the scale factors and $\underline{\mathbf{e}}_i$ the corresponding unit vector. By definition

$$g_{ij} = h_i h_j \underline{\mathbf{e}}_i.\underline{\mathbf{e}}_j$$

Thus $g_{ij} = 0$ for $i \neq j$ and is equal to h_{ii}^2 for $i = j$, and $g^{ii} = 1/h_{ii}^2$. Also for the orthogonal coordinate system, $g_{12} = 0$; $g_{11} = h_1^2$; $g_{22} = h_2^2$ and $\sqrt{g} = h_1 h_2$.

Thus the Laplacian operator for a three-dimensional mapping is given by

$$\nabla^2 = \frac{1}{h_1h_2h_3}\left\{\frac{\partial}{\partial\xi}\left[\frac{h_2h_3}{h_1}\frac{\partial}{\partial\xi}\right] + \frac{\partial}{\partial\eta}\left[\frac{h_3h_1}{h_2}\frac{\partial}{\partial\eta}\right] + \frac{\partial}{\partial\zeta}\left[\frac{h_1h_2}{h_3}\frac{\partial}{\partial\zeta}\right]\right\}.$$

For a two-dimensional map this simplifies to

$$\nabla^2 = \frac{1}{h_1h_2}\left\{\frac{\partial}{\partial\xi}\left(\frac{h_2}{h_1}\frac{\partial}{\partial\xi}\right) + \frac{\partial}{\partial\eta}\left(\frac{h_1}{h_2}\frac{\partial}{\partial\eta}\right)\right\}.$$

In a general mapping problem the singularity appears wherever det $[g]$ vanishes. This can become a problem for a general non-orthogonal transformation. But for orthogonal maps since $g_{12} = 0$, this problem is completely circumvented.

Furthermore, note that if $h_1 = h_2$, then the corresponding map is called the *conformal map*. If one can obtain a conformal map then it is ideal because in such a case we have an aspect ratio of one, thereby the solution process is highly stabilized and it becomes highly convergent. But the equality of scale factors also puts a high premium on the number of grid lines, and hence may not be suitable for viscous flow problems where we know the flow gradients are different in the stream-wise and transverse direction. Furthermore, conformal maps are only possible in two dimensions. Some of these problems are not present in orthogonal grids and their use is recommended whenever possible.

9.5 Generalised Coordinate Transformation

The use of above ideas necessitates that we transform the relevant equations of motion from the physical plane to the transformed plane. The equation of motion is transformed from the physical space (x, y, z) to the computational space (ξ, η, ζ) by the following relations:

$$\tau = t \tag{9.1a}$$

$$\xi = \xi(t, x, y, z) \tag{9.1b}$$

$$\eta = \eta(t, x, y, z) \tag{9.1c}$$

$$\zeta = \zeta(t, x, y, z) \tag{9.1d}$$

The chain rule of partial differentials relate the derivatives in the physical and transformed plane by

$$\frac{\partial}{\partial t} = \frac{\partial}{\partial \tau} + \xi_t \frac{\partial}{\partial \xi} + \eta_t \frac{\partial}{\partial \eta} + \zeta_t \frac{\partial}{\partial \zeta} \tag{9.2a}$$

$$\frac{\partial}{\partial x} = \xi_x \frac{\partial}{\partial \xi} + \eta_x \frac{\partial}{\partial \eta} + \zeta_x \frac{\partial}{\partial \zeta} \tag{9.2b}$$

$$\frac{\partial}{\partial y} = \xi_y \frac{\partial}{\partial \xi} + \eta_y \frac{\partial}{\partial \eta} + \zeta_y \frac{\partial}{\partial \zeta} \tag{9.2c}$$

$$\frac{\partial}{\partial z} = \xi_z \frac{\partial}{\partial \xi} + \eta_z \frac{\partial}{\partial \eta} + \zeta_z \frac{\partial}{\partial \zeta}. \tag{9.2d}$$

9.6 Equations for the Metrics

Detailed expressions for various metrics and transformed equations can be obtained from Hoffmann and Chang (1998). From equations (9.2*a*) through (9.2*d*), the value of the metrics ξ_t, η_t, ζ_t, ξ_x, η_x, ζ_x, ξ_y, η_y, ζ_y, ξ_z, η_z and ζ_z must be obtained analytically or numerically.

From equations (9.1*a*) to (9.1*d*),

$$dt = d\tau, \tag{9.3}$$

$$dx = x_\tau d\tau + x_\xi d\xi + x_\eta d\eta + x_\zeta d\zeta \tag{9.4a}$$

$$dy = y_\tau d\tau + y_\xi d\xi + y_\eta d\eta + y_\zeta d\zeta \tag{9.4b}$$

$$dz = z_\tau d\tau + z_\xi d\xi + z_\eta d\eta + z_\zeta d\zeta. \tag{9.4c}$$

These equations can also be expressed in a matrix form as

$$\begin{bmatrix} dt \\ dx \\ dy \\ dz \end{bmatrix} = \begin{bmatrix} 1 & 0 & 0 & 0 \\ x_\tau & x_\xi & x_\eta & x_\zeta \\ y_\tau & y_\xi & y_\eta & y_\zeta \\ z_\tau & z_\xi & z_\eta & z_\zeta \end{bmatrix} \begin{bmatrix} d\tau \\ d\xi \\ d\eta \\ d\zeta \end{bmatrix}. \tag{9.5}$$

Reversing the role of the independent and dependent variables, one can write

$$d\tau = dt \tag{9.6a}$$

$$d\xi = \xi_t dt + \xi_x dx + \xi_y dy + \xi_z dz \tag{9.6b}$$

$$d\eta = \eta_t dt + \eta_x dx + \eta_y dy + \eta_z dz \tag{9.6c}$$

$$d\zeta = \zeta_t dt + \zeta_x dx + \zeta_y dy + \zeta_z dz \tag{9.6d}$$

which are also expressed as

$$\begin{bmatrix} d\tau \\ d\xi \\ d\eta \\ d\zeta \end{bmatrix} = \begin{bmatrix} 1 & 0 & 0 & 0 \\ \xi_t & \xi_x & \xi_y & \xi_z \\ \eta_t & \eta_x & \eta_y & \eta_z \\ \zeta_t & \zeta_x & \zeta_y & \zeta_z \end{bmatrix} \begin{bmatrix} dt \\ dx \\ dy \\ dz \end{bmatrix}. \tag{9.7}$$

Comparing Equations (9.5) and (9.7), one obtains

$$\begin{bmatrix} 1 & 0 & 0 & 0 \\ \xi_t & \xi_x & \xi_y & \xi_z \\ \eta_t & \eta_x & \eta_y & \eta_z \\ \zeta_t & \zeta_x & \zeta_y & \zeta_z \end{bmatrix} = \begin{bmatrix} 1 & 0 & 0 & 0 \\ x_\tau & x_\xi & x_\eta & x_\zeta \\ y_\tau & y_\xi & y_\eta & y_\zeta \\ z_\tau & z_\xi & z_\eta & z_\zeta \end{bmatrix}^{-1}.$$

This allows one to relate the various partial derivatives of the transformation as

$$\xi_x = J(y_\eta z_\zeta - y_\zeta z_\eta) \tag{9.8a}$$

$$\xi_y = J(x_\zeta z_\eta - x_\eta z_\zeta) \tag{9.8b}$$

$$\xi_z = J(x_\eta y_\zeta - x_\zeta y_\eta) \tag{9.8c}$$

$$\eta_x = J(y_\zeta z_\xi - y_\xi z_\zeta) \tag{9.8d}$$

$$\eta_y = J(x_\xi z_\zeta - x_\zeta z_\xi) \tag{9.8e}$$

$$\eta_z = J(x_\zeta y_\xi - x_\xi y_\zeta) \tag{9.8f}$$

$$\zeta_x = J(y_\xi z_\eta - y_\eta z_\xi) \tag{9.8g}$$

$$\zeta_y = J(x_\eta z_\xi - x_\xi z_\eta) \tag{9.8h}$$

$$\zeta_z = J(x_\xi y_\eta - x_\eta y_\zeta) \tag{9.8i}$$

$$\xi_t = -(x_\tau \xi_x + y_\tau \xi_y + z_\tau \xi_z) \tag{9.8j}$$

$$\eta_t = -(x_\tau \eta_x + y_\tau \eta_y + z_\tau \eta_z) \tag{9.8k}$$

$$\zeta_t = -(x_\tau \zeta_x + y_\tau \zeta_y + z_\tau \zeta_z). \tag{9.8l}$$

Substituting Equations (9.8a) through (9.8i) into Equations (9.8j), (9.8k) and (9.8l), one obtains

$$\xi_t = J[x_\tau(y_\zeta z_\eta - y_\eta z_\zeta) + y_\tau(x_\eta z_\zeta - x_\zeta z_\eta) + z_\tau(x_\zeta y_\eta - x_\eta y_\zeta)], \tag{9.9a}$$

$$\eta_t = J[x_\tau(y_\eta z_\zeta - y_\zeta z_\xi) + y_\tau(x_\zeta z_\xi - x_\xi z_\zeta) + z_\tau(x_\xi y_\zeta - x_\zeta y_\xi)], \tag{9.9b}$$

$$\zeta_t = J[x_\tau(y_\eta z_\xi - y_\xi z_\eta) + y_\tau(x_\xi z_\eta - x_\eta z_\xi) + z_\tau(x_\eta y_\xi - x_\xi y_\eta)]. \tag{9.9c}$$

where J is the Jacobian of the transformation defined by

$$J = \frac{\partial(\xi, \eta, \zeta)}{\partial(x, y, z)} = \frac{1}{x_\xi(y_\eta z_\zeta - y_\zeta z_\eta) - x_\eta(y_\xi z_\zeta - y_\zeta z_\xi) + x_\zeta(y_\xi z_\eta - y_\eta z_\xi)}. \tag{9.10}$$

Example: For a two-dimensional map, evaluate the Laplacian in the transformed plane.

Let $A = \frac{\partial}{\partial x} = \frac{\partial}{\partial \xi}\xi_x + \frac{\partial}{\partial \eta}\eta_x$.

Therefore,

$$\frac{\partial^2}{\partial x^2} = \frac{\partial A}{\partial x} = \xi_x \frac{\partial}{\partial \xi}\left(\xi_x \frac{\partial}{\partial \xi} + \eta_x \frac{\partial}{\partial \eta}\right) + \eta_x \frac{\partial}{\partial \eta}\left(\xi_x \frac{\partial}{\partial \xi} + \eta_x \frac{\partial}{\partial \eta}\right),$$

where

$$\frac{\partial^2}{\partial \xi \partial x} = \frac{\partial}{\partial x}\left(\frac{\partial}{\partial \xi}\right) = \left(\xi_x \frac{\partial}{\partial \xi} + \eta_x \frac{\partial}{\partial \eta}\right)\frac{\partial}{\partial \xi} = \xi_x \frac{\partial^2}{\partial \xi^2} + \eta_x \frac{\partial^2}{\partial \xi \partial \eta},$$

and

$$\frac{\partial^2}{\partial \eta \partial x} = \xi_x \frac{\partial^2}{\partial \xi \partial \eta} + \eta_x \frac{\partial^2}{\partial \eta^2}.$$

Using these two relations we finally can write

$$\frac{\partial^2}{\partial x^2} = \xi_{xx}\frac{\partial}{\partial \xi} + \eta_{xx}\frac{\partial}{\partial \eta} + \xi_x^2 \frac{\partial^2}{\partial \xi^2} + \eta_x^2 \frac{\partial^2}{\partial \eta^2} + 2\xi_x \eta_x \frac{\partial^2}{\partial \xi \partial \eta}.$$

Similarly,

$$\frac{\partial^2}{\partial y^2} = \xi_{yy}\frac{\partial}{\partial \xi} + \eta_{yy}\frac{\partial}{\partial \eta} + \xi_y^2 \frac{\partial^2}{\partial \xi^2} + \eta_y^2 \frac{\partial^2}{\partial \eta^2} + 2\xi_y \eta_y \frac{\partial^2}{\partial \xi \partial \eta}.$$

Finally we can therefore write down the Laplacian operator as

$$\nabla^2 = [\xi_{xx} + \xi_{yy}]\frac{\partial}{\partial \xi} + [\eta_{xx} + \eta_{yy}]\frac{\partial}{\partial \eta} + [\xi_x^2 + \xi_y^2]\frac{\partial^2}{\partial \xi^2} + [\eta_x^2 + \eta_y^2]\frac{\partial^2}{\partial \eta^2} + 2[\xi_x \eta_x + \xi_y \eta_y]\frac{\partial^2}{\partial \xi \partial \eta}.$$

9.7 Navier-Stokes Equation in Transformed Plane

Here, by Navier-Stokes equation we will imply the complete set of equations for mass, momentum and energy as given by Equation (2.36) in the Cartesian frame. If we want to solve the Navier–Stokes equation in the transformed plane, we need to transform the equations also to this plane. This is derived next. Here we will rewrite the conservation equation in a slightly different form, as given by the nondimensional form in the Cartesian frame,

$$\frac{\partial Q}{\partial t}+\frac{\partial E}{\partial x}+\frac{\partial F}{\partial y}+\frac{\partial G}{\partial z}=\left\{\frac{\partial E_v}{\partial x}+\frac{\partial F_v}{\partial y}+\frac{\partial G_v}{\partial z}\right\}\frac{1}{Re},$$

where the state vector and non-dimensional flux terms are given in terms of the density, total internal energy, pressure and velocity components in the Cartesian frame as

$$\begin{aligned}
Q &= \lfloor \rho \;\; \rho u \;\; \rho v \;\; \rho w \;\; \rho e_t \rfloor^T \\
E &= \lfloor \rho u \;\; \rho u^2+p \;\; \rho uv \;\; \rho uw \;\; (\rho e_t+p)u \rfloor^T \\
F &= \lfloor \rho v \;\; \rho uv \;\; \rho v^2+p \;\; \rho vw \;\; (\rho e_t+p)v \rfloor^T \\
G &= \lfloor \rho w \;\; \rho uw \;\; \rho vw \;\; \rho w^2+p \;\; (\rho e_t+p)w \rfloor^T \\
E_v &= \lfloor 0 \;\; \tau_{xx} \;\; \tau_{xy} \;\; \tau_{xz} \;\; R \rfloor^T \\
F_v &= \lfloor 0 \;\; \tau_{xy} \;\; \tau_{yy} \;\; \tau_{yz} \;\; S \rfloor^T \\
G_v &= \lfloor 0 \;\; \tau_{xz} \;\; \tau_{yz} \;\; \tau_{zz} \;\; T \rfloor^T
\end{aligned}$$

with

$$\begin{aligned}
R &= u\tau_{xx}+v\tau_{xy}+w\tau_{xz}+\frac{\kappa}{P_r(\gamma-1)}\frac{\partial}{\partial x}(a^2) \\
S &= u\tau_{xy}+v\tau_{yy}+w\tau_{yz}+\frac{\kappa}{P_r(\gamma-1)}\frac{\partial}{\partial y}(a^2) \\
T &= u\tau_{xz}+v\tau_{zy}+w\tau_{zz}+\frac{\kappa}{P_r(\gamma-1)}\frac{\partial}{\partial z}(a^2)
\end{aligned}$$

where κ is the coefficient of thermal conductivity,

$P_r = \dfrac{\mu c_p}{\kappa}$ is the Prandtl number,

$\gamma = \dfrac{c_p}{c_v}$, the ratio of specific heat,

$a = \sqrt{\gamma RT}$ is the speed of sound,

$e_t = e_{\text{int}} + \dfrac{u^2+v^2+w^2}{2}$ and $e_{\text{int}} = C_v T$.

The Cartesian derivatives are replaced by Equation (9.2a) through (9.2d) to yield

$$\begin{aligned}
&\partial_\tau Q + \xi_t \partial_\xi Q + \eta_t \partial_\eta Q + \zeta_t \partial_\zeta Q + \xi_x \partial_\xi E + \eta_x \partial_\eta E + \zeta_x \partial_\zeta E \\
&\quad + \xi_y \partial_\xi F + \eta_y \partial_\eta F + \zeta_y \partial_\zeta F + \xi_z \partial_\xi G + \eta_z \partial_\eta G + \zeta_z \partial_\zeta G \\
&\quad = \frac{1}{Re}\{\xi_x \partial_\xi E_v + \eta_x \partial_\eta E_v + \zeta_x \partial_\zeta E_v + \xi_y \partial_\xi F_v + \eta_y \partial_\eta F_v \\
&\qquad + \zeta_y \partial_\zeta F_v + \xi_z \partial_\xi G_v + \eta_z \partial_\eta G_v + \zeta_z \partial_\zeta G_v\}.
\end{aligned} \tag{9.11}$$

In this equation we have denoted $\partial_\xi \equiv \frac{\partial}{\partial \xi}$, $\partial_\tau \equiv \frac{\partial}{\partial \tau}$, etc., for ease of representation.

This equation is in non-conservation form and is not considered suitable for capturing shocks. This form is also known as chain rule form, i.e., even though none of the flow variables and their functions appear as coefficients, the metrics do. To produce the strong conservation law form we have to manipulate the equation. If the above Equation (9.11) is first divided by J and then the chain rule used on all the terms, such as

$$\frac{1}{J}\xi_x \frac{\partial E}{\partial \xi} = \frac{\partial}{\partial \xi}\left[\frac{\xi x}{J} E\right] - E\frac{\partial}{\partial \xi}\left(\frac{\xi x}{J}\right),$$

then we can collect all the terms into two groups forming

$$T_1 + T_2 = 0, \tag{9.12}$$

where

$$\begin{aligned}
T_1 = {}& \partial_\tau\left(\frac{Q}{J}\right) + \partial_\xi\left\{\frac{[\xi_t Q + \xi_x E + \xi_y F + \xi_z G - \frac{1}{Re}(\xi_x E_v + \xi_y F_v + \xi_z G_v)]}{J}\right\} \\
& + \partial_\eta\left\{\frac{[\eta_t Q + \eta_x E + \eta_y F + \eta_z G - \frac{1}{Re}(\eta_x E_v + \eta_y F_v + \eta_z G_v)]}{J}\right\} \\
& + \partial_\zeta\left\{\frac{[\zeta_t Q + \zeta_x E + \zeta_y F + \zeta_z G - \frac{1}{Re}(\zeta_x E_v + \zeta_y F_v + \zeta_z G_v)]}{J}\right\},
\end{aligned} \tag{9.13}$$

and

$$\begin{aligned}
T_2 = {}& -\left(E - \frac{Ev}{Re}\right)\left[\frac{\partial}{\partial \xi}\left(\frac{\xi_x}{J}\right) + \frac{\partial}{\partial \eta}\left(\frac{\eta_x}{J}\right) + \frac{\partial}{\partial \zeta}\left(\frac{\zeta_x}{J}\right)\right] - \left(F - \frac{Fv}{Re}\right) \\
& \left[\frac{\partial}{\partial \xi}\left(\frac{\xi_y}{J}\right) + \frac{\partial}{\partial \eta}\left(\frac{\eta_y}{J}\right) + \frac{\partial}{\partial \zeta}\left(\frac{\zeta_y}{J}\right)\right] - \left(G - \frac{Gv}{Re}\right)\left[\frac{\partial}{\partial \xi}\left(\frac{\xi_z}{J}\right) + \frac{\partial}{\partial \eta}\left(\frac{\eta_z}{J}\right) + \frac{\partial}{\partial \zeta}\left(\frac{\zeta_z}{J}\right)\right] \\
& - Q\left[\frac{\partial}{\partial \tau}\left(\frac{1}{J}\right) + \frac{\partial}{\partial \xi}\left(\frac{\xi_t}{J}\right) + \frac{\partial}{\partial \eta}\left(\frac{\eta_t}{J}\right) + \frac{\partial}{\partial \zeta}\left(\frac{\zeta_t}{J}\right)\right].
\end{aligned} \tag{9.14}$$

The following four sets of terms

$$\frac{\partial}{\partial\tau}\left(\frac{1}{J}\right)+\frac{\partial}{\partial\xi}\left(\frac{\xi_t}{J}\right)+\frac{\partial}{\partial\eta}\left(\frac{\eta_t}{J}\right)+\frac{\partial}{\partial\zeta}\left(\frac{\zeta_t}{J}\right)$$

$$\frac{\partial}{\partial\xi}\left(\frac{\xi_x}{J}\right)+\frac{\partial}{\partial\eta}\left(\frac{\eta_x}{J}\right)+\frac{\partial}{\partial\zeta}\left(\frac{\zeta_x}{J}\right)$$

$$\frac{\partial}{\partial\xi}\left(\frac{\xi_y}{J}\right)+\frac{\partial}{\partial\eta}\left(\frac{\eta_y}{J}\right)+\frac{\partial}{\partial\zeta}\left(\frac{\zeta_y}{J}\right)$$

$$\frac{\partial}{\partial\xi}\left(\frac{\xi_z}{J}\right)+\frac{\partial}{\partial\eta}\left(\frac{\eta_z}{J}\right)+\frac{\partial}{\partial\zeta}\left(\frac{\zeta_z}{J}\right)$$

are invariants of the transformation and can be shown to be zero analytically. For example, using the metric definitions in Equations (9.8*a*) to (9.8*l*) for the second invariant equation above, one gets,

$$\begin{aligned}&\frac{\partial}{\partial\xi}(y_\eta z_\zeta - z_\eta y_\zeta)+\frac{\partial}{\partial\eta}(z_\xi y_\zeta - y_\xi z_\zeta)+\frac{\partial}{\partial\zeta}(y_\xi z_\eta - z_\xi y_\eta)\\ &\quad = y_{\eta\xi}z_\zeta - y_\eta z_{\eta\xi} - y_{\zeta\xi}z_\eta - y_\zeta z_{\eta\xi} + z_{\xi\eta}y_\zeta + z_\xi y_{\zeta\eta} - y_{\xi\eta}z_\zeta - y_\xi z_{\eta\zeta}\\ &\qquad + y_{\xi\zeta}z_\eta + y_\xi z_{\eta\zeta} - z_{\xi\zeta}y_\eta - z_\xi y_{\zeta\eta} \equiv 0.\end{aligned} \tag{9.15}$$

Since analytic differentiation is commutative, all of the above terms adds up to zero, as shown above. This then eliminates the T_2 terms of Equation (9.12) completely and the resulting equations are in strong conservation-law form and can be written as

$$\partial_\tau\hat{Q}+\partial_\xi\hat{E}+\partial_\eta\hat{F}+\partial_\zeta\hat{G}=\frac{1}{Re}\{\partial_\xi(\hat{E}_v)+\partial_\eta(\hat{F}_v)+\partial_\zeta(\hat{G}_v)\},$$

where

$$\hat{Q}=\frac{Q}{J}$$

$$\hat{E}=\frac{1}{J}\lfloor \rho U \quad \rho uU+\xi_x p \quad \rho vU+\xi_y p \quad \rho wU+\xi_z p \quad (e_t+p)U-\xi_t p\rfloor^T$$

$$\hat{F}=\frac{1}{J}\lfloor \rho V \quad \rho uV+\eta_x p \quad \rho vV+\eta_y p \quad \rho wV+\eta_z p \quad (e_t+p)V-\eta_t p\rfloor^T$$

$$\hat{G}=\frac{1}{J}\lfloor \rho W \quad \rho uW+\zeta_x p \quad \rho vW+\zeta_y p \quad \rho wW+\zeta_z p \quad (e_t+p)W-\zeta_t p\rfloor^T$$

$$\begin{aligned}\hat{E}_v &= \frac{1}{J}(\xi_x E_v+\xi_y F_v+\xi_z G_v)=\frac{1}{J}\lfloor 0 \quad (\xi_x\tau_{xx}+\xi_y\tau_{xy}+\xi_z\tau_{xz})\\ &\quad (\xi_x\tau_{xy}+\xi_y\tau_{yy}+\xi_z\tau_{zy}) \quad (\xi_x\tau_{xz}+\xi_y\tau_{yz}+\xi_z\tau_{zz}) \quad (\xi_x R+\xi_y S+\xi_z T)\rfloor^T\end{aligned}$$

$$\begin{aligned}\hat{F}_v &= \frac{1}{J}(\eta_x E_v+\eta_y F_v+\eta_z G_v)=\frac{1}{J}\lfloor 0 \quad (\eta_x\tau_{xx}+\eta_y\tau_{yx}+\eta_z\tau_{zx})\\ &\quad (\eta_x\tau_{xy}+\eta_y\tau_{yy}+\eta_z\tau_{zy}) \quad (\eta_x\tau_{xz}+\eta_y\tau_{yz}+\eta_z\tau_{zz}) \quad (\eta_x R+\eta_y S+\eta_z T)\rfloor^T\end{aligned}$$

$$\hat{G}_v = \frac{1}{J}(\zeta_x E_v + \zeta_y F_v + \zeta_z G_v)$$
$$= \frac{1}{J}\lfloor 0 \;\; (\zeta_x \tau_{xx} + \zeta_y \tau_{yx} + \zeta_z \tau_{zx}) \;\; (\zeta_x \tau_{xy} + \zeta_y \tau_{yy} + \zeta_z \tau_{zy})$$
$$(\zeta_x \tau_{xz} + \zeta_y \tau_{yz} + \zeta_z \tau_{zz}) \;\; (\zeta_x R + \zeta_y S + \zeta_z T)\rfloor^T.$$

The contravariant velocities (U, V, W) are for the transformed plane given by

$$U = \xi_t + \xi_x u + \xi_y v + \xi_z w$$
$$V = \eta_t + \eta_x u + \eta_y v + \eta_z w$$
$$W = \zeta_t + \zeta_x u + \zeta_y v + \zeta_z w.$$

9.8 Linearization of Fluxes

In order to solve Equation (9.16) numerically, a linearization procedure in time is introduced and all flux vectors are expressed in terms of the state vector Q. For example, using Taylor series expansion,

$$\hat{E}^{n+1} = \hat{E}^n + \frac{\partial \hat{E}^n}{\partial \hat{Q}} \Delta \hat{Q} + 0(\Delta\tau^2). \tag{9.17}$$

Thus, terms such as $\frac{\partial \hat{E}}{\partial \hat{Q}}$ and similar terms need to be evaluated. They are also defined as the Jacobian matrices (not to be confused with the Jacobian of transformation) that are to be computed. Apart from $\hat{E}^{n+1}$ the other five Jacobian matrices are evaluated for the remaining two inviscid flux terms and the three viscous flux terms. Since flux vectors are (5×1) vectors, each one of the Jacobians are (5×5) matrices. For example, the inviscid Jacobian $\frac{\partial \hat{E}}{\partial \hat{Q}}$ is given by

$$\frac{\partial \hat{E}}{\partial \hat{Q}} = \frac{\partial(\hat{E}_1, \hat{E}_2, \hat{E}_3, \hat{E}_4, \hat{E}_5)}{\partial(\hat{Q}_1, \hat{Q}_2, \hat{Q}_3, \hat{Q}_4, \hat{Q}_5)} = \begin{bmatrix} \frac{\partial \hat{E}_1}{\partial \hat{Q}_1} & \frac{\partial \hat{E}_1}{\partial \hat{Q}_2} & \frac{\partial \hat{E}_1}{\partial \hat{Q}_3} & \frac{\partial \hat{E}_1}{\partial \hat{Q}_4} & \frac{\partial \hat{E}_1}{\partial \hat{Q}_5} \\ \frac{\partial \hat{E}_2}{\partial \hat{Q}_1} & \frac{\partial \hat{E}_2}{\partial \hat{Q}_2} & \frac{\partial \hat{E}_2}{\partial \hat{Q}_3} & \frac{\partial \hat{E}_2}{\partial \hat{Q}_4} & \frac{\partial \hat{E}_2}{\partial \hat{Q}_5} \\ \frac{\partial \hat{E}_3}{\partial \hat{Q}_1} & \frac{\partial \hat{E}_3}{\partial \hat{Q}_2} & \frac{\partial \hat{E}_3}{\partial \hat{Q}_3} & \frac{\partial \hat{E}_3}{\partial \hat{Q}_4} & \frac{\partial \hat{E}_3}{\partial \hat{Q}_5} \\ \frac{\partial \hat{E}_4}{\partial \hat{Q}_1} & \frac{\partial \hat{E}_4}{\partial \hat{Q}_2} & \frac{\partial \hat{E}_4}{\partial \hat{Q}_3} & \frac{\partial \hat{E}_4}{\partial \hat{Q}_4} & \frac{\partial \hat{E}_4}{\partial \hat{Q}_5} \\ \frac{\partial \hat{E}_5}{\partial \hat{Q}_1} & \frac{\partial \hat{E}_5}{\partial \hat{Q}_2} & \frac{\partial \hat{E}_5}{\partial \hat{Q}_3} & \frac{\partial \hat{E}_5}{\partial \hat{Q}_4} & \frac{\partial \hat{E}_5}{\partial \hat{Q}_5} \end{bmatrix}.$$

Now the flux vectors E, F and G can be represented in terms of the components of Q according to

$$E = \begin{bmatrix} \rho u \\ \rho u^2 + p \\ \rho uv \\ \rho uw \\ (\rho e_t + p)u \end{bmatrix} = \begin{bmatrix} Q_2 \\ \dfrac{Q_2^2}{Q_1} + (\gamma - 1)\left[Q_5 - \dfrac{1}{2}\left(\dfrac{Q_2^2}{Q_1} + \dfrac{Q_3^2}{Q_1} + \dfrac{Q_4^2}{Q_1}\right)\right] \\ \dfrac{Q_2 Q_3}{Q_1} \\ \dfrac{Q_2 Q_4}{Q_1} \\ \left[\gamma Q_5 - \dfrac{\gamma - 1}{2}\left(\dfrac{Q_2^2}{Q_1} + \dfrac{Q_3^2}{Q_1} + \dfrac{Q_4^2}{Q_1}\right)\right]\dfrac{Q_2}{Q_1} \end{bmatrix},$$

$$F = \begin{bmatrix} \rho v \\ \rho uv \\ \rho v^2 + p \\ \rho vw \\ (\rho e_t + p)v \end{bmatrix} = \begin{bmatrix} Q_3 \\ \dfrac{Q_2 Q_3}{Q_1} \\ \dfrac{Q_3^2}{Q_1} + (\gamma - 1)\left[Q_5 - \dfrac{1}{2}\left(\dfrac{Q_2^2}{Q_1} + \dfrac{Q_3^2}{Q_1} + \dfrac{Q_4^2}{Q_1}\right)\right] \\ \dfrac{Q_3 Q_4}{Q_1} \\ \left[\gamma Q_5 - \dfrac{\gamma - 1}{2}\left(\dfrac{Q_2^2}{Q_1} + \dfrac{Q_3^2}{Q_1} + \dfrac{Q_4^2}{Q_1}\right)\right]\dfrac{Q_3}{Q_1} \end{bmatrix},$$

$$G = \begin{bmatrix} \rho w \\ \rho wu \\ \rho wv \\ \rho w^2 + p \\ (\rho e_t + p)w \end{bmatrix} = \begin{bmatrix} Q_4 \\ \dfrac{Q_2 Q_4}{Q_1} \\ \dfrac{Q_3 Q_4}{Q_1} \\ \dfrac{Q_4^2}{Q_1} + (\gamma - 1)\left[Q_5 - \dfrac{1}{2}\left(\dfrac{Q_2^2}{Q_1} + \dfrac{Q_3^2}{Q_1} + \dfrac{Q_4^2}{Q_1}\right)\right] \\ \left[\gamma Q_5 - \dfrac{\gamma - 1}{2}\left(\dfrac{Q_2^2}{Q_1} + \dfrac{Q_3^2}{Q_1} + \dfrac{Q_4^2}{Q_1}\right)\right]\dfrac{Q_4}{Q_1} \end{bmatrix}.$$

Since

$$\hat{E}_1 = \frac{1}{J}[\xi_t Q_1 + \xi_x E_1 + \xi_y F_1 + \xi_z G_1],$$

using the above representations we can also write

$$\hat{E}_1 = \frac{1}{J}[\xi_t Q_1 + \xi_x Q_2 + \xi_y Q_3 + \xi_z Q_4].$$

Now the first element of the Jacobian matrix is given by,

$$\frac{\partial \hat{E}_1}{\partial \hat{Q}_1} = \frac{\partial\{\frac{1}{J}[\xi_t Q_1 + \xi_x Q_2 + \xi_y Q_3 + \xi_z Q_4]}{\partial\{\frac{Q_1}{J}\}}$$
$$= \frac{\partial(\xi_t Q_1 + \xi_x Q_2 + \xi_y Q_3 + \xi_x Q_4)}{\partial Q_1} = \xi_t.$$

The remaining elements of the first row are

$$\frac{\partial \hat{E}_1}{\partial \hat{Q}_2} = \xi_x, \quad \frac{\partial \hat{E}_1}{\partial \hat{Q}_3} = \xi_y, \quad \frac{\partial \hat{E}_1}{\partial \hat{Q}_4} = \xi_x \text{ and } \frac{\partial \hat{E}_1}{\partial \hat{Q}_5} = 0.$$

The elements of the second, third, fourth and fifth rows are determined in a similar fashion. The final expressions for the Jacobian are given by

$$\frac{\partial \hat{E}}{\partial \hat{Q}} = [a_{ij}] = \begin{bmatrix} a_{11} & a_{12} & a_{13} & a_{14} & a_{15} \\ a_{21} & a_{22} & a_{23} & a_{24} & a_{25} \\ a_{31} & a_{32} & a_{33} & a_{34} & a_{35} \\ a_{41} & a_{42} & a_{43} & a_{44} & a_{45} \\ a_{51} & a_{52} & a_{53} & a_{54} & a_{55} \end{bmatrix},$$

where,

$$a_{11} = \xi_t, \quad a_{12} = \xi_x, \quad a_{13} = \xi_y, \quad a_{14} = \xi_z, \quad a_{15} = 0$$
$$a_{21} = -u(\xi_x u + \xi_y v + \xi_z w) + \xi_x \left[\tfrac{1}{2}(\gamma - 1)(u^2 + v^2 + w^2)\right]$$
$$a_{22} = \xi_t + \xi_x(2 - \gamma)u + (\xi_x u + \xi_y v + \xi_z w)$$
$$a_{23} = \xi_y u - (\gamma - 1)\xi_x v$$
$$a_{24} = \xi_z u - (\gamma - 1)\xi_x w$$
$$a_{25} = (\gamma - 1)\xi_x$$
$$a_{31} = -v(\xi_x u + \xi_y v + \xi_z w) + \xi_y \left[\tfrac{1}{2}(\gamma - 1)(u^2 + v^2 + w^2)\right]$$
$$a_{32} = \xi_x v - (\gamma - 1)\xi_y u$$
$$a_{33} = \xi_t + \xi_y(2 - \gamma)v + (\xi_x u + \xi_y v + \xi_z w)$$
$$a_{34} = \xi_z v - (\gamma - 1)\xi_y w$$
$$a_{35} = (\gamma - 1)\xi_y$$
$$a_{41} = -w(\xi_x u + \xi_y v + \xi_z w) + \xi_z \left[\tfrac{1}{2}(\gamma - 1)(u^2 + v^2 + w^2)\right]$$
$$a_{42} = \xi_x w - (\gamma - 1)\xi_z u$$
$$a_{43} = \xi_y w - (\gamma - 1)\xi_z v$$

$$a_{44} = \xi_t + \xi_z(2-\gamma)w + (\xi_x u + \xi_y v + \xi_z w)$$
$$a_{45} = (\gamma - 1)\xi_z$$
$$a_{51} = (\xi_x u + \xi_y v + \xi_z w)[-\gamma e_t + (\gamma - 1)(u^2 + v^2 + w^2)]$$
$$a_{52} = \xi_x \left[\gamma e_t - \tfrac{1}{2}(\gamma - 1)(u^2 + v^2 + w^2)\right] - (\gamma - 1)[\xi_x u + \xi_y v + \xi_z w]u$$
$$a_{53} = \xi_y \left[\gamma e_t - \tfrac{1}{2}(\gamma - 1)(u^2 + v^2 + w^2)\right] - (\gamma - 1)[\xi_x u + \xi_y v + \xi_z w]v$$
$$a_{54} = \xi_z \left[\gamma e_t - \tfrac{1}{2}(\gamma - 1)(u^2 + v^2 + w^2)\right] - (\gamma - 1)[\xi_x u + \xi_y v + \xi_z w]w$$
$$a_{55} = \xi_t + \gamma[\xi_x u + \xi_y v + \xi_z w].$$

Similarly, one can obtain the following:

$$\frac{\partial \hat{F}}{\partial \hat{Q}} = [b_{ij}] = \begin{bmatrix} b_{11} & b_{12} & b_{13} & b_{14} & b_{15} \\ b_{21} & b_{22} & b_{23} & b_{24} & b_{25} \\ b_{31} & b_{32} & b_{33} & b_{34} & b_{35} \\ b_{41} & b_{42} & b_{43} & b_{44} & b_{45} \\ b_{51} & b_{52} & b_{53} & b_{54} & b_{55} \end{bmatrix},$$

where,

$$b_{11} = \eta_t, \quad b_{12} = \eta_x, \quad b_{13} = \eta_y, \quad b_{14} = \eta_z, \quad b_{15} = 0$$
$$b_{21} = -u(\eta_x u + \eta_y v + \eta_z w) + \eta_x \left[\tfrac{1}{2}(\gamma - 1)(u^2 + v^2 + w^2)\right]$$
$$b_{22} = \eta_t + \eta_x(2-\gamma)u + (\eta_x u + \eta_y v + \eta_z w)$$
$$b_{23} = \eta_y u - (\gamma - 1)\eta_x v$$
$$b_{24} = \eta_z u - (\gamma - 1)\eta_x w$$
$$b_{25} = (\gamma - 1)\eta_x$$
$$b_{31} = -v(\eta_x u + \eta_y v + \eta_z w) + \eta_y \left[\tfrac{1}{2}(\gamma - 1)(u^2 + v^2 + w^2)\right]$$
$$b_{32} = \eta_x v - (\gamma - 1)\eta_y u$$
$$b_{33} = \eta_t + \eta_y(2-\gamma)v + (\eta_x u + \eta_y v + \eta_z w)$$
$$b_{34} = \eta_z v - (\gamma - 1)\eta_y w$$
$$b_{35} = (\gamma - 1)\eta_y$$
$$b_{41} = -w(\eta_x u + \eta_y v + \eta_z w) + \eta_z \left[\tfrac{1}{2}(\gamma - 1)(u^2 + v^2 + w^2)\right]$$
$$b_{42} = \eta_x w - (\gamma - 1)\eta_z u$$
$$b_{43} = \eta_y w - (\gamma - 1)\eta_z v$$
$$b_{44} = \eta_t + \eta_z(2-\gamma)w + (\eta_x u + \eta_y v + \eta_z w)$$
$$b_{45} = (\gamma - 1)\eta_z$$

$$b_{51} = (\eta_x u + \eta_y v + \eta_z w)[-\gamma e_t + (\gamma - 1)(u^2 + v^2 + w^2)]$$
$$b_{52} = \eta_x \left[\gamma e_t - \tfrac{1}{2}(\gamma - 1)(u^2 + v^2 + w^2)\right] - (\gamma - 1)[\eta_x u + \eta_y v + \eta_z w]u$$
$$b_{53} = \eta_y \left[\gamma e_t - \tfrac{1}{2}(\gamma - 1)(u^2 + v^2 + w^2)\right] - (\gamma - 1)[\eta_x u + \eta_y v + \eta_z w]v$$
$$b_{54} = \eta_z \left[\gamma e_t - \tfrac{1}{2}(\gamma - 1)(u^2 + v^2 + w^2)\right] - (\gamma - 1)[\eta_x u + \eta_y v + \eta_z w]w$$
$$b_{55} = \eta_t + \gamma[\eta_x u + \eta_y v + \eta_z w],$$

and

$$\frac{\partial \hat{G}}{\partial \hat{Q}} = [c_{ij}] = \begin{bmatrix} c_{11} & c_{12} & c_{13} & c_{14} & c_{15} \\ c_{21} & c_{22} & c_{23} & c_{24} & c_{25} \\ c_{31} & c_{32} & c_{33} & c_{34} & c_{35} \\ c_{41} & c_{42} & c_{43} & c_{44} & c_{45} \\ c_{51} & c_{52} & c_{53} & c_{54} & c_{55} \end{bmatrix},$$

where,

$$c_{11} = \zeta_t,\quad c_{12} = \zeta_x,\quad c_{13} = \zeta_y,\quad c_{14} = \zeta_z,\quad c_{15} = 0$$
$$c_{21} = -u(\zeta_x u + \zeta_y v + \zeta_z w) + \zeta_x \left[\tfrac{1}{2}(\gamma - 1)(u^2 + v^2 + w^2)\right]$$
$$c_{22} = \zeta_t + \zeta_x(2 - \gamma)u + (\zeta_x u + \zeta_y v + \zeta_z w)$$
$$c_{23} = \zeta_y u - (\gamma - 1)\zeta_x v$$
$$c_{24} = \zeta_z u - (\gamma - 1)\zeta_x w$$
$$c_{25} = (\gamma - 1)\zeta_x$$
$$c_{31} = -v(\zeta_x u + \zeta_y v + \zeta_z w) + \zeta_y \left[\tfrac{1}{2}(\gamma - 1)(u^2 + v^2 + w^2)\right]$$
$$c_{32} = \zeta_x v - (\gamma - 1)\zeta_y u$$
$$c_{33} = \zeta_t + \zeta_y(2 - \gamma)v + (\zeta_x u + \zeta_y v + \zeta_z w)$$
$$c_{34} = \zeta_z v - (\gamma - 1)\zeta_y w$$
$$c_{35} = (\gamma - 1)\zeta_y$$
$$c_{41} = -w(\zeta_x u + \zeta_y v + \zeta_z w) + \zeta_z \left[\tfrac{1}{2}(\gamma - 1)(u^2 + v^2 + w^2)\right]$$
$$c_{42} = \zeta_x w - (\gamma - 1)\zeta_z u$$
$$c_{43} = \zeta_y w - (\gamma - 1)\zeta_z v$$
$$c_{44} = \zeta_t + \zeta_z(2 - \gamma)w + (\zeta_x u + \zeta_y v + \zeta_z w)$$
$$c_{45} = (\gamma - 1)\zeta_z$$
$$c_{51} = (\zeta_x u + \zeta_y v + \zeta_z w)[-\gamma e_t + (\gamma - 1)(u^2 + v^2 + w^2)]$$
$$c_{52} = \zeta_x \left[\gamma e_t - \tfrac{1}{2}(\gamma - 1)(u^2 + v^2 + w^2)\right] - (\gamma - 1)[\zeta_x u + \zeta_y v + \zeta_z w]u$$

$$c_{53} = \zeta_y\left[\gamma e_t - \tfrac{1}{2}(\gamma-1)(u^2+v^2+w^2)\right] - (\gamma-1)[\zeta_x u + \zeta_y v + \zeta_z w]v$$
$$c_{54} = \zeta_z\left[\gamma e_t - \tfrac{1}{2}(\gamma-1)(u^2+v^2+w^2)\right] - (\gamma-1)[\zeta_x u + \zeta_y v + \zeta_z w]w$$
$$c_{55} = \zeta_t + \gamma[\zeta_x u + \zeta_y v + \zeta_z w].$$

The viscous Jacobians can also be similarly derived. The derivation is straightforward but lengthy, and is not given here. However there are some difference between the inviscid and viscous Jacobians. Since the viscous flux vectors involve gradients of the dependent variables and the Jacobian of transformation, they are in turn functions of independent variables. Thus J remains embedded in the viscous Jacobians unlike the inviscid Jacobian where J cancels out upon evaluation of the Jacobian terms.

Having derived the relevant expressions for the transformed equation for the Navier-Stokes equation in the transformed plane, we mention in passing that some additional simplifications can be brought about to this form of Navier-Stokes equation.

9.9 Thin Layer Navier-Stokes and Parabolised Navier-Stokes Equation

For high Reynolds number flows, the viscous effects are confined near rigid boundaries and in the wake regions. Typically, in 3D finite difference computations one usually takes very fine meshes across the shear layer while in the stream-wise direction very few grid points are taken. Even though the full NS equation may be used, the viscous terms associated with derivatives along the body is not computationally resolved because of the particular choice of the grid system. In fact, for attached and mildly separated flows these terms are negligible. In the thin layer Navier-Stokes (TLNS) equation, this term is dropped and the assumption is akin to what is performed in boundary layer analysis. However, in TLNS the normal momentum equation is retained so that the pressure can vary across the shear layer. The following steps are taken in implementing TLNS approximation:

(*a*) All body surfaces are mapped on to a coordinate surface, for example on to η = const. surface.

(*b*) All the viscous derivatives in the ξ and ζ directions are neglected and all the convection terms retained.

The final form of the TLNS equation is then given by

$$\frac{\partial \hat{Q}}{\partial \tau} + \frac{\partial \hat{E}}{\partial \xi} + \frac{\partial \hat{F}}{\partial \eta} + \frac{\partial \hat{G}}{\partial \zeta} = \frac{1}{Re}\frac{\partial \hat{F}_v}{\partial \eta}.$$

Additionally, for steady flows one can also make the governing PDE into a parabolic equation by making following changes: The stream-wise viscous stresses are neglected and the convective flux vector $\hat{E}$ modified such that $\frac{\partial \hat{E}}{\partial \hat{Q}}$ has only positive eigen values. The parabolic nature of the PDE allows stable marching in the stream-wise direction. It can be shown that the parabolized Navier Stokes (PNS) equations retain their parabolic nature as long as the stream-wise velocity u is positive. We will not discuss about this any further.

9.10 Grid Generation

Transformed plane Equations (9.16) or the TLNS equation given in the previous section requires grid metric and grid Jacobian evaluation. While it is possible to generate either structured or unstructured grids, here we will discuss only structured grids which provide maximum resolution for a given hardware platform.

Any successful structured grid should have the following essential properties:

(*a*) The mapping between the transformed and the physical plane must be one-to-one, i.e. the grid lines of the same family in the physical plane should not cross each other.

(*b*) The grid distribution must be smooth. Any discontinuities of the grid metrics or Jacobian will lead to serious errors in the solution, or at some time this may also lead to solution divergence (or unphysical solutions).

(*c*) At the same time we must have a grid system that is capable of resolving all the scales, i.e., we should be able to control the grid point distribution.

(*d*) Since the solutions have a very large gradient near the body, we must have very high accuracy for the solution. In general the grid points are usually clustered close to the body to resolve the viscous layer. But it is the non-orthogonal terms of the transformed equation that gives rise to large error. This can be avoided by employing either an orthogonal grid or at least use a grid system that is orthogonal near the body.

(*e*) Also, for fast convergence of any iterative schemes employed for the solution of linear algebraic equations, we must have a cell aspect ratio that is not too large.

9.11 Types of Grids

Usually there are three specific types of structured grids used for computation. These are classified depending on the type of topology expressed by one of the constant grid lines. They are explained below.

O-type grid: Here one of the grid lines resembles the general shape of the letter O; such grids are typically used for multiply-connected body as shown in Fig. 9.4.

Fig. 9.4 *An O-type orthogonal grid generated by solving hyperbolic partial differential equation.*

C-type grid: Here one of the grid lines resembles the letter C. C-grids are preferred when one is interested in investigating the wake of a body with fine grids in the wake. Notice that in O-grids there are large number of points ahead of the body where the flow is computed, but where hardly anything happens! This is specifically avoided in C-type grid. A typical example is shown in Fig. 9.5.

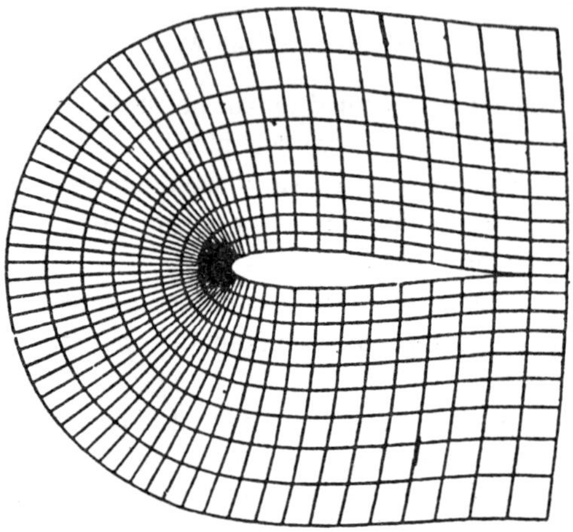

Fig. 9.5 *A C-type grid generated by algebraic method.*

H-type grid: Here the grid lines form a lattice network similar to the stream function contour lines and the velocity potential contour lines of the irrotational potential flow. Such grids have the good attributes of a C-type grid. They are mostly used in internal flows, as in turbomachines.

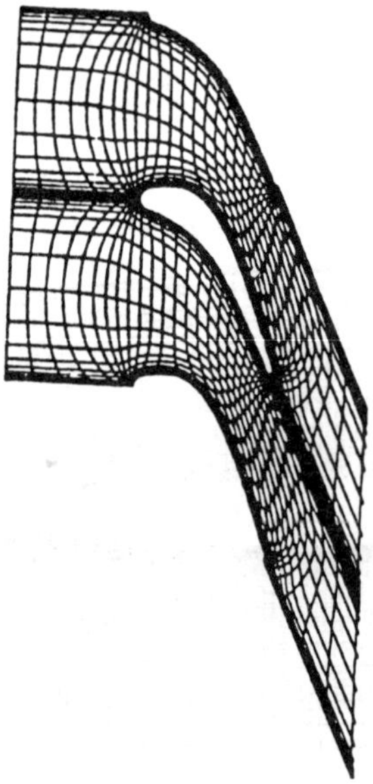

Fig. 9.6 *An H-type grid generated for a turbine nozzle by solving elliptic partial differential equation.*

9.12 Grid Generation Methods

There are various methods of grid generations starting from the algebraic method to grids generated by solving partial differential equations. We discuss some representative methods only.

9.13 Algebraic Method (Eiseman, 1985)

This technique interpolates the boundary data to generate interior grids. The explicit interpolation may be in one- or multi-dimensions. The distribution of grid points in the interior is mainly governed by the stretching functions on the boundaries. The distributions of points along the boundary of the domain is handled effectively by defining normalized 1D stretching functions along boundary segments, typically corresponding to each side of computational domain. In Fig. 9.7, such stretching along boundary segments EA, AB, BC and CD are to be provided for effective grid point control in the interior of the domain. Boundary stretching functions can be effectively used in generating interior points by algebraic techniques subsequently.

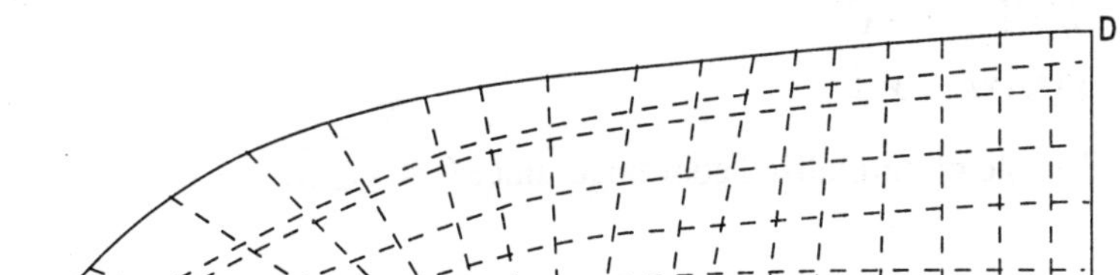

Fig. 9.7 *Grid generation by algebraic method using stretching function on the boundaries.*

9.14 One-Dimensional Stretching Functions

One-dimensional (1D) stretching is used for distributing points along a particular boundary such that a specific region of the domain can be resolved accurately. For the geometry shown below, it is appropriate to introduce 1D stretching function on AE and CD so that grids are clustered close to ABC.

For simple geometries one can combine 1D boundary stretching with a simple shearing transformation to establish interior points such that

$$x = (1 - \bar{\mu})\text{EA}^x(v_k) + \bar{\mu}\text{CD}^x(v_k),$$
$$y = (1 - \bar{\mu})\text{EA}^y(v_k) + \bar{\mu}\text{CD}^y(v_k),$$

with

$$\bar{\mu} = \frac{\mu - \mu_1}{\mu_{\max} - \mu_1};$$

where μ_1 corresponds to EA and $\mu_{\max}$ correspond to CD.

Along EA we can define a normalized independent variable

$$\eta^* = \frac{\eta - \eta_A}{\eta_E - \eta_A}$$

such that each η value corresponds to $v = v_k$ line. Therefore η^* varies between 0 and 1 as $\eta_A \leq \eta \leq \eta_E$. An effective stretching reported in Eiseman (1985) is

$$s = P\eta^* + (1 - P)\left\{1 - \frac{\tanh[Q(1 - \eta^*)]}{\tanh Q}\right\}$$

where P and Q provide grid point control, with P providing the slope of the distribution close to the body ($\eta^* = 0$) and Q the damping factor controlling the departure of linear variation of s versus η^*. Small values of Q causes small departures from linearity. When P is close to 1, the departure from linearity will be small and will occur only when η^* is close 1. Once s is obtained it is used to specify the distribution of x and y. For example, one can define on EA,

$$x = x_A + f(s)[x_E - x_A],$$
$$y = y_A + g(s)[y_E - y_A].$$

s can also play the role of normalized coordinate since $0 \leq s \leq 1$.

9.15 Two-Boundary Technique

We will explain this with reference to the channel shown below in Fig. 9.8.

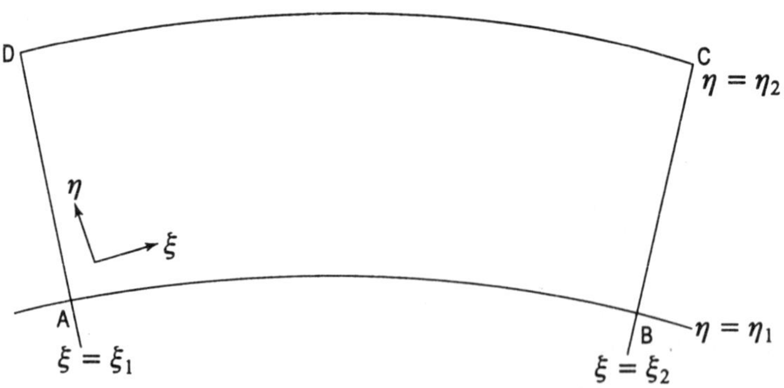

Fig. 9.8 *Algebraic grid generation by two-boundary technique.*

We define

$$\eta^* = \frac{\eta - \eta_1}{\eta_2 - \eta_1}$$

so that a normalized coordinate along AD and BC is given by

$$s_{AD} = P_1\eta^* + (1 - P_1)\left\{1 - \frac{\tanh[Q(1-\eta^*)]}{\tanh Q}\right\}$$

$$s_{BC} = P_2\eta^* + (1 - P_2)\left\{1 - \frac{\tanh[Q(1-\eta^*)]}{\tanh Q}\right\}.$$

To obtain s between surfaces AD and BC, a simple linear interpolation is recommended. Thus

$$s = s_{AD} + \xi^*(s_{BC} - s_{AD}) \quad \text{with} \quad \xi^* = \frac{\xi - \xi_1}{\xi_2 - \xi_1}.$$

In a similar way the distribution of grid points along the surfaces AB and CD are controlled by 1D stretching functions r_{AB} and r_{DC}. If r_{AB} and r_{DC} are interpreted as normalized coordinates measured along the surface, i.e., $r = \xi^*$, x_{AB} and y_{AB} follow directly from

$$x_{AB} = x_A + r_{AB}[x_B - x_A]$$

$$y_{AB} = y_A + r_{AB}[y_B - y_A].$$

Similarly,

$$x_{DC} = x_D + r_{DC}(x_C - x_D)$$

$$y_{DC} = y_D + r_{DC}(y_C - y_D).$$

Once the boundary points are specified, the interior points are interpolated by

$$x(\xi, \eta) = (1 - s)x_{AB} + s x_{DC} = [1 - s(\eta)]x_{AB}(\xi) + s(\eta)x_{DC}(\xi),$$

$$y(\xi, \eta) = (1 - s)y_{AB} + s y_{DC} = [1 - s(\eta)]y_{AB}(\xi) + s(\eta)y_{DC}(\xi).$$

The above interpolation can give undesirable grids if the corresponding boundary point (x_{AB}, y_{AB}) and (x_{DC}, y_{DC}) are out of alignment. This may produce grids adjacent to the boundary with large distortion. This could be avoided by the following prescription given in Eiseman (1985),

$$x(\xi, \eta) = \mu_1(s)x_{AB}(r_{AB}) + \mu_2(s)x_{DC}(r_{DC}) + T_1\mu_3\left(\frac{dy_{AB}}{dr_{AB}}\right) + T_2\mu_4\left(\frac{dy_{DC}}{dr_{DC}}\right),$$

$$y(\xi, \eta) = \mu_1 y_{AB} + \mu_2 y_{DC} - T_1\mu_3\frac{dx_{AB}}{dr_{AB}} - T_2\mu_4\frac{dx_{DC}}{dr_{DC}},$$

where

$$\mu_1 = 2s^3 - 3s^3 + 1; \qquad \mu_2 = -2s^3 + 3s^2$$

$$\mu_3 = s^3 - 2s^2 + s; \qquad \mu_4 = s^3 - s^2.$$

This grid is locally orthogonal at the boundaries AB and CD. The parameters T_1 and T_2 control how far into the grid interior orthogonality is enforced. Large values of T_1 and T_2 cause grid lines to cross.

9.16 Partial Differential Equation Techniques

The grid generation techniques that use PDEs can be based on any type of PDEs, namely, parabolic, elliptic and hyperbolic method. Here we will discuss only the elliptic and hyperbolic methods of grid generation.

Elliptic grid generators

This method is well suited to the situation when all the boundaries are prescribed and one is interested in generating interior grid points subject to the specific requirements of grid point control. It can be well understood when we look at the property of solution to Laplace's equation. We know that the solution provides unique solution to each and every point, i.e., if we generate grids as a solution to Laplace's equation then the mapping will be one-to-one without any grid lines crossing the other. Such a property of uniqueness is also retained for Poisson equation if the right hand side is not a large negative source term. The grid generating equation is then given for a two-dimensional grid generation problem by,

$$\xi_{xx} + \xi_{yy} = P(\xi, \eta)$$

$$\eta_{xx} + \eta_{yy} = Q(\xi, \eta)$$

where P and Q are the source terms to control the grid spacing. These equations can be rewritten in the transformed plane as

$$ax_{\xi\xi} - 2bx_{\xi\eta} + cx_{\eta\eta} = -\frac{1}{J_1^2}(Px_\xi + Qx_\eta) = -J^2(Px_\xi + Qx_\eta),$$

$$ay_{\xi\xi} - 2by_{\xi\eta} + cy_{\eta\eta} = -\frac{1}{J_1^2}(Py_\xi + Qy_\eta) = -J^2(Py_\xi + Qy_\eta),$$

where J is the Jacobian of transformation as given in Equation (9.10) suitably changed for the two-dimensional problem. Similarly J_1 is given by $\frac{\partial(x,y)}{\partial(\xi,\eta)}$. The functions P and Q are selected according to the requirements. It might be desirable to cluster grid lines closer to a particular coordinate lines or in the vicinity of a particular point. Thompson et al.(1974) suggested the following choice of functions for P and Q:

$$P(\xi, \eta) = -\sum_{l=1}^{L} a_l \text{Sign}(\xi - \xi_l) e^{(-c_l|\xi - \xi_l|)} - \sum_{m=1}^{M} b_m \text{Sign}(\xi - \xi_m) e^{\{-d_m[(\xi-\xi_m)^2 + (\eta-\eta_m)^2]^{\frac{1}{2}}\}}$$

$$Q(\xi, \eta) = -\sum_{l=1}^{L} a_l \text{Sign}(\eta - \eta_l) e^{-c_l|\eta - \eta_l|} - \sum_{m=1}^{M} b_m \text{Sign}(\xi - \xi_m) e^{-d_m\sqrt{(\xi-\xi_m)^2 + (\eta-\eta_m)^2}}$$

The first exponential term attracts the lines of *constant*-ξ to the vicinity of ξ_l. The second exponential term attracts the near-by grids lines to the specific point (ξ_m, η_m). The multiplicative constants a and b are the amplification factors and c and d the decay factors. They are usually chosen by trial and error.

9.17 Hyperbolic Grid Generation Method

The elliptic grid generation method works for closed domain, i.e. where both the inner and outer boundaries are prescribed. However in many problems of external aerodynamics the positioning of the outer boundary is at the discretion of the user, where some asymptotic form of boundary conditions are supported. In such cases one can generate a grid which start off from the body and marches outwards. This is essentially the idea behind the hyperbolic grid generation methods. Usually, grid generation by hyperbolic method also produces orthogonal grid. Thus this gives us the following benefits. Firstly, we can march step by step, thereby we control the spacing between successive grid lines. Secondly, it produces an orthogonal grid and thus we have less numerical error. As such the grid generation process is very fast as compared to elliptic grid generator. The only disadvantage of the method is that it does not work for any arbitrary shape. Also, the body must have a well defined normal at each and every point. This therefore requires that the slope discontinuities have to be smoothed out over a very small extent. Also, the generating grids around highly cambered shape requires careful choice of spacing. Figure 9.10 shows the grid generated around an aerofoil. There are small fillet fitted at the trailing edge which are not visible to the naked eye. Even in solving NS equation such small fillets are of the dimension of viscous dissipation length scale and the flow would not "see" such rapid changes. The method and the related steps are given in Nair & Sengupta (1998), which was used to generate grids suitable for viscous flow calculations for NACA series of aerofoils, with and without camber.

9.18 Orthogonal Grid Generation for Navier-Stokes Computations

One of the primary activities in computational fluid dynamics is obtaining a suitable grid for the solution of the governing differential equation. In the present study, our concern is with orthogonal structured grid generation in two dimensions. Orthogonal grids are preferred for the following reasons:

(*a*) Simpler transformed equations and their differenced form.
(*b*) Accurate and easy implementation of boundary conditions.
(*c*) Smaller numerical errors of the solution for governings differential equations by faster methods. The accuracy and speed of methods could be due to the mapping.

Orthogonal grid is the optimum choice between complete non-orthogonal grids and conformal mapping. Orthogonal grids have been produced by solving either elliptic partial differential equations [Ryskin and Leal (1983), Eca (1996)] or hyperbolic differential equations [Steger and Chaussee (1980), Nair and Sengupta (1998), Cordova and Barth (1988)].

Ryskin and Leal (1983) and Duraiswami and Prosperetti (1982) and Eca (1996) have solved the covariant Laplace equations

$$\left.\begin{aligned}\frac{\partial}{\partial\xi}\left(f\frac{\partial x}{\partial\xi}\right)+\frac{\partial}{\partial\eta}\left(\frac{1}{f}\frac{\partial x}{\partial\eta}\right)=0\\ \frac{\partial}{\partial\xi}\left(f\frac{\partial y}{\partial\xi}\right)+\frac{\partial}{\partial\eta}\left(\frac{1}{f}\frac{\partial y}{\partial\eta}\right)=0\end{aligned}\right\} \tag{9.18}$$

for transformation from the physical (x,y)-plane to the mapped (ξ, η)-plane. In the above equation, f is the distortion function, given by

$$f=\frac{h_{22}}{h_{11}}=\frac{\sqrt{x_\eta^2+y_\eta^2}}{\sqrt{x_\xi^2+y_\xi^2}}. \tag{9.19}$$

In Ryskin and Leal (1983) the f distribution was interpolarated as a bivariate distribution; Duraiswami and Prosperetti (1982) used a specific method of obtaining a class of f functions where it is expressed as a product of two independent functions of the transformed coordinates with conditions imposed on them for proper grid generation. However, a general distortion function (as used here) according to these authors is *highly desirable to achieve a more localized control on the density of the coordinate lines.* Eca (1996) used a procedure to iteratively obtain f along with the main grid generation iteration for the resultant nonlinear system. The disadvantage of the methods used by Ryskin and Leal (1983) and Duraiswami and Prosperetti (1982) is that the grid point distribution on the defining boundaries is made to change as part of the iterative method. Note that Equation (9.18) is obtained by utilising the definition of the distortion function (Equation (9.19)) and the orthogonality constraint

$$g_{12}=x_\xi x_\eta+y_\xi y_\eta=0. \tag{9.20}$$

It has been pointed out in Duraiswami and Prosperetti (1982) that any solution of Equation (9.18) alone does not necessarily satisfy the orthogonality condition given by Equation (9.20)—a fact that can be demonstrated easily.

The hyperbolic methods, in this context, are superior in producing orthogonal grids because they use Equation (9.20) as one of the governing grid generation differential equations. In fact, Steger & Chaussee (1980) have clearly shown that the grid generation process of solving Equation (9.20) and one of the equations of Equation (9.24) constitutes a hyperbolic problem, while solving both of Equations (9.24) is an ill-posed problem. In general, hyperbolic methods are attractive for external flow problems where the shape of the outer boundary is not a constraint. However, they have also been successfully implemented for internal flow problems by Jeng et al (1995), Cordova and Barth (1982) by relaxing orthogonality.

In the present method, Equation (9.20) along with one of the Beltrami equations as given in Equation (9.24) (referred to as generalized Cauchy–Riemann equation in Ryskin and Leal (1983)) are solved with arc-length prescribed in the normal direction. This method of using constant arc-length is similar to the suggestion of Dwyer (1994) to overcome the problem of grid shock formation in hyperbolic grid generation over strong concave surfaces.

The general mapping in two dimensions is given by

$$\xi = \xi(x, y), \qquad \eta = \eta(x, y). \tag{9.21}$$

The metrics of the transformation is given by

$$g = \begin{vmatrix} g_{11} & g_{12} \\ g_{21} & g_{22} \end{vmatrix}. \tag{9.22}$$

The scale factors of the transformation are given by

$$h_{ij} = \sqrt{g_{ij}}. \tag{9.23}$$

The orthogonality condition given by Equation (9.20) can be combined with the definition of the Jacobian to yield the Beltrami equations

$$x_\eta = f y_\xi, \qquad y_\eta = -f x_\xi. \tag{9.24}$$

It is easy to see that Equation (9.18) follows straight from Equation (9.24) upon eliminating x and y, respectively.

Let us demonstrate this, where O-type grids are produced using the finite difference method. Here, Equation (9.20) and any one of Equation (9.24) are solved along with the prescribed arc-length in the hyperbolic direction by taking the following steps:

(*a*) The points on the body are distributed along $\eta = 0$ line which fixes the scale factor h_{11} distribution on this line.

(*b*) The grid line increment ΔS_η in the η-direction is prescribed and this fixes the other scale factor h_{22}, given by

$$h_{22} = \frac{\Delta S_\eta}{\Delta \eta}. \tag{9.25}$$

Thus, the distortion function f can be evaluated on $\eta = 0$ line.

(*c*) Equation (9.20) and either part of Equation (9.24) are then integrated by one step. This produces (x, y) co-ordinates of the $\eta = \Delta\eta$ line.

(*d*) Steps (a)–(c) are repeated to obtain the successive grid lines.

The present procedure for the evaluation of f not only makes the grid generation process very fast, it also gives a rational basis for the choice of the function f. This process of fixing the distortion function makes the governing grid generation equations linear and hence the grid generation process is very fast. Furthermore, the ability to prescribe the grid line spacing in the normal direction is a very useful feature of the generated grids for viscous flow calculations.

The above mentioned method has been implemented on various aerofoils. In Fig. 9.9 the physical and computational planes are illustrated. For a typical plane O-type grid the physical body is represented as ABC, where the cuts AF and CD are introduced starting from the trailing edge so that the computational domain is simply connected. Dirichlet conditions are prescribed on ABC and the grid spacing in the wake of the aerofoil is achieved by distributing points on the cuts AF and CD. The first grid line is located as close as 0.0001 c of the aerofoil. The successive grid lines are expanded smoothly by increasing the spacing by a constant value.

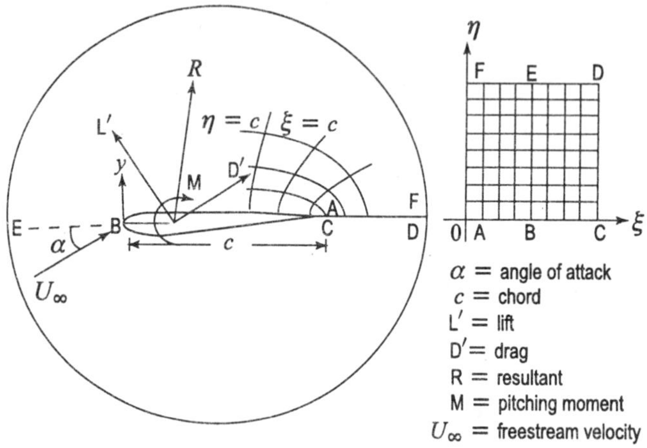

Fig. 9.9 *The physical and computational domains for the O-type grid.*

The first 100 lines in the hyperbolic direction of the grid generated for NACA 0015 are shown in Fig. 9.10. The slope discontinuity at the trailing edge is avoided by fitting a parabola in the open trailing edge, with continuous slope everywhere. There are 233 points in the ξ-direction, 301 lines in the η-direction, and the outer boundary is 14.76 chords away. The included angle between the constant ξ- and η- lines are virtually orthogonal everywhere except at three points along the

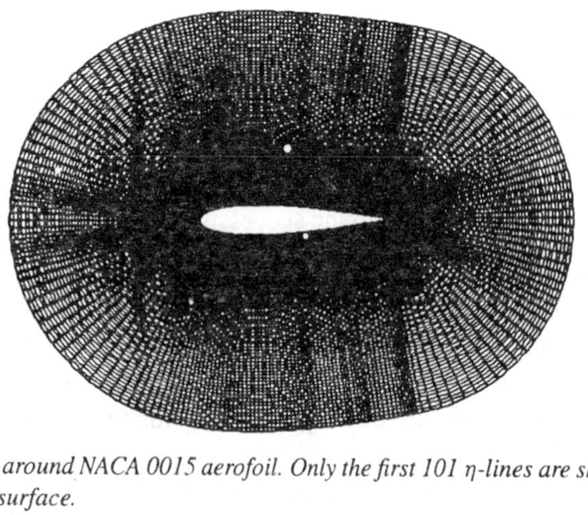

Fig. 9.10 *Grid generated around NACA 0015 aerofoil. Only the first 101 η-lines are shown. There are 233 points on the aerofoil surface.*

second η-line, where the worst departure from the orthogonality is given by the included angle of 89.476°. The included angle is calculated from

$$\theta = \cos^{-1}\left[\frac{g_{12}}{h_{11}h_{22}}\right]. \tag{9.26}$$

In calculating the metrics, second-order central differences are used. The major advantage of the present method is that the addition and deletion of the points have only local effects. The orthogonality property of the generated grids is better than the earlier efforts by Steger and Chaussee (1980), Duraiswami and Prosperetti (1982) and Eca (1980). For a cambered turbine blade, Steger and Chaussee (1980) have reported the worst non-orthogonality as $\approx$ 20%; Duraiswami and Prosperetti (1982) reported 1.11° as the maximum departure from orthogonality; and Eca (1980) reports this as 4.41° for the NACA 0015 aerofoil. In Fig. 9.11 the distortion function has been plotted as a function of η and ξ for NACA 0015. This figure shows that the present method uses f, which is quite different from the analytic expression used in Duraiswami and Prosperetti (1982). In fact, the local control of grid lines as seen in the present method precludes such simple representation of f. Although f depends strongly on ξ and η near the trailing edge, later on it is a weak function of η.

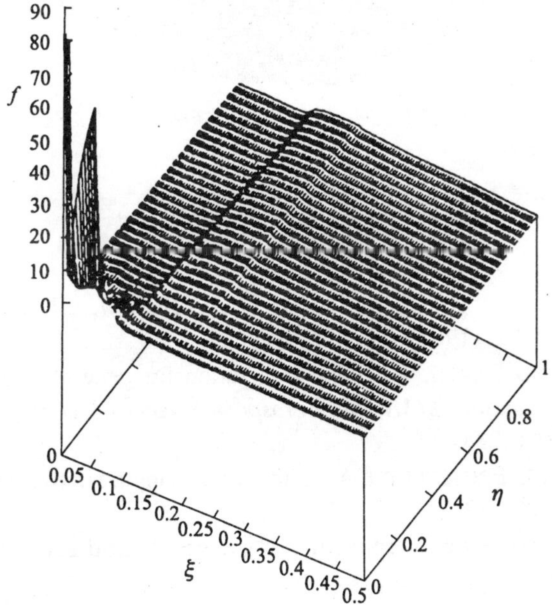

Fig. 9.11 *The f distribution in the (ξ, η) plane. Note the strong dependence of f on (ξ, η) near the trailing edge only.*

The grid for a thicker (18%) aerofoil with camber (NACA 23018) is shown in Fig. 9.12. The first 101 η- lines of the generated grids are shown, and the effect of the large camber on generated grids produces grid clustring on the lower side near the leading edge, due to the shape of the camber line. The worst non-orthogonality here is obtained near the trailing edge with a value of $\theta = 89.33°$. Once again, the open trailing is closed by an analytic curve affecting only the last 0.01% of the chord near the trailing edge.

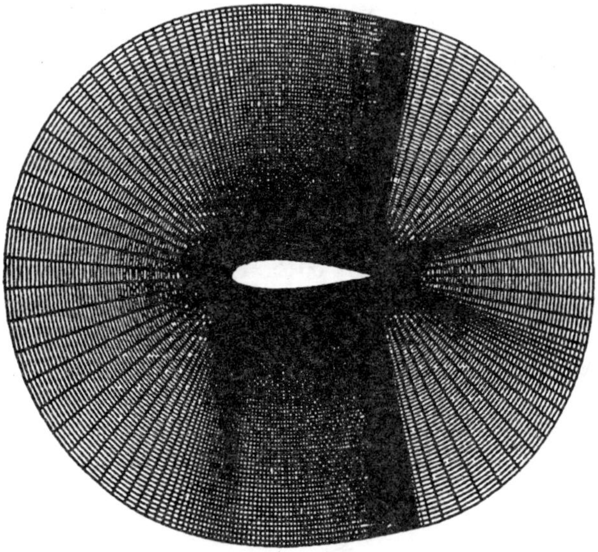

Fig. 9.12 *Grid generated around NACA 23018 aerofoil. Only the first 101 η-lines are shown. There are 229 points on the aerofoil surface.*

References

[1] ARIS, R., *Tensors and Basic Equations of Fluid Mechanics*, Prentice Hall, Englewood Cliffs, New Jersy, USA (1962).

[2] CORDOVA, J.Q. AND BARTH, T.J., Grid generation for general two-dimensional regions using hyperbolic equation, *AIAA 26th Aerospace Sciences Meeting*, Reno Nevada, *Paper No. AIAA-88-0520*, (1988).

[3] DURAISWAMI, R. AND PROSPERETTI, A., Orthogonal mapping in two dimensions, *J. Comput. Phys.*, **98**, 254–268 (1982).

[4] DWYER, H.A., A geometric interpretation of hyperbolic grid generation, *Comput. Fluids*, **23**, 737–748 (1994).

[5] ECA, L., 2D orthogonal grid generation with boundary point distribution control, *J. Comput. Phys.*, **125**, 440–453 (1996).

[6] EISEMAN, P.R., Grid generation for fluid mechanics computations, *Annual Review of Fluid Mechanics*, **17**, pp 487–522 (1985).

[7] HOFMANN, K.A. & CHIANG, S.T., *Computational Fluid Dynamics for Engineers*, **Vol 1**, Engineering Educational System, Austin, Texas, USA (1998).

[8] JENG, J.N., SHU, Y.L. AND LIN, W.W., Grid generation for internal flow problem by methods using hyperbolic equations, *Numer. Heat Trans. B*, **27**, 43–61 (1982).

[9] NAIR, M.T. & SENGUPTA, T.K., Orthogonal grid generation for Navier-Stokes computations, *Int. J. Num. Methods in Fluids*, **28**, pp 215–224 (1998).

[10] RYSKIN, G. AND LEAL, L.G., Orthogonal mapping, *J. Comput. Phys.*, **50**, 71–100 (1983).

[11] STEGER, J.L. AND CHAUSSEE, D.S., Generation of body-fitted coordinates using hyperbolic partial differential equations, *Siam J. Sci. Stat. Comput.*, **1**, 431–437 (1980).

[12] THOMPSON, J.F., THAMES, F.C. AND MASTIC, C.W., Automatic numerical generation of body fitted curvilinear coordinate systems for fields containing any number of arbitrary two-dimensional bodies, *J. Comput. Phys.*, **15**, pp 299–319 (1974).

10 Spectral Analysis of Numerical Schemes and Aliasing Error

The basis of spectral analysis of numerical methods is to study the associated physical events as a general wave process described aptly by Laplace–Fourier transforms for general aperiodic behaviour of the fluid dynamical system. In Chapter 4 we defined a space–time dependent function $f(x, t)$ in terms of its Laplace–Fourier transform as

$$f(x,t) = \int\int F(\omega, k) e^{i(kx-\omega t)} d\omega dk. \tag{10.1}$$

We have also shown that if 'f' is governed by 1D wave equation with information propagating along the positive x-axis direction with a phase speed c, then the governing equation is given by

$$\frac{\partial f}{\partial t} + c\frac{\partial f}{\partial x} = 0 \quad \text{with } c > 0. \tag{10.2}$$

The corresponding *physical dispersion* relation is given by

$$\omega = kc. \tag{10.3}$$

While c denotes the phase speed, it is the group velocity (V_g) which determines the speed at which the energy propagates and hence it is the more meaningful parameter physically; for equation (10.2) this is given by

$$V_g = \frac{d\omega}{dk} = c. \tag{10.4}$$

If one wants to analyse numerical schemes that are to be used for the solution of Navier-Stokes equation, then one must investigate the efficacy of discretization of each and every term in the governing differential equation. We have seen that there are methods that treat spatial and temporal discretization separately to arrive at difference equations that are linear algebraic equations. Irrespective of the route taken to discretize, the final difference equation should be such that apart from it being stable and consistent (as a consequences it is also convergent by Lax's theorem), it must also preserve the correct space–time behaviour simultaneously for all wave number–circular frequency combinations. In the spectral plane this is given by the dispersion relation. Any numerical scheme (irrespective of its being defined in physical or transformed plane) that is used to solve the governing differential equation satisfying the physical dispersion relation in a numerical sense is called Dispersion Relation Preservation (DRP) scheme. In DNS applications, spurious dispersion may be misinterpreted as turbulence. Thus this aspect of CFD studies should be undertaken carefully. Why is it specifically important for DNS?

We realise that for high Reynolds number flows the energy spectrum is broadband. To represent this flow field, the adopted numerical scheme must be able to resolve this wide spectrum of length and time scales. Thus, one can immediately envisage the requirement for developing high resolution schemes which are accurate up to a wide range of wave numbers and circular frequencies. While, for incompressible flows one can generate higher order schemes of any order, for the compressible flow it is not so easy due to presence of various discontinuities. Because of discontinuities and the associated Gibbs' phenomena, there will necessarily be oscillations and over- and under-shoot of the solution in the near vicinity of such discontinuities. And in the presence of non-linearity in governing differential equation, this can lead to additional numerical instability. Usage of higher than second order reconstruction schemes are therefore fraught with danger. Thus, Harten (1983) described any scheme as high resolution scheme for compressible flows, if the numerical method is at least second order accurate in the smooth part of the flow. The reason behind this statement is the near-unanimous agreement among researchers about the need to add lowest order (second order) dissipation to "capture" shocks and reduce Gibbs' phenomena. However, for incompressible flows there are no such restrictions. Hence the "high resolution" schemes for incompressible flows are truly very high order schemes. In general, we require not only very high resolution schemes, we also require that each and every resolved wave number must propagate its phase and energy at the right speed. A numerical scheme designed without much attention to this aspect will show dispersion phenomenon, i.e., different wave numbers will propagate at different speeds even though the corresponding physical system is non-dispersive—as in the case of one-dimensional wave equation given by equation (10.2). This shows why one requires a DRP scheme.

The reason that the energy is broadband is due to the various physical processes. For example, for three-dimensional flows it is the nonlinear vortex-stretching term (the first term on the right hand side of equation (2.48)) that is responsible for the energy to cascade to smaller length scales. It is also now understood that apart from this mechanism, response of fluid dynamical system to the omnipresent background noises are highly dispersive. This dispersion of the response itself is equivalent to exciting smaller scales as has been demonstrated in Sengupta et al. (2003) as the other mechanism by which energy appears at the smaller scales. It has been proven that this dispersion is due to the linear instability mechanism and can be present in 2D flows as well! In any case, it is important to realise that the nonlinear term in Euler / Navier-Stokes equation should be modelled correctly, which is the source of appearance of energy at small scales by both these mechanisms. The error in modelling of this term can arise from two sources: It can be due to (i) the accuracy of the discretization of the derivative terms, and (ii) aliasing error. The role of aliasing error of discretization is not well appreciated or understood at present. While aliasing may not be a problem for low Reynolds number flow problem or for low order methods (which we will discuss in this chapter), it is a major problem for higher order accurate methods and at higher Reynolds number. It will also be shown that aliasing can be an undesirable outcome of computation in the transformed plane—even for a linear problem! These error sources can be analysed effectively using spectral analysis tools.

10.1 Error associated with Discretization of First Derivatives

Let us consider the spatial derivative of the function given by Equation (10.1). Analytically,

$$\frac{\partial f}{\partial x} = \int\int ikF(\omega, k)e^{i(kx-\omega t)}dkd\omega. \tag{10.5}$$

When this is evaluated numerically, we can write out an equivalent spectral representation for the derivative as

$$\frac{\partial f}{\partial x} = \int\int ik_{\text{eq}}F(\omega, k)e^{i(kx-\omega t)}dkd\omega. \tag{10.6}$$

k_{eq} is different for different discretization scheme. Let us evaluate some of them for some finite differencing schemes.

10.2 Second Order Central Differencing (CD_2) Scheme

If we represent the first derivative in a uniform grid in the physical plane by

$$\left.\frac{\partial f}{\partial x}\right|_m = \frac{f_{m+1} - f_{m-1}}{2\Delta x\ .}, \tag{10.7}$$

then since $x = m\Delta x$, using Equation (10.1) in Equation (10.7) we get

$$\begin{aligned}\frac{\partial f}{\partial x} &= \int\int F(\omega, k)\frac{e^{ik(m+1)\Delta x} - e^{ik(m-1)\Delta x}}{2\Delta x}e^{-i\omega t}dkd\omega \\ &= \int\int F\frac{e^{ik\Delta x} - e^{-ik\Delta x}}{2\Delta x}e^{i(kx-\omega t)}dkd\omega.\end{aligned} \tag{10.8}$$

Comparing the expressions given in Equations (10.6) and (10.8), we get

$$(k_{\text{eq}})_{\text{CD}_2} = \frac{\sin k\Delta x}{\Delta x}. \tag{10.9}$$

10.3 Discrete Computing and Nyquist Criterion

At this point we should discuss the range of k that is resolved for a given choice of Δx. We know that any wave can be represented by three quantities—the amplitude, wavelength and the phase shift. Thus to fix three unknowns, we require at least three points for a pure monochromatic wave. For a grid size Δx, one can resolve a wave of wavelength $2\Delta x$ which intersects the x-axis at three points.

Thus the minimum wavelength that is resolved by such a grid is given by

$$\lambda_{\min} = 2\Delta x.$$

Therefore,

$$k_{\max} = \frac{2\pi}{\lambda_{\min}} = \frac{\pi}{\Delta x}. \tag{10.10}$$

In digital signal processing, the above relates the maximum circular frequency with the data sampling rate and is known as Nyquist criterion (see for example Press et al. (1989)).

10.4 Spectral Accuracy of Differentiation

Thus one can investigate the effectiveness of the differencing strategy given by Equation (10.9) when the argument varies in the range $0 \leq k\Delta x \leq \pi$. It is well known that the spectral method has the best accuracy among all discrete schemes. Hence we will calibrate all other methods with respect to spectral methods and the effectiveness of derivative evaluation will be represented by plotting k_{eq}/k versus $k\Delta x$. This is shown in Fig. 10.1 for the CD_2 scheme.

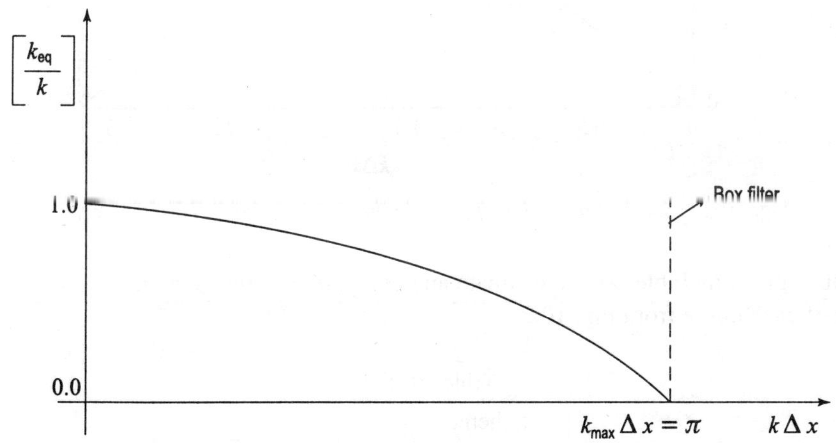

Fig. 10.1 *Spectral analysis of CD_2 scheme. Dotted line is for spectral method using the same uniform grid.*

One of the most attractive features of this analysis tool is the view it affords of the ability of different discretization schemes on a scale-by-scale basis.

Ideally, one would like to have a numerical method for which k_{eq}/k is equal to one, as shown by the dotted line. However, one notices that at higher wave numbers the derivative components are severely attenuated. Thus any discretization scheme, while evaluating derivatives, imposes a *filter* of its own on the unknown. Remembering this will assist us in understanding how the Large Eddy Simulation (LES) works. This attribute of filtering due to discrete evaluation shows that even the

Nyquist criterion is an over-estimate to define the cut-off wave number for discrete computations. Thus one should define the cut-off wave number as

$$k_c = \alpha \frac{\pi}{\Delta x}. \qquad (10.11)$$

Different discretization schemes have different values of α. In Fig. 10.2 is shown the plot of $k_{eq}\Delta x$ versus $k\Delta x$ for various schemes taken from Sengupta & Nair (1999). This figure can be used to estimate the value of α.

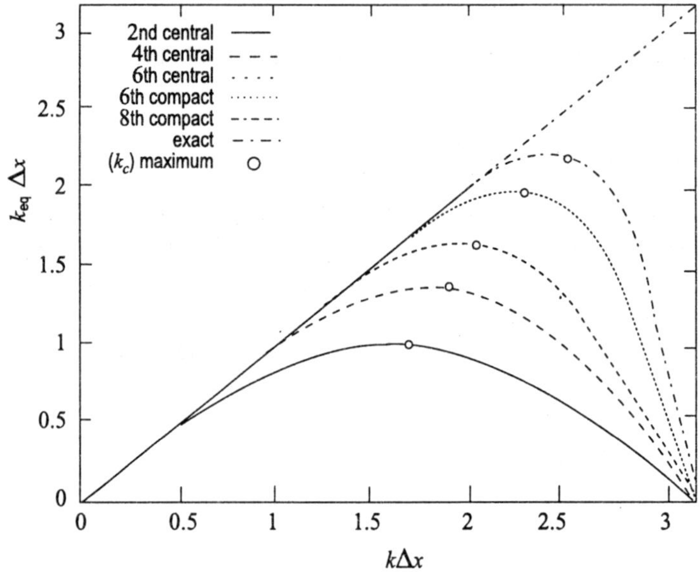

Fig. 10.2 *Spectral resolution of central schemes shown as k_{eq}/k versus $k\Delta x$.*

The values given in Table 10.1 to estimate and compare different schemes are obtained as the maximum of each curve from Fig. 10.2.

Table 10. 1

Scheme	α
CD_2	$1/\pi$
CD_4	0.43678
CD_6	0.50478
Compact-6th	0.63324
Compact-8th	0.67936

The last two entries in the above table correspond to the compact differences schemes (as in Lele (1992)) that will be described in Chapter 11.

10.5 Fourth Order Central Difference Scheme (CD_4)

Instead of using Equation (10.7), if we use the following representation for the first derivative,

$$\left.\frac{\partial f}{\partial x}\right|_m = \frac{f_{m-2} - 8f_{m-1} + 8f_{m+1} - f_{m+2}}{12\Delta x}, \tag{10.12}$$

then the corresponding discretization is fourth order accurate. Its k_{eq} can be obtained from

$$= \int\int F \frac{e^{-2ik\Delta x} - 8e^{-ik\Delta x} + 8e^{ik\Delta x} - e^{2ik\Delta x}}{12\Delta x} e^{i(kx-\omega t)} dk d\omega,$$

and on simplification this yields

$$(k_{eq})_{CD_4} = \frac{(4 - \cos k\Delta x)\sin k\Delta x}{3\Delta x}. \tag{10.13}$$

Consistency condition: Note the fact that both the Equations (10.9) and (10.13) show that as $k\Delta x \rightarrow 0$, the ratio, $\left(k_{eq}/k\right)$ is unity. This must be so for a proper discretization scheme, because the limit $k\Delta x \rightarrow 0$ implies approaching the continuum limit from the discrete limit. This is what we have already defined as the *consistency* property in Chapter 6. The above is the same statement in the spectral plane. For the sixth order central difference scheme, look at Exercise 10.1.

10.6 Role of Upwinding

So far we have been looking at central schemes. However, in computing real flows one often uses upwind schemes, the reason being similar to what has been said about numerical stability while discussing the solution methods for hyperbolic equations in Chapter 8.

10.6.1 First order upwind schemes (UD_1)

Consider for the moment the first order upwind differencing for the first derivative appropriate to Equation (10.2), i.e., for $c > 0$,

$$\left.\frac{\partial f}{\partial x}\right|_m = \frac{f_m - f_{m-1}}{\Delta x} = \int F \frac{1 - e^{-ik\Delta x}}{\Delta x} e^{i(kx-\omega t)} dk d\omega. \tag{10.14}$$

Therefore,

$$(k_{eq})_{UD_1} = \frac{\sin k\Delta x}{\Delta x} - i\frac{1 - \cos k\Delta x}{\Delta x}. \tag{10.15}$$

The real part of k_{eq} is the same as the one obtained for CD_2—as one would expect. The imaginary part is due to the implicit second order derivative term added for artificial dissipation. The imaginary part of the above scheme and other upwind schemes are shown in Fig. 10.3.

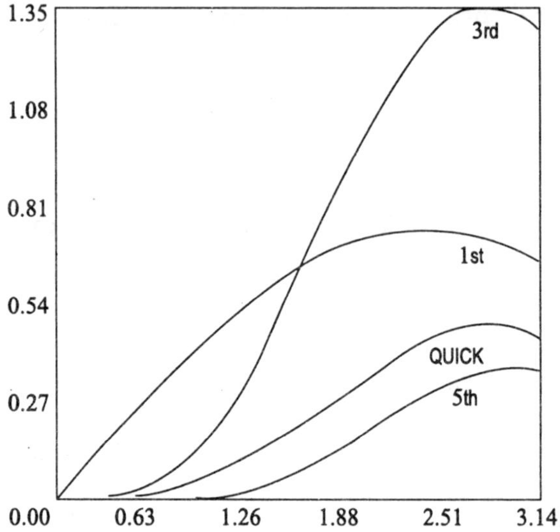

Fig. 10.3 *The imaginary part of the k_{eq}/k versus $k\Delta x$ for different explicit upwind schemes.*

The figure establishes the fact that apart from filtering—as represented by the real part of Equation (10.15)—the imaginary part further attenuates the component across all wave numbers. In fact this added distortion of all wave number components must be the reason for which this scheme should be used with care for any computations. The maximum dissipation is at $k\Delta x = 2.3 - 2.4$, much above the cut-off wave number (k_c) of the scheme. Thus we can state that the numerical dissipation is a monotonically increasing function of the wave number of interest.

10.6.2 Third order upwind scheme (UD_3):

As we have seen above, that the first order upwind scheme is equivalent to a second order central differencing scheme plus an artificial dissipation term that is nothing but the second order derivative term itself. Proceeding the same way, a third order upwind scheme can be constructed from a blend of CD_4 and the fourth derivative term. The CD_4 scheme is given by

$$\left(\frac{\partial f}{\partial x}\bigg|_m\right)_{CD_4} = \frac{f_{m-2} - 8f_{m-1} + 8f_{m+1} - f_{m+2}}{12\Delta x}. \tag{10.16}$$

Following the same method used in Equation (5.35) to evaluate the sixth derivatives, one can obtain

$$\frac{\partial^4 f}{\partial x^4} = \frac{f_{m+2} - 4f_{m+1} + 6f_m - 4f_{m-1} + f_{m-2}}{(\Delta x)^4}, \tag{10.17}$$

so that one can construct the third order upwind scheme as,

$$\left(\frac{\partial f}{\partial x}\right)_{UD_3} = \left(\frac{\partial f}{\partial x}\right)_{CD_4} \pm \beta \frac{\Delta x^3}{4!}\frac{\partial^4 f}{\partial x^4}. \tag{10.18}$$

Note that by choosing the parameter β in Equation (10.18) one can control the amount of dissipation. The choice of the sign of the dissipation term is guided by feedback stabilization. This is discussed next.

10.7 Numerical Stability and Concept of Feedback

At this stage the reader should be made aware of the role of upwinding in providing numerical stability. For any evolution equation of the type

$$\frac{\partial u}{\partial t} = L(u), \tag{10.19}$$

irrespective of the nature of the operator on the right hand side of above, after discretizing the spatial derivative(s), one can write the above for the lth node as

$$\frac{\partial u}{\partial t} = \sum_{k=0}^{N} a_k u_{l\pm k} + b_l. \tag{10.20}$$

It is to be noted that even if $L(x)$ is nonlinear, we quasi-linearize such terms and Equations (10.20) is a valid expression.

Also, note that the value of N is dictated by the order of the discretization of the spatial derivatives. The higher the order of such discretization, the higher would be the value of N. Also, depending on the type of equation and its discretization, there might not be a term involving u_l on the right hand side of Equation (10.20). For fluid dynamical applications we have convection terms in $L(u)$ as exemplified by Equation (10.2). If we employ central differencing, as given by Equations (10.7) and (10.12), then $a_0 \equiv 0$. For such a case the time evaluation is independent of u_l, and then we have the corresponding neutral numerical stability for the error derived from such quasi-linearization considerations discussed above. This will show up as oscillatory error variation in space with wavelength twice the gird spacing ($2\Delta x$). This error is then at the Nyquist wave number. In literature this is referred to as odd–even oscillation, i.e., as if the odd and even point values are decoupled. However, due to aliasing the error spectrum will fill up and may even lead to numerical instability across all wave numbers.

To control such oscillations or numerical instability, one must add dissipation. The rationale behind such a move is to view Equation (10.20) in a way that the error (e_l) is governed by

$$\frac{\partial e_l}{\partial t} = -\alpha e_l + p \qquad \text{where } \alpha > 0 \tag{10.21}$$

such that $e_l \sim Ke^{-\alpha t}$. Note that Equation (10.21) is typical of a feedback system provided $\alpha > 0$. When we use first order upwinding, this condition of $\alpha > 0$ is equivalent to adding $\frac{\partial^2 u}{\partial x^2}$. For the third order upwind scheme, it is equivalent to subtracting $\frac{\partial^4 u}{\partial x^4}$ and so on for other higher order upwind schemes.

Based on our discussion on classification of PDEs, we see that wave propagation should be viewed as passage of information in fluid flow. Thus the sign of the dissipation term in Equation (10.18) for application in incompressible fluid flow should be decided by the local convection velocity. For compressible flows or for problems of aeroacoustics, this has to be determined by finding the eigenvalues of the Jacobian matrices, and it is possible to have omnidirectional anisotropic propagation of information. Also, note that in NS equation the physical dissipation is due to the second derivative terms and hence the added dissipation terms of Equation (10.18) does not alter physical dissipation, especially at lower wave numbers, unlike the case for first order upwinding. Secondly, note that even in spectral methods artificial dissipation is added occasionally to control aliasing error. Since the added dissipation terms are higher order dissipation – they are called *hyperviscosity* terms.

Historically, higher order upwinding was originally suggested by Leonard (1978) as a means of overcoming odd–even coupling problem. The idea was simply that as Reynolds number increases, the central differencing of the convection term produces a solution that shows sustained oscillation. This is due to the neutral stability inherent with all forms of central differencing. Leonard's idea of introducing numerical dissipation was to provide a negative feedback to the system of equation which will then reduce the amplitude of oscillation with time/iteration as discussed above. Later, in the 80s, Kawamura and his associates also started using third order upwinding scheme (Kawamura et al. (1985)) as given by Equation (10.21) with $\beta = 6$, i.e., they introduce numerical dissipation which is six times the leading truncation error term. The imaginary part of $\left(k_{eq}/k\right)_{UD_3}$ is given by

$$\text{Imag}\left\{\left(\frac{k_{eq}}{k}\right)_{UD_3}\right\} = \frac{3\cos 2k\Delta x - 12\cos k\Delta x + 9}{6k\Delta x} \tag{10.22}$$

and already shown in Fig. 10.3. Fifth order upwind scheme case is given as an exercise.

So far we have discussed the discretization of the convection term. Similarly, for solving Navier-Stokes equation one has to discretize the dissipation term. The role of dissipation discretization in NS equation is discussed later. If proper care is not exercised in choosing the grid parameters, then the higher wave number components of the physical dissipation is switched completely. This loss of dissipation can lead to severe problems in DNS and LES.

10.8 Spectral Stability Analysis

We have already discussed this topic while studying the von Neumann stability analysis for numerical methods while solving parabolic PDEs. There we used the normal mode approach; here we will extend it to more general PDEs.

For example, for the parabolic PDE

$$f_t = f_{xx}, \tag{10.23}$$

if we use the spectral representation of f as given by Equation (10.1), then using this in Equation (10.23) will yield the following relation between the wave number k, and the circular frequency ω.

$$\omega = ik^2. \tag{10.24}$$

In Equation (4.4), we referred to such a relation as the dispersion relation. Therefore the phase speed and group velocity of the system are given by

$$c = \frac{\omega}{k} = -ik, \tag{10.25a}$$

$$V_g = \frac{\partial \omega}{\partial k} = -2ik. \tag{10.25b}$$

The very fact that both of these are complex implies that parabolic PDEs, like the heat equation, do not exhibit 'wave like' behaviour—as would be expected. Also, for Equation (10.23) if we use an explicit method like FTCS, the finite difference equation is given by

$$\frac{f_m^{n+1} - f_m^n}{\Delta t} = \frac{f_{m+1}^n - 2f_m^n + f_{m-1}^n}{\Delta x}. \tag{10.26}$$

As before, if we use $r = \frac{\Delta t}{\Delta x^2}$ and split the time (indicated by n) and space dependence of f as

$$f(x,t) = \int F^n(k)e^{ikx}dk, \tag{10.27}$$

then substitution of Equation (10.27) in Equation (10.26) yields the following:

$$\frac{F^{n+1} - F^n}{\Delta t} = F^n \frac{e^{ik\Delta x} - 2 + e^{-ik\Delta x}}{\Delta x^2}. \tag{10.28}$$

Defining $G = F^{n+1}/F^n$ as the amplification factor, we get from the above the equivalent numerical amplification rate as

$$G(k) = 1 + r(2\cos k\Delta x - 2) = 1 - 4r\sin^2(k\Delta x/2).$$

Since $0 \leq k\Delta x \leq \pi$, $G(k)$ varies between 1 and $(1-4r)$. Thus the choice of r determines whether we have instability or not. Also, the instability can appear only when $(1 - 4r) \leq -1$, i.e., $r \geq \frac{1}{2}$ —a result we obtained from von Neumann stability analysis. So what is new?

Here it is possible to estimate the theoretical amplification factor. Using the operator notations used in Equations (5.21a)–(5.21f), we note that the amplification factor operator is given by

$$f(x, t + \Delta t) = \int GF(k,t)e^{ikx}dk.$$

Therefore, define $\triangle_t$ as the forward difference operator such that

$$\triangle_t f = f(x, t + \triangle t) - f(x, t) = \int (G - 1) F(k, t) e^{ikx} dk. \tag{10.29}$$

By using the definition of the function f as given by Equation (10.1),

$$\triangle_t f = \int\int \widehat{F} e^{i(kx - \omega t)} [e^{-i\omega \triangle t} - 1] dk d\omega$$

where $F(k, t) = \int \widehat{F} e^{-i\omega t} d\omega$.

Hence,

$$G(\omega) = 1 + e^{-i\omega\triangle t} - 1 = e^{-i\omega\triangle t} \tag{10.30a}$$

$$= 1 - i\omega\triangle t - \frac{\omega^2 \triangle t^2}{2!} + \ldots \tag{10.30b}$$

These are the theoretical estimates of the exact amplification factor for any differential equation. Different differential equations have different dispersion relations which can be substituted in either Equation (10.30*a*) or (10.30*b*) for the theoretical amplification factor of the differential equation. For example, for the heat equation we can use the dispersion relation given in Equation (10.24) to get the actual amplification rate of the heat equation as

$$G_{\text{act}}(k) = e^{-k^2 \triangle t}.$$

Now for small time steps we can obtain the approximate estimate of the actual amplification rate by expanding the exponential terms and retaining terms of up to $O(\triangle t)$ as,

$$G_{\text{act}}(k) \simeq 1 - k^2 \triangle t. \tag{10.31}$$

How does this relate to various amplification factors? Let us look at the numerical amplification factor of the explicit method given by

$$G_{\text{exp}}(k) = 1 - \frac{4\triangle t}{\triangle x^2} \sin^2 \frac{k\triangle x}{2}. \tag{10.32}$$

Now it is only for $k\triangle x \to 0$,

$$\text{Lim}_{k\triangle x \to 0} G_{\text{exp}}(k) \simeq 1 - \frac{4\triangle t}{\triangle x^2} \left(\frac{k\triangle x}{2}\right)^2 = 1 - k^2 \triangle t.$$

Thus the numerical amplification of the explicit method given by Equation (10.32) matches with the exact amplification rate given by Equation (10.31) only for small wave numbers, i.e., in the infra-red range. Thus if we try to solve a problem where energy resides at the high wave number range, then the numerical solution will be erroneous as compared to the exact solution of such a simple differential equation! This extra insight of the performance of numerical methods is made possible only by spectral analysis.

Now let us focus our attention on the wave equation (a hyperbolic PDE) as given by Equation (10.2). If we use FTCS scheme for this equation and substitute Equation (10.27) in Equation (10.2), we get,

$$\frac{F^{n+1} - F^n}{\Delta t} = -\frac{c}{2\Delta x}[e^{ik\Delta x} - e^{-ik\Delta x}]F^n.$$

Therefore the amplification factor of the explicit second order central difference scheme in solving Equation (10.2) is given by

$$G_{\text{CD}_2} = 1 - iN_c \sin k\Delta x, \tag{10.33}$$

where N_c is the CFL number.

Since $|G_{\text{CD}_2}| \geq 1$ for all $k\Delta x$, the FTCS scheme is unstable. But is it really so? From Equation (10.30a) we see that the exact amplification factor in this case is given by

$$G = e^{-i\omega\Delta t}. \tag{10.34}$$

If we linearise the expression we get the complex amplification rate as

$$G(\omega) \simeq 1 - i\omega\Delta t. \tag{10.35}$$

This shows that the theoretical amplification rate should approach unity as $\Delta t \to 0$ from the unstable side because $|G(\omega)| = (1 + \omega^2\Delta t^2)^{1/2}$. Hence any discrete-time numerical method is *theoretically* unstable! This has prompted researchers to state that Euler time marching as inherently unstable, which is also known from the analytic solution which simply states that the initial solution convects with the phase speed. However, the above expression also states all computational methods which are in the limit neutrally stable, will be marginally unstable for any $\Delta t \neq 0$.

For example, if we use Euler time marching, we are in essence using the amplification factor given by Equation (10.35) only in the limit of $k\Delta x$, $\omega\Delta t \to 0$. However, note that in Equation (10.33) we should first use the dispersion relation, and then use the equivalent wave number of the discretization scheme used for the spatial derivative. Thus we can write the numerical amplification factor of any numerical method that uses Euler time marching scheme for the wave equation as

$$G \simeq 1 - ick_{eq}\Delta x. \tag{10.36}$$

If we use the fourth order accurate space discretization scheme along with the Euler time marching scheme, then the equivalent wave number that was derived in Equation (10.13) can be used in Equation (10.36) to obtain the numerical amplification factor as

$$G_{\text{CD}_4} = 1 - \frac{iN_c}{3}\sin k\Delta x(4 - \cos k\Delta x). \tag{10.37}$$

Here also, $|G_{\text{CD}_4}| \geq 1$ and is unstable across all wave numbers. But if we choose a very small CFL number, then the numerical amplification factors (as given by Equations (10.33) and (10.37)) have the right analytical behaviour for the central schemes when the time step chosen (or the CFL number) is very small. But in real computations this is impractical because all computations are going to be performed with finite time steps and as shown in Equation (10.35)—the exact amplification factor would be unstable corresponding to these time steps. So a practical method has to be chosen that should be *computationally stable* and has an amplification factor as close to one as possible. Thus in a computational environment, we should try to achieve the theoretical amplification factor $G = 1$ for the limit $\Delta t \to 0$, while performing calculations with finite Δt. This is achieved by using upwinding schemes with very small CFL number.

If we use first order upwinding schemes for Equation (10.2), then we already have obtained the equivalent wave number in Equation (10.15) as

$$(k_{eq})_{UD_1} = \frac{\sin k\Delta x}{\Delta x} - i\frac{1-\cos k\Delta x}{\Delta x}.$$

Hence the numerical amplification factor of this scheme is given by

$$\left.\begin{aligned} G_{UD_1} &= 1 - iN_c\left\{\frac{\sin k\Delta x}{\Delta x} - i\frac{1-\cos k\Delta x}{\Delta x}\right\} \\ &= \{1 - N_c(1-\cos k\Delta x)\} - iN_c \sin k\Delta x \end{aligned}\right\}. \quad (10.38)$$

For the third order upwind scheme, the imaginary part of k_{eq} is already obtained in Equation (10.22). Hence the amplification factor is given by,

$$\left.\begin{aligned} G_{UD_3} &= 1 - \tfrac{iN_c}{3}\sin k\Delta x(4-\cos k\Delta x) - \tfrac{N_c}{2}[\cos 2k\Delta x - 4\cos k\Delta x + 3] \\ &= 1 - \tfrac{iN_c}{3}\sin k\Delta x(4-\cos k\Delta x) - \tfrac{N_c}{2}(\cos k\Delta x - 1)^2 \end{aligned}\right\}. \quad (10.39)$$

The plots of $|G_{UD_3}|$ versus $k\Delta x$ for $N_c = 0.001, 0.01$ and 0.1 are shown in Fig. 10.4.

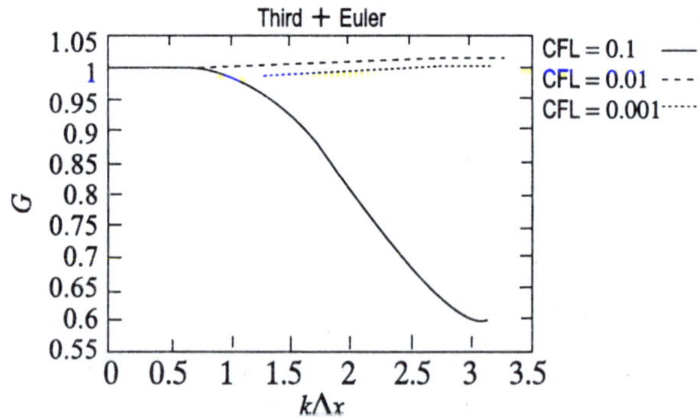

Fig. 10.4 *The amplification factor for 1D wave equation when discretized by 3rd order spatial upwind scheme and Euler time integration scheme for the indicated CFL number in the figure.*

It is clear that with higher values of CFL the method shows excessive damping at high wave numbers and have significant departure from the actual amplification rate and hence would be of no value.

We can, thus, conclude that while the central schemes have the correct trend as far as the theoretical amplification factor is concerned, they are practically unusable for long-time integration purposes. The upwind schemes offer a practical alternative but the first order upwinding scheme is far too dissipative! The higher order upwind methods hold the maximum promise in this regard as these methods are computationally stable, while the amplification factors are very close to unity for low CFL number – as one would expect analytically.

So far we have been looking at only first order accurate time discretization schemes, namely the Euler time marching scheme. Next we will investigate some higher order time marching schemes that have been used by researchers for DNS and LES studies; see for e.g. Kim & Moin (1985) for DNS and Schumann (1987) for LES and many other such efforts.

10.8.1 Mid-point leap frog scheme

For example in Geophysical Fluid Dynamics (GFD) and meteorology, time integration is often performed using mid-point leap frog method. For the wave equation given by (10.2), the corresponding finite difference equation is obtained by central differencing both the terms as given below:

$$\frac{f_m^{n+1} - f_m^{n-1}}{2\Delta t} = -\frac{c}{2\Delta x}(f_{m+1}^n - f_{m-1}^n). \tag{10.40}$$

Hence, following our notations there would be two amplification factors given by the following quadratic equation

$$G - \frac{1}{G} = -N_c(e^{ik\Delta x} - e^{-ik\Delta x}), \tag{10.41}$$

whose roots are given by

$$G_{1,2} = -iN_c \sin k\Delta x \pm \sqrt{1 - N_c^2 \sin^2 k\Delta x} \tag{10.42}$$

The interesting feature of this scheme is that $|G_{1,2}| = 1$ for all wave numbers irrespective of the choice of CFL number. Contrast this with the results obtained for the first order upwinding given in Equation (8.5*b*) which would mimic the actual amplification factor for all length scales for either $N_c = 1$ or 0. This feature leads to very accurate simulation of weather forecast codes. However, the simulations tends to de-correlate in long-time integration sequences because of aliasing error, phase error and the effect of nonlinearity in the governing equation.

10.8.2 Second order Adams-Bashforth scheme

The second time integration scheme to be discussed is due to Adams & Bashforth (see e.g. Lilly (1965) for this and other time integration schemes). It is interesting to note that the original scheme was given in Bashforth and Adams (1883) and the current name for the method is traced to Lilly (1965), and here the time integration proceeds via

$$f_m^{n+1} = f_m^n + \left[\frac{3}{2}\frac{\partial f}{\partial t}\bigg|_m^n - \frac{1}{2}\frac{\partial f}{\partial t}\bigg|_m^{n-1}\right]\Delta t + O(\Delta t^3). \tag{10.43}$$

If we use this method to solve Equation (10.2), then the above can also be written as

$$f_m^{n+1} = f_m^n + c\frac{\Delta t}{2}\left[-3\frac{\partial f}{\partial x}\bigg|_m^n + \frac{\partial f}{\partial x}\bigg|_m^{n-1}\right]. \tag{10.44}$$

If we express the space derivative by

$$\frac{\partial f}{\partial x}\bigg|_m^n = \int\int ik_{\text{eq}}F(k,\omega)e^{i(kx-\omega n\Delta t)}dkd\omega, \tag{10.45}$$

then from Equation (10.44) we get the following dispersion relation:

$$e^{-i\omega\Delta t} = 1 + i\frac{c\Delta t}{2}k_{\text{eq}}(\cos\omega\Delta t + i\sin\omega\Delta t - 3). \tag{10.46}$$

The amplification factor is obtained from Equation (10.44) as

$$G = 1 + i\frac{c\Delta t}{2}k_{\text{eq}}\left(\frac{1}{G} - 3\right). \tag{10.47}$$

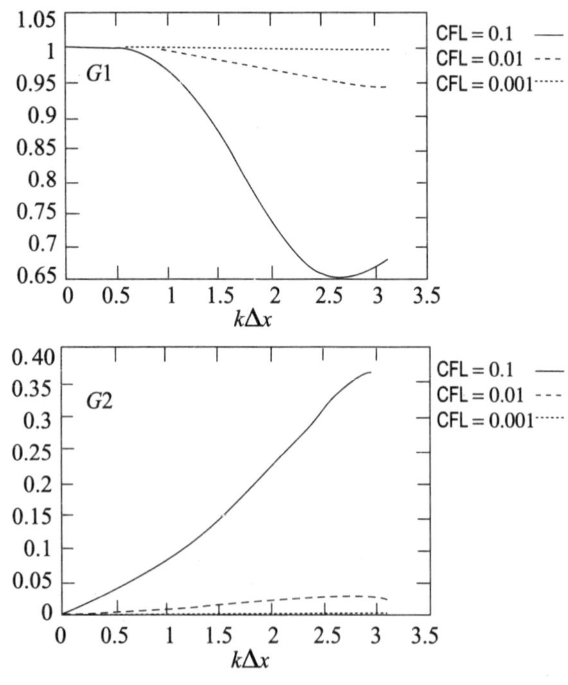

Fig. 10.5 *The amplification factor for 1D wave equation discretized by 3rd order spatial scheme and Adams-Bashforth time integration scheme.*

This is once again a three time-level integration scheme and would have two amplification factors which are shown in Fig. 10.5 for the case where third order upwind scheme for the space derivative has been used.

While G_1 is similar to the amplification factor obtained for the Euler time integration scheme, it is the second amplification factor G_2 which shows severe attenuation property across all wave numbers. For the overall solution, the time dependence will be governed by two functions. The first functions is dictated by the amplification factor G_1 and in the GFD literature is called the *physical mode*. The second function is dictated by G_2 and is called the *computational mode*. However there appears to be some misinterpretation about the behaviour of these modes. For example, Haltiner & Williams (1980) and Durran (1999) seem to suggest that the behaviour of *computational mode* "tends to damp, which is favourable".

This can indeed be favourable for a stable fluid dynamical system—as the case may be for laminar flows. However, for transitional or turbulent flows, the *physical mode* may itself have inherent physical instabilities, and will also be damped by the *computational mode*. This will be an unwarranted scenario. It is to be noted that the DNS results presented in Kim & Moin (1985) have used this time integration strategy. It can also be shown that the *computational mode* carries a part of the initial condition which is eventually damped. This loss of information can produce spurious results.

10.8.3 Multi-stage Runge-Kutta scheme

The above mentioned problem with the Adams-Bashforth scheme is not encountered by various multi-step Runge-Kutta integrators like RK_2 and RK_3 schemes that have been used for performing DNS by Le & Moin (1991) for incompressible flows and for high resolution computing of compressible flows by Shu & Osher (1998, 1989). To summarize, the finding of this subsection is the following: Any scheme should be tested first with some simple model equation to test the fundamental properties that any numerical method must possess before it is used for solving complex equations.

The two-and three-stage Runge-Kutta schemes, also known as Total Variation Diminishing schemes, in advancing from the nth time step to $(n+1)$th time step solution for the equation $u_t = L(u)$ are given below.

2nd order Runge Kutta method (RK_2 scheme)

$$\left.\begin{aligned} u^{(1)} &= u^{(n)} + \Delta t L(u^n) \\ u^{(n+1)} &= \frac{u^{(n)}}{2} + \frac{u^{(1)}}{2} + \frac{\Delta t}{2} L(u^{(1)}) \end{aligned}\right\} \tag{10.48}$$

3rd order Runge Kutta method (RK_3 scheme)

$$\left.\begin{aligned} u^{(1)} &= u^{(n)} + \Delta t L(u^{(n)}) \\ u^{(2)} &= \frac{3}{4}u^{(n)} + \frac{1}{4}u^{(1)} + \frac{\Delta t}{4} L(u^{(1)}) \\ u^{(n+1)} &= \frac{u^{(n)}}{3} + \frac{2u^{(2)}}{3} + \frac{2\Delta t}{3} L(u^{(2)}) \end{aligned}\right\} \tag{10.49}$$

10.9 Long-Time Integration and Dispersion Relation Preservation (DRP) Property

It has been noticed often that the nonlinear instability experienced at high Reynolds number cannot be suppressed by lowering the CFL number, i.e., by reducing the time step for integrating convection–diffusion equation. Even when such numerical instabilities are avoided, one is still faced with difficulties associated with solutions which seem to match well with experiments, but which as time progresses, seem to decorrelate. This brings us to the important criterion for DNS which relates to the dispersion relation that the governing equation has in the continuum sense, and which the discretized equation must satisfy. One of the problems associated with space–time propagation of numerical solution is that the various wave numbers and circular frequencies that are excited must have an appropriate relation between them. This has been termed as the dispersion relation in Equation (4.4) for one-dimensional wave equation.

It is clearly evident that for linear wave equation the phase speed and the group velocity of the disturbance are the same, i.e., the system is *non-dispersive*. For Euler and Navier-Stokes equation this wave property of the solution is built-in because of the nonlinear convection term. However, one does not know a priori whether the system is *non-dispersive* or not. Without providing any proof we state that the general fluid dynamical systems are *dispersive*. For any chosen numerical method, to solve these equations, the selected space and time step has to be such that for one-dimensional wave equation the non-dispersive nature is preserved. There is no reason to believe that any numerical scheme that is incapable of solving the simple wave equation, will perform correctly in solving Euler / Navier-Stokes equation. At the same time, one is not making a claim that correctly solving the one-dimensional wave equation will ensure proper solution for Euler/Navier-Stokes equation either. Thus the requirement of a numerical scheme to solve wave equations correctly tantamounts to satisfying a *necessary* condition that is not a *sufficient* condition though. The numerical dispersion relation for one-dimensional wave equation is given by

$$\omega_{\text{eq}} = ck_{\text{eq}}. \tag{10.50}$$

For example, if we choose third order upwinding scheme to spatially discretize Equation (10.2) and the Euler time marching scheme, then the numerical dispersion relation of this method is given by equating the real part of Equation (10.50) and is given by

$$\frac{\sin \omega\Delta t}{\Delta t} = \frac{c}{3\Delta x}(4 - \cos k\Delta x) \sin k\Delta x. \tag{10.51}$$

We propose here to quantify the effect of dispersion by comparing the group velocity of the numerical dispersion relation with the theoretical estimate of the same for the linear wave equation. The numerical group velocity above is obtained for the scheme presented in Equation (10.50) as

$$\frac{d\omega}{dk} = \frac{c}{3}\frac{4\cos k\Delta x - \cos 2k\Delta x}{\cos \omega\Delta t}. \tag{10.52}$$

Since the actual group velocity is known in this case, we can compare the performance of the numerical method for different choice of the space and time steps by plotting the area in the

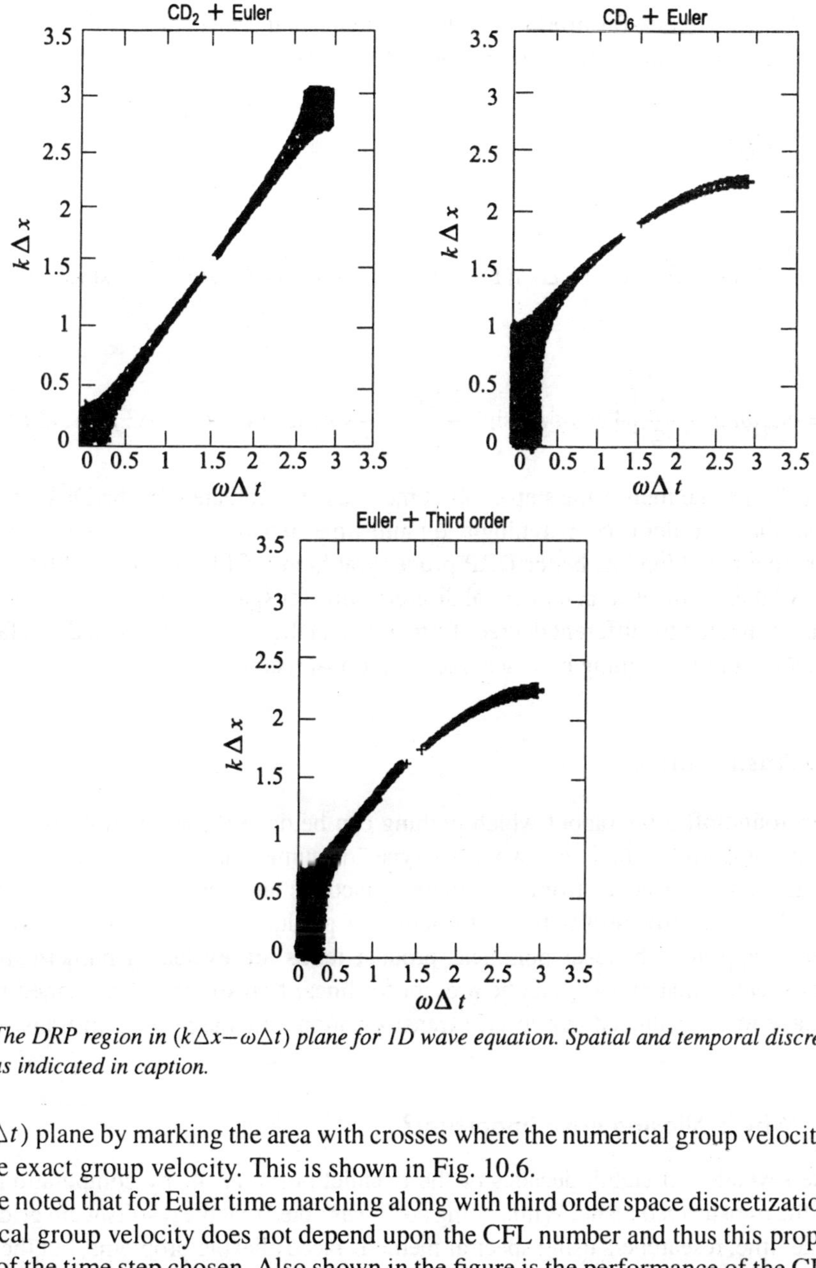

Fig. 10.6 *The DRP region in $(k\Delta x - \omega\Delta t)$ plane for 1D wave equation. Spatial and temporal discretizations are as indicated in caption.*

$(k\Delta x - \omega\Delta t)$ plane by marking the area with crosses where the numerical group velocity is within $\pm 5\%$ of the exact group velocity. This is shown in Fig. 10.6.

It is to be noted that for Euler time marching along with third order space discretization scheme, the numerical group velocity does not depend upon the CFL number and thus this property is not a function of the time step chosen. Also shown in the figure is the performance of the CD_2 scheme along with Euler time marching scheme. Comparison of these two schemes clearly show that there is a very narrow range of space and time step that can be taken for the CD_2 scheme as compared to the third order upwind scheme. When the fifth order upwind scheme is considered it is seen that even larger space steps can be chosen where the dispersion errors are minimum.

If one chooses Adams-Bashforth time integration scheme then the DRP region becomes a function of the CFL number; the numerical group velocity of such a scheme is given by

$$\frac{d\omega}{dk} = \frac{c}{2}\frac{N}{D}, \tag{10.53}$$

where

$$N = \frac{1}{3}(4\cos k\Delta x - \cos 2k\Delta x)(1 - \cos\omega\Delta t) + 4\sin k\Delta x \sin^2\frac{k\Delta x}{2}\sin\omega\Delta t$$

and

$$D = \cos\omega\Delta t - \frac{N_c}{2}\left[4\cos\omega\Delta t\sin^4\frac{k\Delta x}{2} + \frac{1}{3}\sin k\Delta x(4 - \cos k\Delta x)\sin\omega\Delta t\right].$$

In Fig. (10.7) some multi-time step explicit methods are evaluated for the DRP property.

It is clear that, if Euler time marching and multi-time step numerical schemes are both stable, then Euler time marching has better DRP property at higher CFL number and hence should be preferred. All these indicate that numerical discretization for space and time restricts the allowable grid size and time step by different degrees further when DRP is considered, and this factor should not be lost sight of in designing high accuracy numerical methods.

10.10 Aliasing Error

Apart from round-off error (about which nothing can be done (?) apart from increasing the bit precision of computing), there are two other types of numerical errors: truncation and aliasing errors. Truncation error arises from discretizing functions and derivatives and was described in Chapter 5. Aliasing error results from evaluation of product terms in a finite grid. In general, aliasing error appears whenever nonlinear product terms are evaluated numerically, but it is important to realise that aliasing can be present for linear terms in the transformed plane where products have to be evaluated due to grid transformation terms appearing in transformed plane equations.

10.10.1 Why is aliasing error important?

During the seventh and eighth decades of the twentieth century, many computations were performed in the transformed plane. While doing so, it was noted that the results were grid dependent. At the same time, researchers using spectral methods faced aliasing problems, but there was little appreciation that such problems could bedevil other discrete computations as well. The reason that aliasing problem was not noted was due to the fact that rarely DNS at high Reynolds numbers was performed using these discrete methods. As higher order methods came into existence, researchers started looking at unsteady problems at high Reynolds numbers, and computational results revealed the lurking presence of aliasing error.

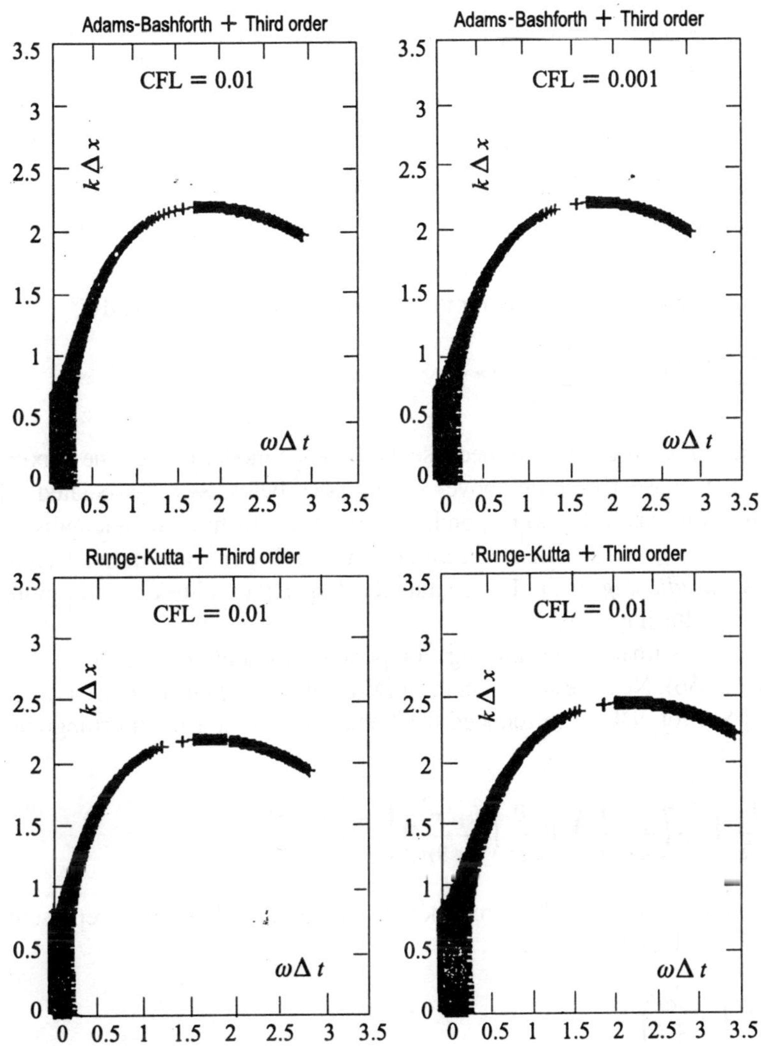

Fig. 10.7 *The DRP region in $(k\Delta x-\omega\Delta t)$ plane for 1D wave equation for third order upwind spatial with Adams-Bashforth and RK_2 time discretization.*

It also became clear to some researchers that the grid dependence of CFD solution in the transformed plane is related to the aliasing issues. Here a clear exposition is given for the same. However, it must be pointed out that in discrete computations that use a non-spectral method, this problem was diagnosed for the numerical estimation of the nonlinear convection terms by Phillips (1959), Arakawa (1966), Orszag (1971), and in recent times by Kravchenko & Moin (1997), Sengupta & Pin (2002), Sengupta & Nair (1999).

For example, consider the product of two terms in the physical space as

$$w(x_j) = u(x_j)v(x_j). \tag{10.54}$$

Consider the finite Fourier series representation of the above terms as given in a uniform grid of size Δx,

$$\left.\begin{aligned} w(x_j) &= \sum_{k=-\frac{N}{2}}^{\frac{N}{2}-1} \hat{w}_k e^{ik(j\Delta x)} \\ u(x_j) &= \sum \hat{u}_n e^{in(j\Delta x)} \\ v(x_j) &= \sum \hat{v}_m e^{im(j\Delta x)} \end{aligned}\right\} \text{where } \Delta x = \frac{2\pi}{N}. \tag{10.55}$$

Forming the direct product and obtaining the Fourier transform of w yields

$$\hat{w}_k = \sum_{n+m=k} \hat{u}_n \hat{v}_m + \sum_{n+m=k\pm N} \hat{u}_n \hat{v}_m. \tag{10.56}$$

Note that the second set of term arises because the $(n+m)$ modes exceed the maximum resolved mode N, i.e., the phase increases above $+\pi$ or goes below $-\pi$ (associated with the $2\Delta x$ wavelength). In these cases the corresponding harmonic will have wavelengths corresponding to $[2\pi - (m+n)\Delta x]$, and it will therefore appear as a lower wave number component. This phenomenon is the *aliasing* error. In general, the larger the value of k the more will be the contribution of the aliasing error.

So far we have estimated the aliasing components for nonlinear product terms in Equations (10.54) to (10.56). Next we show that in CFD calculations such error also arises in evaluating linear terms. In Section 9.4 we introduced the Laplacian operator in an orthogonal grid system by

$$\nabla^2 = \frac{1}{h_1 h_2}\left\{\frac{\partial}{\partial \xi}\left(\frac{h_1}{h_2}\frac{\partial}{\partial \xi}\right) + \frac{\partial}{\partial \eta}\left(\frac{h_1}{h_2}\frac{\partial}{\partial \eta}\right)\right\}. \tag{10.57}$$

In general, the scale factors h_1 and h_2 are functions of ξ and η. Therefore a representative term of Equation (10.57) can be written as

$$d(\xi) = \frac{\partial}{\partial \xi}\left(f\frac{\partial \omega}{\partial \xi}\right). \tag{10.58}$$

It is clearly evident that although f is a known function of ξ, evaluation of $d(\xi)$ involves taking a product of f and $\frac{\partial \omega}{\partial \xi}$, and that will cause aliasing.

Let us represent

$$f(\xi) = \int_{-k_m}^{k_m} F(k')e^{ik'\xi}dk', \tag{10.59}$$

where

$$k_m = \frac{\pi}{\Delta \xi} \tag{10.60}$$

from the Nyquist criterion defined in Equation (10.10). Present analysis derives the benefit of working in the transformed plane where we have a uniform grid, and the tools of Fourier series and transforms are directly applicable.

Similarly, let us also define

$$\omega(\xi) = \int_{-k_m}^{k_m} \Omega(k) e^{ik\xi} dk \tag{10.61}$$

and

$$d(\xi) = \int_{-k_m}^{k_m} D(\bar{k}) e^{i\bar{k}\xi} d\bar{k}. \tag{10.62}$$

Using these in Equation (10.58) results in

$$\int_{-k_m}^{k_m} D(\bar{k}) e^{i\bar{k}\xi} d\bar{k} = \int \int_{-k_m}^{k_m} -(k^2 + kk')F(k')\Omega(k) e^{i(k+k')\xi} dk' dk. \tag{10.63}$$

Thus equating the same phase components on either side of Equation (10.63) results in

$$D(\bar{k}) = -\int_{-k_m}^{k_m} (k^2 + kk')F(k')\Omega(k) dk \tag{10.64}$$

where

$$\bar{k} = (k + k'). \tag{10.65}$$

Hence

$$D(\bar{k}) = -\int_{-k_m}^{k_m} \{k^2 + k(\bar{k} - k)\} F(\bar{k} - k)\Omega(k) dk. \tag{10.66}$$

Note that the equality in Equation (10.63) is obtained for the condition given by Equation (10.65). However, there are other combinations of k and k' that exceed beyond k by $|k_m|$; at the same time it is possible that the phase on either side of Equation (10.63) matches. Since this exceeds $|k_m|$, the phase folds it back within Nyquist limit. This folding back of spectral components due to phase equality is called the aliasing error. It is to be emphasized that such folding back is necessary, due to the fact that one is restricted for the limits in the Fourier integrals in Equations (10.59) to (10.62) due to finite discretization fixed by the Nyquist criterion.

Thus the aliasing error arise when

(a) both k and k' are positive and $(k + k')$ is greater than k_m,
(b) both k and k' are negative and $(k + k')$ is less than $-k_m$.

Both these conditions are possible if and only if $k * k' > 0$.

This is indicated in Fig. 10.8, where the hatched areas are the combinations for which aliasing will be caused.

For case (a) that corresponds to the top right triangular area in Fig. 10.8,

$$k + k' = \bar{k} + 2k_m \tag{10.67}$$

because working in a fixed grid restrains one within the wave number range fixed by Nyquist criteria given by the condition of Equations (10.60). Thus in Fig. 10.9, the point A for which $(k + k')$ exceed k_m, will be folded back to a wave number at A′, such that Equation (10.67) holds.

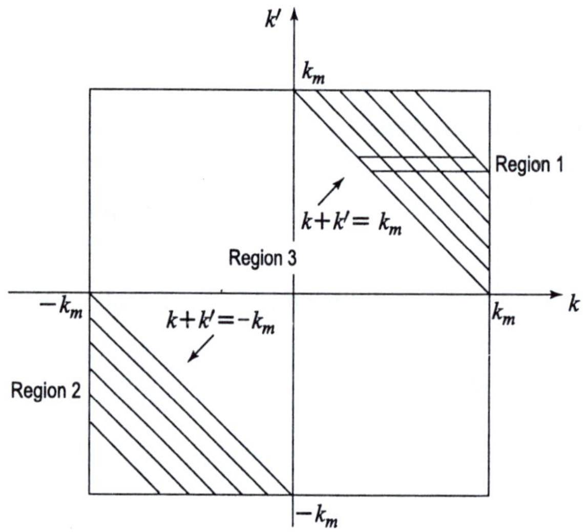

Fig. 10.8 *Spectral domain of interest for finite grid computations. The shaded areas contribute to aliasing problem in evaluating product terms.*

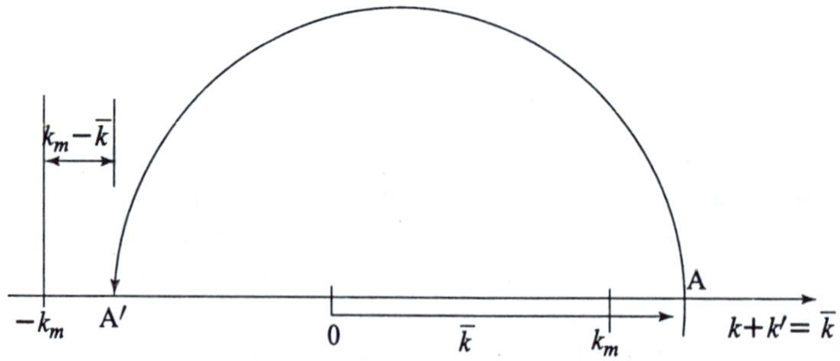

Fig. 10.9 *The point A is aliased to A′ due to finite grid computation that fixes the maximum wave number.*

Since

$$\bar{k} = -k_{\max} + (k + k' - k_{\max}),$$

it is easy to see that

$$i(k+k')\xi = i\left(\bar{k}+\frac{2\pi}{\Delta\xi}\right)\xi,$$

where $\xi = m\Delta\xi$; therefore $e^{i(k+k')\xi} = e^{i\bar{k}\xi + i2\pi i m}$.

As $e^{2\pi i m} \equiv 1$, hence the phases are matched for these combinations of k and k'.

For case (b), $(k+k')$ is negative and beyond $-k_m$, as indicated in Fig. 10.10.

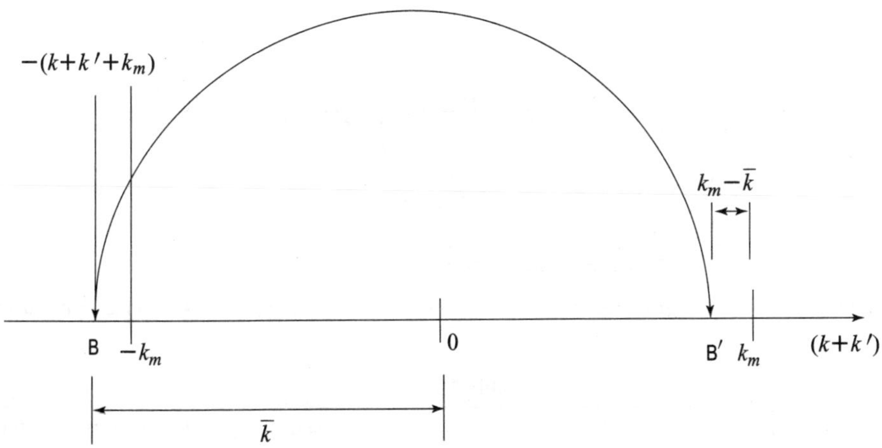

Fig. 10.10 *The point B is aliased to B′ due to finite grid computation that fixes the maximum wave number.*

Hence the point B is aliased on to B′ such that

$$-(k+k'+k_{\max}) = k_{\max}-\bar{k},$$

or

$$\bar{k} = k+k'+2k_m. \tag{10.68}$$

Thus from Equations (10.67) and (10.68), aliasing will occur when

$$k+k' = \bar{k}\pm 2k_m. \tag{10.69}$$

The above discussion graphically displays that evaluation of linear terms in the physical plane—as shown for the term given in equation (10.58)—will give rise to aliasing in the spectral plane. For Equation (10.63), we can calculate the aliasing contribution to $D(\bar{k})$ from

$$D_{\text{alias}}(\bar{k}) = -\int_{-k_m}^{k_m} k(\bar{k}\pm 2k_m)F(\bar{k}\pm 2k_m - k)\Omega(k)dk. \tag{10.70}$$

Next we demonstrate through an example, as to why it is said that the effect of aliasing is more at high wave numbers.

Example: Consider the case where k and k only take discrete integer values (i.e., when f and ω are strictly periodic functions expressed by Fourier series) with k_m equal to 5. Then, $(k+k')$

can take values between -10 to $+10$. Out of these values, only the values within ± 5 are the valid ones while the ones outside the limit will be aliased in $-k_m \leq k \leq k_m$.

The aliasing event for the case is shown in Fig. 10.11, where it is clearly shown that the values of $(k + k^{'})$ and the correspondence of the points where the aliased components migrate—shown by arrowheads. Let us take stock of the situation in a tabular form as given in Table 10.2.

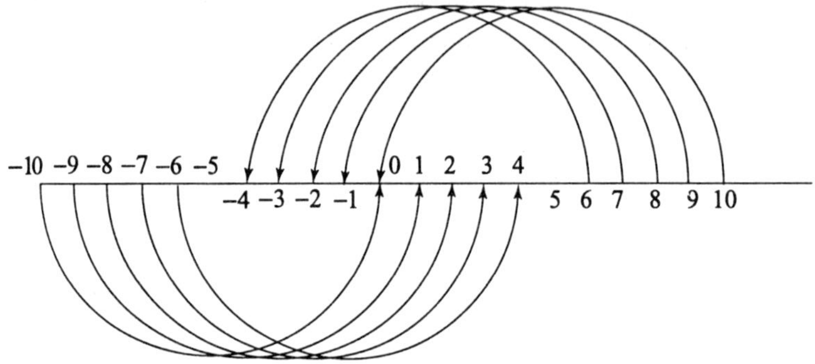

Fig. 10.11 *Schematic diagram showing aliasing from outside the range (–5, +5) to the resolved scale.*

Table 10. 2

$(k + k^{'})$ values for which aliasing occurs	No. of occurrences (the events)	Aliased to $\bar{k}$	The weight of integral in equation (10.64) $= k(k + k^{'})$
-10	$1\,(-5-5)$	0	50
-9	$2\,(-5-4, -4-5)$	1	$45 + 36 = 81$
-8	$3(-5-3, -4-4, -3-5)$	2	$40 + 32 + 24 = 96$
-7	$4(-5-2, -4-3, 3-4, -2-5)$	3	$35 + 28 + 21 + 14 = 98$
-6	$5(-5-1, -4-2, -3-3, -2-4, -1-5)$	4	$30 + 24 + 18 + 12 + 6 = 90$
6	$5\,(5+1, 4+2, 3+3, 2+4, 1+5)$	-4	90
7	$4\,(5+2, 4+3, 3+4, 2+4)$	-3	98
8	$3\,(5+3, 4+4, 3+5)$	-2	96
9	$2(5+4, 4+5)$	-1	81
10	$1\,(5+5)$	0	50

We note the following from the table. The first column contains the values of $(k + k')$ and this varies between -10 to 10. In the second column we note the number of times such an event of the first column occurs. For example, $(k + k')$ can take a value of -10 for a single combination of k and k', while $k + k' = 7$ can occur in four possible ways indicated by the entries in the second column within parenthesis. The third column indicates where the combined $|(k + k')|$ aliases to when its value exceeds 5 (for this case). For example, $|k + k'| = 10$ are aliased to $\bar{k} = 0$, as indicated in Fig. 10.11. The fourth column is self-explanatory.

The number of occurrences for aliasing increases with wave number $(\bar{k})$. However, the spectral weights are not a monotonic function as seen from the fourth column of Table 10.2. This means that if we consider the most restrictive case of $F(k')$, $\Omega(k)$ being constant, then the maximum aliasing will not be at $\bar{k} = \pm 4$ but at $\bar{k} = 0$. Needless to say that in practical cases the function $f(\xi)$ will have a transform that decreases with k'. A similar trend may be predicted for $\Omega(k)$ such that the maximum aliasing will be towards the larger values of wave numbers (as can be concluded from an inspection of Table 10.2). A priori, one cannot say exactly at which wave number one will have maximum aliasing.

10.11 Estimating the Aliased Component for Equation (10.60): An Example

We have purposely taken this case to demonstrate the effect of aliasing on a term which is essentially linear in nature, but due to product evaluation gives rise to aliasing.

For the quantity represented by Equation (10.63), this is given by

$$d(\xi) = \int \int_{-k_m}^{k_m} -(k^2 + kk')F(k')\Omega(k)e^{i(k+k')\xi}dkdk'.$$

We note that the integral is evaluated in $(k-k')$ plane in the rectangular box shown in Fig. 10.8. The aliasing contribution comes from the shaded regions. The limits for Region 1 and Region 2 are constituted respectively by the included lines $k + k' = \pm k_m$.

For calculation in the spectral plane, we can rewrite Equation (10.63) as

$$d(\xi) = -\int \int k_{\text{exact}}^2 F(k')\Omega(k)e^{i(k+k')\xi}dk'dk, \tag{10.71}$$

where $\xi = m\Delta\xi$ and $k_{\text{exact}}^2 = k^2 + kk'$.

To calculate the aliased component we will perform the integral along the horizontal strips indicated in Fig. 10.8 such that k ranges from $(k_m - k')$ to k_m on this strip. If we define this aliased component as $d_{sp}^a(\xi)$, then

$$d_{sp}^a(\xi) = \int_0^{k_m} \int_{k_m-k'}^{k_m} (k^2 + kk')\Omega(k)F(k')e^{i(k+k')\xi}dkdk'. \tag{10.72}$$

For an analytical evaluation of the integral in Equation (10.72), let us consider a shot noise or white noise process for which $\Omega(k) \equiv 1$.

Also, let

$$
\begin{aligned}
g_{sp}(k', \xi) &= \int_{k_m-k'}^{k_m} (k^2 + kk')e^{ik\xi} dk \\
&= (k^2 + kk')\frac{e^{ik\xi}}{i\xi}\bigg|_{k_m-k'}^{k_m} - \int (2k + k')\frac{e^{ik\xi}}{i\xi} dk \\
&= \frac{e^{ik\xi}}{\xi}\left[\frac{2i}{\xi^2} + \frac{(2k + k')}{\xi} - i(k^2 + kk')\right]_{k_m-k'}^{k_m} \\
&= \frac{(-1)^m}{\xi}\left[\frac{2i}{\xi^2} + \frac{(2k_m + k')}{\xi} - ik_m(k_m + k')\right] \\
&\quad - \frac{(-1)^m e^{-ik'\xi}}{\xi}\left[\frac{2i}{\xi^2} + \frac{(2k_m - k')}{\xi} - ik_m(k_m - k')\right].
\end{aligned}
$$

Hence after simplification, for $\xi = m\Delta\xi$,

$$
\begin{aligned}
g_{sp}(k', \xi) = \frac{(-1)^m}{\xi}\bigg[&\frac{2i}{\xi^2}(1 - e^{-ik'\xi}) + \frac{1}{\xi}(2k_m(1 - e^{-ik'\xi}) \\
&+ k'(1 + e^{-ik'\xi})) - ik_m(k_m(1 - e^{ik'\xi}) + k'(1 + e^{-ik'\xi}))\bigg].
\end{aligned} \tag{10.73}
$$

Using Equation (10.73) in (10.72) one obtains the aliasing error contribution coming from Region 1 of Fig. 10.8 for the shot noise process as

$$
d_{sp}^a(\xi) = \int_0^{k_m} g_{sp}(k', \xi)F(k')e^{ik'\xi} dk'. \tag{10.74}
$$

This is the aliased component for the evaluation of term given by Equation (10.58) for spectral calculations. Note that the expressions given in Equation (10.74) have to be numerically evaluated, because g_{sp} is a function of ξ. If g_{sp} were not a function of ξ the integral could have been evaluated by FFT routines. It is also possible to estimate the aliased component for the evaluation of Equation (10.58) if we estimate $\frac{\partial}{\partial\xi}(f\frac{\partial\omega}{\partial\xi})$ by a finite difference procedure. For example, if we use second order central differencing scheme, then

$$
d(\xi) = -\int\int k_{\text{eq}}^2 \Omega(k)F(k')e^{i(k+k')\xi} dk dk', \tag{10.75}
$$

where

$$
k_{\text{eq}}^2 = \frac{2}{\Delta\xi^2}\left[\cos\left(\frac{k'\Delta\xi}{2}\right) - \cos\left(k + \frac{k'}{2}\right)\Delta\xi\right]. \tag{10.76}
$$

Proof of equation (10.76)

The expression (10.58) is discretized in the following manner:

$$\frac{\partial}{\partial \xi}\left(f\frac{\partial \omega}{\partial \xi}\right)_m = \frac{1}{\Delta\xi^2}[f_{m-1/2}\omega_{m-1} - (f_{m-1/2} + f_{m+1/2})\omega_m + f_{m+1/2}\omega_{m+1}].$$

Using the representations (10.59) and (10.61) in the above, one can write

$$\frac{\partial}{\partial \xi}\left(f\frac{\partial \omega}{\partial \xi}\right)_m = \frac{1}{\Delta\xi^2}\int\int F(k')\Omega(k)\Big[e^{-ik'\frac{\Delta\xi}{2}-ik\Delta\xi} - \left(e^{-\frac{ik'\Delta\xi}{2}} + e^{\frac{ik'\Delta\xi}{2}}\right)$$
$$+ e^{\frac{ik'\Delta\xi}{2}+ik\Delta\xi}\Big]e^{i(k+k')\xi}dkdk'.$$

Comparing with Equation (10.71), we can write

$$d(\xi) = -\int\int k_{\text{eq}}^2 F(k')\Omega(k)e^{i(k+k')\xi}dkdk'.$$

Simplifying the expression within the square bracket, one gets

$$k_{\text{eq}}^2 = \frac{2}{\Delta\xi^2}\left[\cos\left(\frac{k'\Delta\xi}{2}\right) - \cos\left(k+\frac{k'}{2}\right)\Delta\xi\right].$$

Having obtained the equivalent expression for $d(\xi)$ as given by Equation (10.75), we can obtain the aliased component coming from Region-1 of Fig. 10.8 for this finite difference discretization (for the shot noise process).

$$d^a_{\text{CD}_2}(\xi) = \int_0^{k_m}\int_{k_m-k'}^{k_m} k_{\text{eq}}^2 F(k')e^{i(k+k')\xi}dkdk'. \tag{10.77}$$

If we define

$$g_2(k',\xi) = \int_{k_m-k'}^{k_m}\frac{2}{\Delta\xi^2}\left[\cos\frac{k'\Delta\xi}{2} - \cos\left(k+\frac{k'}{2}\right)\Delta\xi\right]e^{ik\xi}dk, \tag{10.78}$$

then

$$d^a_{\text{CD}_2}(\xi) = \int_0^{k_m} g_2(k',\xi)F(k')e^{ik'\xi}dk', \tag{10.79}$$

where

$$\frac{\Delta\xi^2}{2}g_2(k',\xi) = \frac{e^{ik\xi}}{i\xi}\cos\frac{k'\Delta\xi}{2} - \cos\frac{k'\Delta\xi}{2}\int e^{ik\xi}\cos(k\Delta\xi)dk$$
$$+ \sin\frac{k'\Delta\xi}{2}\int e^{ik\xi}\sin(k\Delta\xi)dk.$$

If $I_1 = \int e^{ik\xi} \cos(mk)dk$ and $I_2 = \int e^{ik\xi} \sin(mk)dk$, then it is easy to see

$$I_1 \left(1 - \frac{m^2}{\xi^2}\right) = -\frac{e^{ik\xi}}{\xi} \left(\frac{m}{\xi} \sin(mk) + i \cos(mk)\right)$$

and

$$I_2 \left(1 - \frac{m^2}{\xi^2}\right) = \frac{e^{ik\xi}}{\xi} \left(\frac{m}{\xi} \cos(mk) - i \sin(mk)\right).$$

Thus

$$\frac{\Delta\xi^2}{2} g_2(k', \xi) = \left[\frac{e^{ik\xi}}{i\xi} \cos\frac{k'\Delta\xi}{2} + \frac{e^{ik\xi}}{(\xi - \frac{\Delta\xi^2}{\xi})} \left\{ \cos\frac{k'\Delta\xi}{2} \left(\frac{\Delta\xi}{\xi} \sin k'\Delta\xi + i \cos k\Delta\xi\right) \right.\right.$$

$$\left.\left. + \sin\frac{k'\Delta\xi}{2} \left(\frac{\Delta\xi}{\xi} \cos k\Delta\xi - i \sin k\Delta\xi\right) \right\}\right]_{k_m - k'}^{k_m},$$

which upon further simplification yields

$$g_2(k', \xi) = \frac{(-1)^m 2\xi}{\Delta\xi(\xi^2 - \Delta\xi^2)} \left[\frac{\xi^2 - \Delta\xi^2}{\xi} \cos\frac{k'\Delta\xi}{2} + (-1)^m i \cos\frac{k'\Delta\xi}{2}\right.$$

$$+ (-1)^m \frac{\Delta\xi}{\xi} \sin\frac{k'\Delta\xi}{2} - e^{-ik'\xi} \left\{ \left[\frac{\xi^2 - \Delta\xi^2}{\xi} - (-1)^m \frac{\Delta\xi}{\xi} \sin k'\Delta\xi\right.\right.$$

$$\left. + i(-1)^m \cos k'\Delta\xi \right] \cos\frac{k'\Delta\xi}{2} + \left[(-1)^m \frac{\Delta\xi}{\xi} \cos k'\Delta\xi\right.$$

$$\left.\left.\left. + i(-1)^m \sin k'\Delta\xi \right] \sin\frac{k'\Delta\xi}{2} \right\}\right]. \tag{10.80}$$

10.12 Some Numerical Estimates of Aliasing Error

Let us choose two popular grid types used for estimation of aliasing error in a transformed plane using (a) spectral method and (b) second-order central differencing schemes. The stretched grids in the physical x-plane is mapped to a uniform grid in the computational ξ-plane via analytical transformation

$$\xi = \xi(x) \tag{10.81}$$

such that the term in the transformed plane, given by Equation (10.58), is actually a linear term $\frac{\partial^2\omega}{\partial x^2}$ in the physical plane. Let us consider the following analytical grid types.

Grid 1: For one-dimensional stretching the grid spacing of Δx_0 at $x = 0$ increases by d progressively such that the points in the physical plane are distributed in an arithmetic progression. If there are N grid points, then the point distribution in the physical plane is given by

$$x_A(\xi) = (N-1)\xi \left\{\Delta x_0 + \frac{d}{2}[(N-1)\xi + 1]\right\}. \tag{10.82}$$

As $f(\xi) = \dfrac{1}{dx/d\xi}$,

$$\text{so} \quad f_A(\xi) = \frac{1}{(N-1)[\Delta x_0 + d\{\xi(N-1) + \frac{1}{2}\}]}. \tag{10.83}$$

Grid 2: In this case, a grid is chosen for which the grid spacing increases by a constant proportion $r(= \Delta x_j/\Delta x_{j-1})$ with $r > 1$. The resultant points in the physical plane are distributed in a geometric progression. The distribution of points in the physical plane is then given by

$$x_G(\xi) = \frac{\Delta x_0}{r-1}\{r^{(N-1)\xi} - 1\}, \tag{10.84}$$

and the metric of the transformation is given by

$$f_G(\xi) = \frac{r^{-\xi(N-1)}}{\Delta x_0(\frac{N-1}{r-1})\ln r}. \tag{10.85}$$

In Fig. 10.12, f_A and f_G are shown as a function ξ. And in Fig. 10.13, the corresponding Fourier transform $F_A(k^{'})$ and $F_G(k^{'})$ are plotted. It is apparent from Fig. 10.12 that the AP distribution

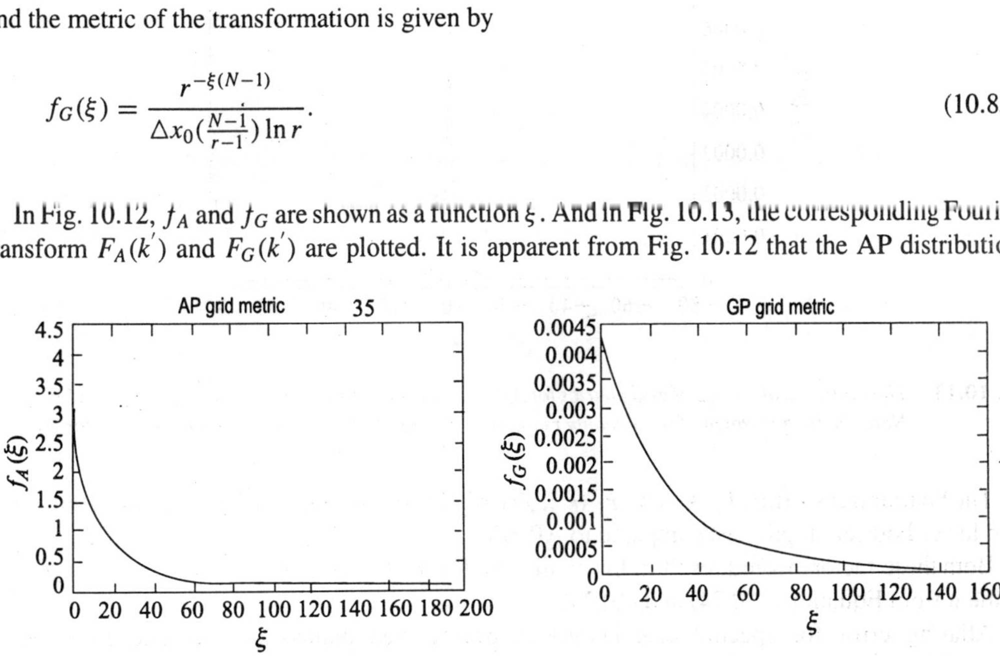

Fig. 10.12 *The grid metrics for AP (top) and GP grids (bottom) in the transformed plane. The abscissa indicates the grid indices. To reach the same final point the AP grid requires more number of points.*

produces a grid whose metric of transformation is much higher compared with the metric for the GP grid. To arrive at the same terminal grid point in the physical plane ($x \simeq 10$), we require 200 points for AP grid as compared to 160 points for GP grid, when the initial grid spacing is kept the same for both the grids.

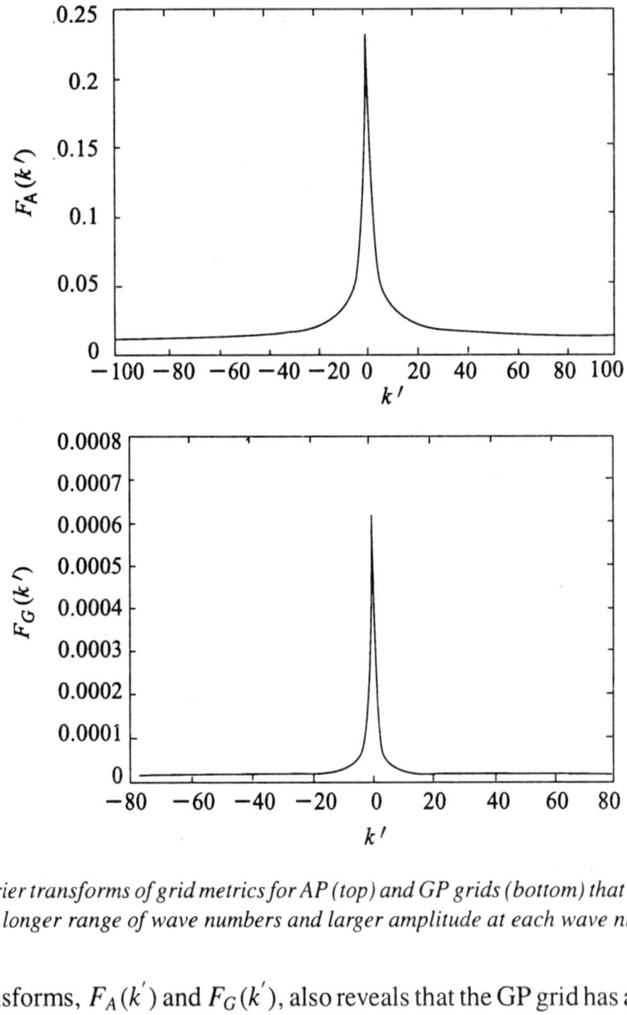

Fig. 10.13 *The Fourier transforms of grid metrics for AP (top) and GP grids (bottom) that are shown in Fig. 10.12. Note the longer range of wave numbers and larger amplitude at each wave numbers for the AP grid.*

The Fourier transforms, $F_A(k')$ and $F_G(k')$, also reveals that the GP grid has a narrower spectrum and lower Fourier amplitudes compared to AP grid.

Both these factors would result in lower aliasing for the GP grid as compared to AP grids – as evident from Equations (10.74) and (10.79).

Aliasing error for spectral calculations in transformed plane—as evaluated from equation (10.74) coming from Region 1 of Fig. 10.8 are displayed in Fig. 10.14 and 10.15 for AP and GP grid respectively. As reasoned above, the AP grid aliasing error is thousand-fold more than that for the GP grid.

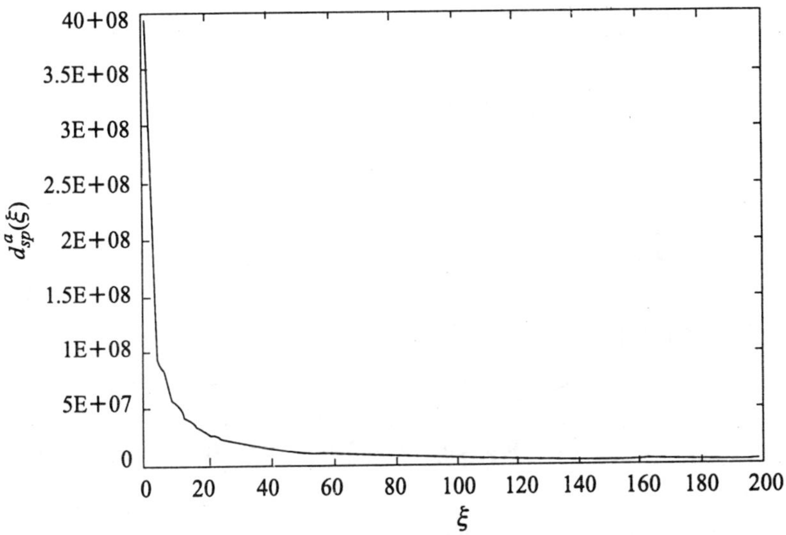

Fig. 10.14 *The aliased component coming from Region 1 of Fig. 10.8 for spectral calculations using the AP grid in Equation (10.74).*

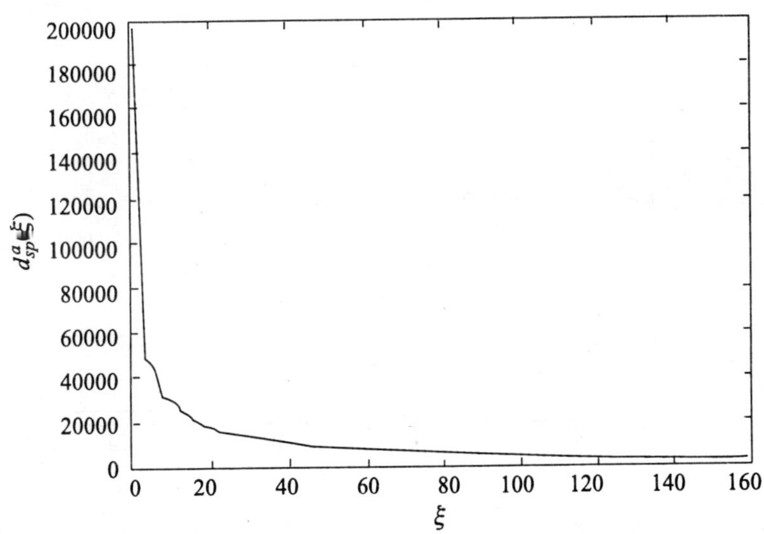

Fig. 10.15 *The aliased component coming from Region 1 of Fig. 10.8 for spectral calculations using the GP grid in Equation (10.74).*

In Figs. 10.16 and 10.17, the aliased components for the quantity given by Equation (10.58) as discretized by second order accurate central differencing (coming from the Region 1 of Fig. 10.8) are shown for AP and GP grids. Once again one notices very small level of aliasing for the GP grid as compared to the AP grid.

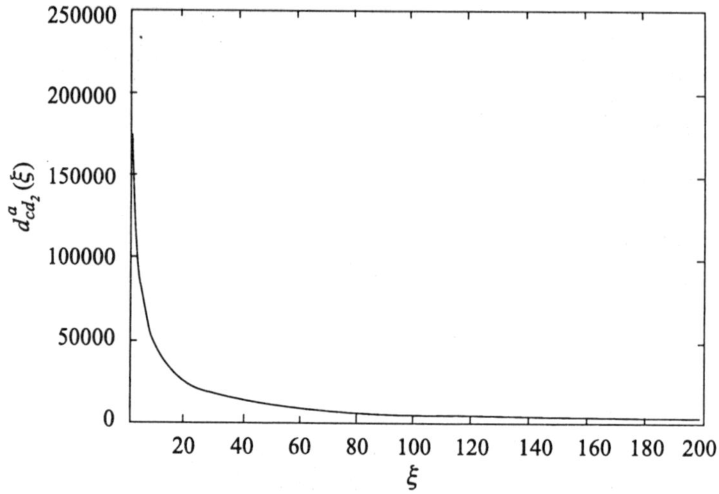

Fig. 10.16 *The aliased component coming from Region-1 of Fig. 10.8 for CD_2 calculations using the AP grid in Equation (10.79).*

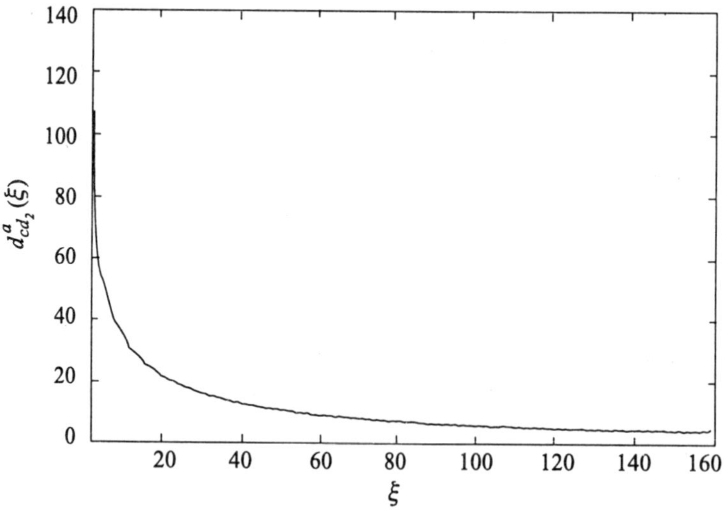

Fig. 10.17 *The aliased component coming from Region 1 of Fig. 10.8 for CD_2 calculations using the GP grid in Equation (10.79).*

The error is thousand-fold lower for the GP grid as compared to the AP grid. Secondly, we also notice another qualitative difference between the spectral calculation and finite difference calculations by comparing the results of Figs. 10.14 and 10.16 for the AP grid. The spectral method has an error that is thousand-fold more as compared to the finite difference method.

This ratio between errors incurred between spectral calculations and finite difference calculations is retained for GP grids – as seen by comparing Figs. 10.15 and 10.17. Thus, it is important to note that the level of aliasing error in discretizing the dissipation terms (a representative term shown in Equation (10.58)) by using spectral method in the transformed plane (as obtained by a AP gird) is million times more than the error committed if the term is computed using second order central differencing in a GP grid. It is for this reason that spectral calculations are never performed in the transformed plane—even when there is the tempting proposition of using a uniform grid. It is not difficult to point out the reason for this behaviour. From Fig. 10.1, it is apparent that the discrete computational methods other than spectral methods act like a low-pass filter. (note that $k_{\text{eq}} < k_{\text{spectral}}$ for estimating the 1st derivative). Thus the quantities in Region 1 and 2 of Fig. 10.8 are underestimated by finite difference schemes as compared to the spectral method.

We would also like to point out an interesting feature of the aliased error. From Equation (10.80), it is noted that the function $g_2(k', \xi)$ monotonically increased with the number of points (i.e. by reducing $\Delta\xi$), and this would imply that usage of a finer grid will give rise to higher aliasing.

This is particularly significant, since we have seen in Chapter 5 that the truncation error reduces with reduction in grid spacing. Thus truncation error and aliasing error places conflicting demand on the grids. Before closing this section we also note that the estimates provided here is for a special shot-noise/white noise process for which $\Omega(k) = 1$ for all wave numbers. However, for physical processes, usually $\Omega(k)$ is band-limited and decreases rapidly with k. Hence the presented estimates are the worst-case scenario – but none the less it is helpful to compare different methods and grid types.

Before we proceed to discuss ways of controlling aliasing, it is appropriate to discuss two grid types we have used so far in our discussions.

To resolve the shear layer forming over solid bodies it is customary to take a GP grid (see e.g. Cebeci & Bradshaw (1977)) to resolve the flow gradient. However, when such a grid was used for solving Navier-Stokes equation using higher order upwinding (as in Kawamura et al. (1985) and Sengupta & Sengupta (1994)), it exhibited strong grid dependence. In fact, the situation was found to be the same for many transformed plane Navier-Stokes solvers. This has led to a shift from transformed plane methods to those methods that work in the physical plane. This behaviour was explained and a solution was provided in Nair et al. (1998). We will discuss this next. In solving Navier-Stokes equation in the (ψ–ω) formulation using higher order upwinding of the convection terms, the vorticity transport equation (VTE) is given in the transformed plane (via an orthogonal map) by

$$h_1 h_2 \frac{\partial \omega}{\partial t} + h_2 u \frac{\partial \omega}{\partial \xi} + h_1 v \frac{\partial \omega}{\partial \eta} = \frac{1}{Re}\left\{\frac{\partial}{\partial \xi}\left(\frac{h_2}{h_1}\frac{\partial \omega}{\partial \xi}\right) + \frac{\partial}{\partial \eta}\left(\frac{h_1}{h_2}\frac{\partial \omega}{\partial \eta}\right)\right\}. \tag{10.86}$$

The convection terms (second and third term on left hand side of Equation (10.86)) are represented by a third order upwind scheme in Kawamura et al. (1985) and Sengupta & Sengupta (1994) as

$$f\left.\frac{\partial \omega}{\partial \xi}\right|_{m,n} = \frac{f_{m,n}}{6\Delta\xi}[\omega_{m+2,n} - 2\omega_{m+1,n} + 9\omega_{m,n} - 10\omega_{m-1,n} + 2\omega_{m-2,n}] \tag{10.87a}$$

$$\text{for } f_{m,n} \geq 0,$$

$$= \frac{f_{m,n}}{6\Delta\xi}[-2\omega_{m+2,n} + 10\omega_{m+1,n} - 9\omega_{m,n} + 2\omega_{m-1,n} - \omega_{m-2,n}] \tag{10.87b}$$

$$\text{for } f_{m,n} < 0.$$

These expressions are the same as the one given in Equation (10.18) with $\beta = 6$. Thus the numerical dissipation added in this scheme is six times the leading truncation error of the Taylor series expansion,

$$f_{m,n}\frac{(\Delta\xi)^3}{4!}\frac{\partial^4 \omega}{\partial \xi^4}. \tag{10.88}$$

Thus the usage of the third order upwinding is equivalent to adding a fourth order *dissipation term*—in the transformed plane. It is to be noted that in the physical plane it will not be equivalent to adding higher order dissipation. What is really added by higher order upwinding in the transformed plane?

Let us consider that we are using a grid transformation given by

$$\left.\begin{aligned} x &= x(\xi, \eta) \\ y &= y(\xi, \eta) \end{aligned}\right\}. \tag{10.89}$$

Thus successive application of differentiation by chain-rule will yield the first two partial derivatives of ω with respect to ξ as,

$$\frac{\partial \omega}{\partial \xi} = \frac{\partial \omega}{\partial x}\left(\frac{\partial x}{\partial \xi}\right) + \frac{\partial \omega}{\partial y}\left(\frac{\partial y}{\partial \xi}\right), \tag{10.90}$$

$$\begin{aligned} \frac{\partial^2 \omega}{\partial \xi^2} &= \frac{\partial^2 \omega}{\partial x^2}\left(\frac{\partial x}{\partial \xi}\right)^2 + \frac{\partial^2 \omega}{\partial y^2}\left(\frac{\partial y}{\partial \xi}\right)^2 \\ &\quad + 2\frac{\partial^2 \omega}{\partial x \partial y}\left(\frac{\partial x}{\partial \xi}\right)\left(\frac{\partial y}{\partial \xi}\right) + \frac{\partial \omega}{\partial x}\frac{\partial^2 x}{\partial \xi^2} + \frac{\partial \omega}{\partial y}\frac{\partial^2 y}{\partial \xi^2}. \end{aligned} \tag{10.91}$$

Thus, in a computation, if one resorts to first order upwinding in the transformed plane, then in the physical plane the leading order truncation error term alters the physical dissipation by the first three terms in Equation (10.91). However, it also alters the physical convection term via the last two terms in Equation (10.91).

If one continues the process of differentiation, then it can be shown—as in Nair et al. (1998), that Equation (10.88) can be rewritten as,

$$= f\frac{\Delta\xi^3}{4}[A + B + C + D + E + F + G], \tag{10.92}$$

where the quantities in the parenthesis are given by

$$A = \frac{\partial^4 x}{\partial \xi^4}\frac{\partial \omega}{\partial x} + \frac{\partial^4 y}{\partial \xi^4}\frac{\partial \omega}{\partial y}, \tag{10.93}$$

$$B = \left[\left(\frac{\partial^2 x}{\partial \xi^2}\right)^2 + 4\frac{\partial x}{\partial \xi}\frac{\partial^3 x}{\partial \xi^3}\right]\frac{\partial^2 \omega}{\partial x^2} + \left[3\left(\frac{\partial^2 y}{\partial \xi^2}\right)^2 + 4\frac{\partial y}{\partial \xi}\frac{\partial^3 y}{\partial \xi^3}\right]\frac{\partial^2 \omega}{\partial y^2}, \tag{10.94}$$

$$C = \left[4\frac{\partial x}{\partial \xi}\frac{\partial^3 y}{\partial \xi^3} + 6\frac{\partial^2 x}{\partial \xi^2}\frac{\partial^2 y}{\partial \xi^2} + 4\frac{\partial y}{\partial \xi}\frac{\partial^3 x}{\partial \xi^3}\right]\frac{\partial^2 \omega}{\partial x \partial y}, \tag{10.95}$$

$$D = 6\left(\frac{\partial x}{\partial \xi}\right)^2\frac{\partial^2 x}{\partial \xi^2}\frac{\partial^3 \omega}{\partial x^3} + 6\left(\frac{\partial y}{\partial \xi}\right)^2\frac{\partial^2 y}{\partial \xi^2}\frac{\partial^3 \omega}{\partial y^3}, \tag{10.96}$$

$$E = 6\left[\left(\frac{\partial x}{\partial \xi}\right)^2\frac{\partial^2 y}{\partial \xi^2} + 2\frac{\partial x}{\partial \xi}\frac{\partial y}{\partial \xi}\frac{\partial^2 x}{\partial \xi^2}\right]\frac{\partial^3 \omega}{\partial x^2 \partial y} + 6\left[\left(\frac{\partial y}{\partial \xi}\right)^2\frac{\partial^2 x}{\partial \xi^2} + 2\frac{\partial x}{\partial \xi}\frac{\partial y}{\partial \xi}\frac{\partial^2 y}{\partial \xi^2}\right]\frac{\partial^3 \omega}{\partial x \partial y^2}, \tag{10.97}$$

$$F = \left(\frac{\partial x}{\partial \xi}\right)^4\frac{\partial^4 \omega}{\partial x^4} + \left(\frac{\partial y}{\partial \xi}\right)^4\frac{\partial^4 \omega}{\partial y^4}, \tag{10.98}$$

$$G = 4\left(\frac{\partial x}{\partial \xi}\right)^2\frac{\partial y}{\partial \xi}\frac{\partial^4 \omega}{\partial x^3 \partial y} + 6\left(\frac{\partial x}{\partial \xi}\frac{\partial y}{\partial \xi}\right)^2\frac{\partial^4 \omega}{\partial x^2 \partial y^2} + 4\frac{\partial x}{\partial \xi}\left(\frac{\partial y}{\partial \xi}\right)^2\frac{\partial^4 \omega}{\partial x \partial y^3}. \tag{10.99}$$

Thus the addition of a fourth order *dissipation term* in the transformed plane is equivalent to various terms that are clubbed together based on their physical meaning. For example, the term A as given by Equation (10.93) alters the physical convection. The term B represents addition of lowest order dissipation. The term D in Equation (10.96) adds dispersion numerically. The terms C and E represent addition of mixed order dissipation and dispersion term. It is only the terms F and G that are truly higher order dissipation terms in the physical plane. Thus, working on a computational plane is equivalent to altering the basic physical phenomenon itself if proper care is not exercised in the choice of the grids.

It is clearly evident that the usage of an orthogonal grid (with $x = x(\xi)$ and $y = y(\eta)$) will radically alter the truncation error terms in Equation (10.93) to (10.99). This is one of the strongest incentive for the choice of an orthogonal grid.

For example, $A = \dfrac{d^4 x}{d\xi^4}\dfrac{\partial \omega}{\partial x}$ and $B = \left[3\left(\dfrac{d^2 x}{d\xi^2}\right)^2 + 4\dfrac{\partial x}{\partial \xi}\dfrac{d^3 x}{d\xi^3}\right]\dfrac{\partial^2 \omega}{\partial x^2}$ and so on.

Even among the orthogonal grids, it is clearly evident that if one chooses an AP grid (as given in Equation (10.82)), then all the partial derivatives of x with respect to ξ of order greater than two are identically zero. Therefore, for such a grid $A \equiv 0$ and the other terms are vastly reduced. However, for a GP grid (as given by Equation (10.84)), none of the grid-metric terms are zero. Hence the results using GP grids are more likely to be grid dependent, than using an AP grid, as noted in Nair et al. (1998).

In short, then one can see the reason why one should adopt higher order methods. From Table 10.1, it is easy to see that a fourth-order method provides 37% more resolution of wave numbers as compared to the second-order methods. Thus to attain the same level of accuracy one can take fewer points for a higher order accurate method. The relatively coarser grids also reduce the aliasing error, as shown above. Are there any other methods for systematically reducing aliasing error? This is explored next.

10.13 Controlling Aliasing Error

The general discussion here and in the cited literature (Phillips (1959), Arakawa (1966) and Orszag (1971)), have revealed that the aliasing error is a high wave number phenomenon. Hence the early efforts in containing aliasing is also through the route of manipulating high wave number components of the solution.

However, Phillips (1959) did not diagnose this as an aliasing problem, but termed it as *nonlinear computational instability*. It was so termed due to the fact that the instability at the smallest resolvable scales (k_m) could not be suppressed by merely reducing the CFL number. For the numerical integration of a barotropic non-divergent atmospheric model, Phillips (1959) periodically removed all components of solution with wavelengths smaller than four times the grid size. This is equivalent to suppressing all wave number components above $k_m/2$. Thus this method of controlling aliasing can be termed as filtering the finite difference results by a box filter.

A similar approach was advocated by Orszag (1971) who also suggested elimination of aliasing in discrete computation by filtering high wave number components. In the same context of controlling aliasing, the small amount of numerical dissipation introduced in the higher order upwind schemes suppressed numerical instabilities caused by aliasing due to poorly resolved high wave number component. Smagorinsky (1993) notes that the ease with which the numerical dissipation can be introduced in spectral calculation (by introducing numerical dissipation terms $\nabla^4\omega$ or $\nabla^8\omega$) analytically, makes it an attractive candidate solution technique over finite difference calculation method.

The added numerical dissipation terms in the spectral calculations are referred to as *hyperviscosity* term (see Sengupta & Nair (1999)). In Sengupta & Nair (1999), various higher order finite difference explicit upwind methods have been spectrally analyzed and the connection between upwinding and Sub-Grid Stress (SGS) models for LES established.

The explicit removal of Fourier components from the solution by Phillips (1959) is equivalent to taking $\alpha = 1/2$ in Table 10.1 over and above what is filtered via the discretization schemes themselves for different operators. This seems to be particularly a severe restriction. In higher order upwind schemes for finite difference calculations or spectral methods using hyperviscosity, the added numerical dissipation performs a similar function—albeit smoothly. In Fig. 10.18, the added dissipation of variable finite difference upwind schemes are shown.

The compact difference (CD) schemes referred to in Fig. 10.18 are introduced in Chapter 11. From the figure it is apparent that the addition of fourth or sixth-order dissipation term will adequately control aliasing problem while leaving the resolved scales (that presumably are not affected by aliasing) unmodified. It is interesting to note that Lesieur & Metais (1996) suggested SGS model (for LES) features that matches with the plotted numerical dissipation in Fig. 10.18.

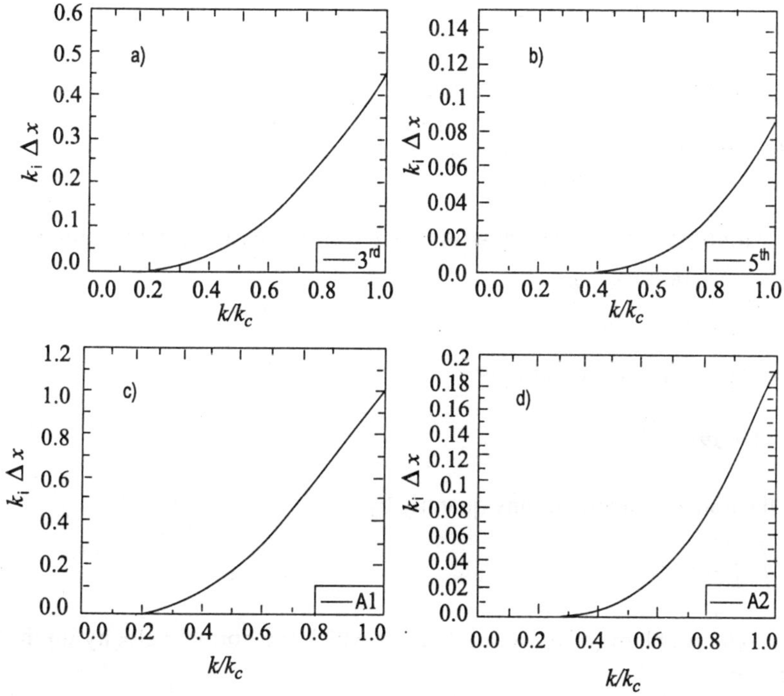

Fig. 10.18 *The numerical dissipation added to control aliasing via upwinding. The abscissa is sacled by the maximum resolved wave number as in Fig. 10.2. (a) 3rd-order upwind scheme; (b) 5th-order upwind schem; (c) compact difference scheme with 4th-order dissipation; (d) compact difference scheme with 6th-order dissipation.*

In finite difference calculations there is another way to control aliasing via the choice of the appropriate equation type. This was originally performed by Arakawa (1966) and later again addressed in Kravchenko & Moin (1997) and the references contained therein. The present analysis is motivated by the above references.

First of all, let us discuss Arakawa's skew-symmetric differencing. Let us consider the 2D barotropic vorticity equation which is nothing but the full VTE without the viscous terms. The full VTE is given by Equation (2.53).

$$\frac{\partial \omega}{\partial t} + V.\nabla\omega = \nu\nabla^2\omega.$$

Now writing the inviscid version of the equation in an orthogonal grid using the relations given in Section 9.4, one gets the governing equation in the transformed $(\xi-\eta)$ plane as

$$h_1 h_2 \frac{\partial \omega}{\partial t} + h_2 u \frac{\partial \omega}{\partial \xi} + h_1 v \frac{\partial \omega}{\partial \eta} = 0, \tag{10.100}$$

since by definition

$$h_2 u = \frac{\partial \psi}{\partial \eta} \quad \text{and} \quad h_1 v = -\frac{\partial \psi}{\partial \xi}.$$

Therefore Equation (10.100) can be rewritten as

$$h_1 h_2 \frac{\partial \omega}{\partial t} + \left(\frac{\partial \psi}{\partial \eta} \frac{\partial \omega}{\partial \xi} - \frac{\partial \psi}{\partial \xi} \frac{\partial \omega}{\partial \eta} \right) = 0.$$

The quantity within parenthesis can be written in terms of the Jacobian operator as

$$J(a, b) = \frac{\partial a}{\partial \xi} \frac{\partial b}{\partial \eta} - \frac{\partial a}{\partial \eta} \frac{\partial b}{\partial \xi}.$$

Thus the barotropic vorticity equation is

$$\frac{\partial \omega}{\partial t} + J(\psi, \omega) = 0. \tag{10.101}$$

If we define the domain integral of any quantity by

$$\langle a \rangle = \int \int a d\xi d\eta,$$

then for a periodic problem (or for a problem for which the normal velocity on the boundary is zero) we can show,

$$\langle J(\psi, \omega) \rangle = \left\langle \frac{\partial}{\partial \xi} \left(\psi \frac{\partial \omega}{\partial \eta} \right) \right\rangle - \left\langle \frac{\partial}{\partial \eta} \left(\psi \frac{\partial \omega}{\partial \xi} \right) \right\rangle = 0.$$

This ensures that

$$\langle \psi J(\psi, \omega) \rangle = \left\langle J \left(\frac{\psi^2}{2}, \omega \right) \right\rangle = 0 \tag{10.102}$$

and

$$\langle \omega J(\psi, \omega) \rangle = \left\langle J \left(\psi, \frac{\omega^2}{2} \right) \right\rangle = 0. \tag{10.103}$$

Now multiplying Equation (10.101) by ω and integrating over the whole domain gives

$$h_1 h_2 \frac{\partial}{\partial t} \left(\left\langle \frac{\omega^2}{2} \right\rangle \right) + \langle \omega J(\psi, \omega) \rangle = 0.$$

Using Equation (10.103) in the above ensures that

$$\left\langle \frac{\partial}{\partial t} \left(\frac{\omega^2}{2} \right) \right\rangle = 0. \tag{10.104}$$

This implies that the total enstrophy $(= \frac{\omega^2}{2})$ does not change with time.

From Appendix A,

$$\nabla.(f\vec{F}) = f\nabla.\vec{F} + \vec{F}.\nabla f$$

If we use $f \equiv \psi$ & $\vec{F} = \frac{\partial}{\partial t}\nabla\psi$, then

$$\nabla.\left(\psi\frac{\partial}{\partial t}\nabla\psi\right) = \psi\nabla.\left(\frac{\partial}{\partial t}\nabla\psi\right) + \frac{\partial}{\partial t}(\nabla\psi).\nabla\psi.$$

Hence

$$\psi\frac{\partial}{\partial t}(\nabla^2\psi) = \nabla.\left(\psi\frac{\partial}{\partial t}\nabla\psi\right) - \frac{\partial}{\partial t}\left(\nabla\psi.\frac{\nabla\psi}{2}\right),$$

and integrating over the domain and using Equations (10.102) and (10.103), one gets

$$\left\langle\frac{\partial}{\partial t}\left(\frac{|\nabla\psi|^2}{2}\right)\right\rangle = 0 \tag{10.105}$$

because $\left\langle\nabla.\left(\psi\frac{\partial}{\partial t}\nabla\psi\right)\right\rangle = 0$ by divergence theorem.

Like enstrophy, the total kinetic energy over the whole domain does not change with time. The invariance of kinetic energy and enstorphy with time allows one to define a time-invariant average wave number, k_{avg}.

$$\text{Since} \quad \langle\nabla\psi.\nabla\psi\rangle = -\langle\psi\nabla^2\psi\rangle,$$

so $\langle\nabla\psi.\nabla\psi\rangle \sim k^2\langle\Psi^2\rangle$.

And therefore, $\langle\omega^2\rangle = \langle(\nabla^2\psi)^2\rangle \sim k^4\langle\Psi^2\rangle$.

$$\text{Hence,} \qquad k_{\text{avg}} \simeq \left[\frac{\langle\omega^2\rangle}{\langle\nabla\psi.\nabla\psi\rangle^2}\right]^{1/2} \tag{10.106}$$

This interesting result also ensures the principle of backscatter for two dimensional flows. As k_{avg} does not change with time, any energy transfer that takes place from small to large wave number must be compensated by an inverse cascade of energy from small scales to large scales.

Any numerical scheme used to discretize the inviscid VTE must satisfy the above physical principle of backscatter. To this end Arakawa (1966) proposed the following discretization for the Jacobian in Equation (10.101).

$$\begin{aligned} J(\psi,\omega) = &\frac{1}{3}\left[\frac{\partial\psi}{\partial\xi}\frac{\partial\omega}{\partial\eta} - \frac{\partial\psi}{\partial\eta}\frac{\partial\omega}{\partial\xi}\right] + \frac{1}{3}\left[\frac{\partial}{\partial\xi}\left(\psi\frac{\partial\omega}{\partial\eta}\right) - \frac{\partial}{\partial\eta}\left(\psi\frac{\partial\omega}{\partial\xi}\right)\right] \\ &+ \frac{1}{3}\left[\frac{\partial}{\partial\eta}\left(\omega\frac{\partial\psi}{\partial\xi}\right) - \frac{\partial}{\partial\xi}\left(\omega\frac{\partial\psi}{\partial\eta}\right)\right] \end{aligned} \tag{10.107}$$

In this representation, Arakawa (1966) suggested using second order central differences for all the terms, including the time derivative. Usually in all meteorological calculations, mid-point leap frog time marching are used for its neutral amplification factor as discussed in Section 10.8.1.

The above representation of the convection terms satisfies Equations (10.102) and (10.103) numerically. Roache (1982) notes that the formal truncation error is $O(\Delta t^2, \Delta\xi^2, \Delta\eta^2)$. It has an amplification factor $|G| = 1$ and together with the conservation properties of vorticity, enstrophy and kinetic energy, makes it specially suited to hydrodynamic stability problems. It is specifically noted that aliasing errors are present but remains bounded, since enstrophy remains bounded. However, it is also stated that the method is suspectible to time-splitting instability. Of course, it is to be expected of any numerical method with $|G| = 1$ for the inviscid equation.

In the context of finite difference methods Orszag (1971) stated that "schemes with no quadratic semi-conservation properties may be unstable due to aliasing error." The term quadratic semi-conservation implies setting up an equivalent ODE for time-marching that should be stable. Arakawa's method is one such method steeped in strong physical principles. Unfortunately, the advantages of such a method cannot be directly extrapolarated to solution methods for primitive variable formulations. This route has been followed by some researchers.

For example Teman (1984) advocated the usage of skew-symmetric form of Navier-Stokes equation, as given in Equations (2.69) and (2.70). This form of equation guarantees quadratic conservation of kinetic energy. Another quadratic semi-conservation form of Navier-Stokes equation, suggested by Heywood et al. (1996) is the one given by

$$\rho\left[\frac{\partial \vec{V}}{\partial t} + \vec{V}.\nabla\vec{V} - (\nabla V).\vec{V}\right] = -\nabla P_{\text{tot}} + \mu\nabla^2 V. \tag{10.108}$$

This is a variation of the rotational form of Navier-Stokes equation given by Equation (2.67) where the following vector identity has been used:

$$\vec{\omega} \times \vec{V} = \vec{V}.\nabla\vec{V} - (\nabla\vec{V}).\vec{V}.$$

It has been shown in Kravchenko & Moin (1997), for the skew-symmetric and rotational forms of Navier-Stokes equation, both spectral and finite differences methods are energy conserving even in the presence of aliasing errors.

In Kravchenko & Moin (1997) the issue of discretizing the convection term in the divergence form and the convective form was considered. If one writes

$$\text{Div} = \frac{d}{dx}(uv) \tag{10.109}$$

and

$$\text{Con} = u\frac{dv}{dx} + v\frac{du}{dx} \tag{10.110}$$

as two alternative forms of expressing the same quantity, then their spectral representation is given by

$$\text{Div}(x) = \int\int_{-k_m}^{k_m} ikU(k_1)V(k_2)e^{i(k_1+k_2)x}dk_1dk_2. \tag{10.111}$$

This can be separated into (following the depicted regions in Fig. 10.8)

$$\text{Div}(x) = \int\int_{\text{Region 3}} ikU(k_1)V(k_2)e^{ikx}dk_1dk_2$$

$$+\int\int_{\text{Region 1\&2}} ikU(k_1)V(k_2)e^{ikx}dk_1dk_2,$$

where the second integral on the right hand side gives rise to aliasing.

Similarly one can show

$$\text{Con}(x) = \int\int_{\text{Region 3}} ikU(k_1)V(k_2)e^{ikx}dk_1dk_2$$

$$+\int\int_{\text{Region 1\&2}} i(k \pm 2k_{\max})U(k_1)V(k_2)e^{ikx}dk_1dk_2. \tag{10.112}$$

It has been shown by Kravchenko & Moin (1997) that the aliasing error component of Div(x) and Con(x) are of opposite signs. This observation is made use of in skew-symmetric form of N-S equation which is an arithmetic average of Div(x) and Con(x). Thus without any specific dealiasing also, the skew symmetric form has the lowest aliasing error.

For turbulent channel flow simulations reported in Kravchenko & Moin (1997) that used spectral, sixth-order compact difference (CD_6) and second order finite difference (FD_2) formulations, the following are the observations as summarized in Table 10.3.

Table 10. 3 *Simulation of turbulent channel flow*

	Aliased calculations			Dealising performed during calculations		
Form of convection terms	Spectral	CD_6	FD_2	Spectral	CD_6	FD_2
Rotational	Soln. Laminarizes	Stable soln.	Stable soln.	Stable soln.	Stable soln.	Stable soln.
Divergence	Unstable soln.	Unst able soln.	Unst able soln.	Unstable soln.	Unstable soln.	Unstable soln.
Skew-symmetric	Stable soln.	Stable soln.	Stable soln.	Stable soln.	Stable soln.	Stable soln.

It is interesting to note that working on a staggered grid is in effect equivalent to a dealising procedure. We have already discussed some de-aliasing procedures. In the following section we discuss a few dealising techniques that are used in the spectral method.

10.14 Aliasing Removal by Zero Padding

Referring to Fig. 10.8, it is noted that the aliasing arises from the triangular Regions 1 and 2 where $k_m < |k_1 + k_2| < 2k_m$. This region can be made smaller by zero padding – as explained with Fig. 10.19.

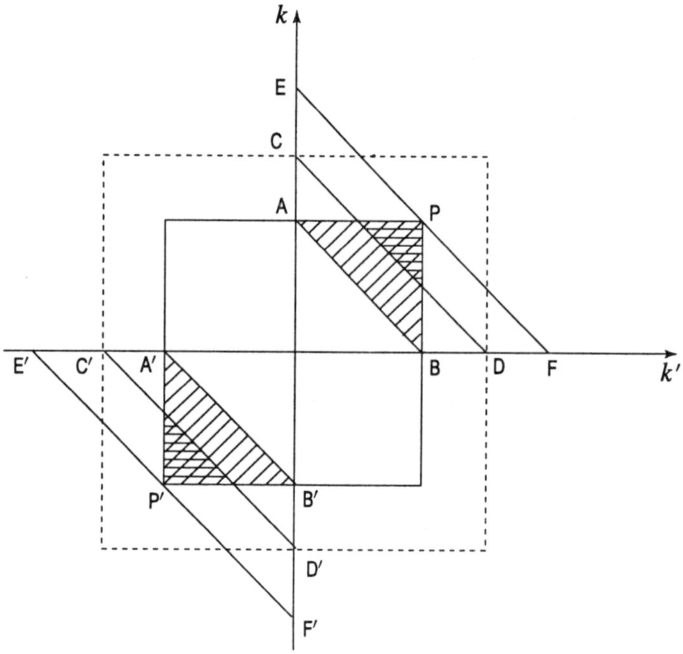

Fig. 10.19 *Removal of aliasing by zero padding. The line CD corresponds to the case when padding is done up to $3k_m/2$ and EF corresponds to padding up to $2k_m$.*

Let us once again consider the evaluation of the term

$$d(\xi) = \frac{\partial}{\partial \xi}\left(f\frac{\partial \omega}{\partial \xi}\right),$$

where we represent the variables by their bilateral Laplace transform

$$f(\xi) = \int_{-k_m}^{k_m} F(k^{'})e^{ik^{'}\xi}dk^{'}$$

and

$$\omega(\xi) = \int_{-k_m}^{k_m} \Omega(k)e^{ik\xi}dk.$$

In Fig. 10.19, the line AB is given by $k + k^{'} = k_m$. For this discretization the area APB contributes to the aliasing. If we pad zeros, i.e. $F(k^{'}) \equiv 0$ for the segment BD and $A'C'$; $\Omega(k) \equiv 0$ for AC and $B'D'$, then the spectral estimation of Equation (10.58) by

$$d(\xi) = \int_{C'}^{D} \int_{D'}^{C} (k^2 + kk^{'})\Omega(k)F(k^{'})e^{i(k+k^{'})\xi}dkdk^{'}$$

will now take place over the bigger square shown by the dotted lines. It is easy to see that the area which will now contribute to aliasing is reduced – as shown by the cross-hatched triangular patches.

Thus by padding more zeros it is possible to shrink this area to zero. For example, if we pad zeros for

$$\left.\begin{aligned} F(k^{'}) &\equiv 0 \quad \text{for } k_m < |k^{'}| \le 2k_m \\ \Omega(k) &\equiv 0 \quad \text{for } k_m < |k| \le 2k_m \end{aligned}\right\}, \tag{10.113}$$

then the estimation of $d(\xi)$ spectrally will be completely dealiased. However, in literature and other texts this has been suggested erroneously. For example, Caunto et al. (1988) suggested that the zero padding should be from $k_m < |k|, |k^{'}| \le \frac{3k_m}{2}$. Of course, depending on the dynamics of the system, if $F(k^{'})$ and $\Omega(k)$ are negligible beyond $|k|, |k^{'}| \ge \frac{k_m}{2}$, this prescription will work. Instead, if one considers a tougher situation, like the one of shot/ white noise processes, then what is suggested in Equation (10.113) will solve the problem correctly.

It is also to be noted that in spectral calculations one employs a box filter, while in all other discrete computations a low pass filter is operational that reduces the Laplace amplitudes with increase in wave number. This fundamental difference between spectral and non-spectral method mitigates the aliasing problem for the latter. In recent years, with advent of higher order methods like compact differencing schemes, it is possible to reach near-spectral accuracy and this is the reason for more and more computations being affected by aliasing that uses accurate methods. Also, note that the aliasing error increases with grid points, making fine grid calculations more susceptible to aliasing problem for non-spectral calculations. For second order central difference scheme, the aliased component is given by the expressions (10.78) and (10.79), and the appearance of $\Delta\xi$ in the denominator will attest to this fact.

In Chapter 5, we have discussed numerical errors and it was stated that truncation error decreases while round-off increases with grid refinement. Hence as computer power increases and more and more 'refined' calculations are done, CFD practitioners will be faced with the task of devising means to control aliasing and round-off errors.

10.14.1 Aliasing Removal by Phase Shifts and Grid-Staggering

Let us consider the product evaluation

$$w(\xi) = u(\xi)v(\xi) \tag{10.114}$$

where

$$u(\xi) = \int_{-k_m}^{k_m} U(k_1)e^{ik_1\xi}dk_1 \tag{10.115}$$

and

$$v(\xi) = \int_{-k_m}^{k_m} V(k_2)e^{ik_2\xi}dk_2. \tag{10.116}$$

If

$$w(\xi) = \int W(k)e^{ik\xi}dk, \tag{10.117}$$

then

$$W(k) = \int w(\xi)e^{-ik\xi}dk. \tag{10.118}$$

Hence

$$\begin{aligned} W(k) = &\int\int_{\text{for } \mathrm{k_1+k_2=k}} U(k_1)V(k_2)dk_1dk_2e^{ik\xi} \\ &+\int\int_{\text{for } \mathrm{k_1+k_2=k\pm 2k_m}} U(k_1)V(k_2)dk_1dk_2e^{ik\xi}. \end{aligned} \tag{10.119}$$

This last quantity in Equation (10.119) is the aliased component.

Now to remove aliasing in the product evaluation of Equation (10.114), in spectral method the variables are phase-shifted as follows.

$$\tilde{u}(\xi) = \int_{-k_m}^{k_m} \tilde{U}(k_1)e^{ik_1(\xi+\triangle)}dk_1 \tag{10.120}$$

$$\tilde{v}(\xi) = \int_{-k_m}^{k_m} \tilde{V}(k_2)e^{ik_2(\xi+\triangle)}dk_2 \tag{10.121}$$

Thus if $\tilde{w} = \tilde{u}\tilde{v}$, then

$$\tilde{W}(k) = \int \tilde{w}(\xi)e^{-ik\xi}dk \tag{10.122}$$

$$= \int\int_{k_1+k_2=k} U(k_1)V(k_2)dk_1dk_2e^{ik\xi} + e^{\pm ik_m\triangle}\int\int_{k_1+k_2=k\pm 2k_{max}} UVdk_1dk_2e^{ik\xi}.$$

Thus if we choose, $k\Delta = k\Delta\xi = \frac{\Pi}{N}$, i.e. shift the variables by half the grid spacing, then,

$$W_{\text{correct}}(k) = \frac{1}{2}[W(k) + \tilde{W}(k)]. \tag{10.123}$$

Therefore, evaluating the product twice, one can obtain the dealiased product. In the spectral space this procedure is easily extended to two or three dimensions.

In finite difference calculations a similar idea is practiced where computations have been performed using staggered grids.

This was first used with MAC method by Harlow & Welch (1965) and in many other latter methods. It is somewhat disconcerting to note that some authors often present results which claim that adopting staggered grid is a must. This is only necessary for non-dissipative numerical methods. The observations of Harlow & Welch (1965) was also reported in Kravchenko & Moin (1997) – as reported in Table 10.2. In the latter reference a second order central difference scheme was noted to solve the convection terms. At the same time Kawamura et al. (1985), while using the MAC method did not use staggered grids. In the third order upwind scheme used in Kawamura et al. (1985), there is sufficient numerical dissipation to remove any energy at high wave number that will accumulate otherwise. This role of general upwind schemes to remove aliasing error is discussed in Sengupta & Nair (1999) and Sengupta & Pin (2002).

An explicit formula for $\frac{\partial}{\partial \xi}\left(f\frac{\partial \omega}{\partial \xi}\right)$ for CD_4 formula

Using the half-node points, one can write down a CD_4 expansion of a first derivative as

$$\left.\frac{\partial g}{\partial \xi}\right|_m^{CD_4} = \frac{1}{6\Delta\xi}[g_{m-1} - 8g_{m-1/2} + 8g_{m+1/2} - g_{m+1}].$$

Hence

$$\left. f\,\frac{\partial \omega}{\partial \xi}\right|_m = \frac{f_m}{6\Delta\xi}[\omega_{m-1} - 8\omega_{m-1/2} + 8\omega_{m+1/2} - \omega_{m+1}].$$

Similarly

$$\frac{\partial}{\partial \xi}\left(f\frac{\partial \omega}{\partial \xi}\right) = \frac{1}{6\Delta\xi}\left[\frac{\partial g_1}{\partial \xi} - 8\frac{\partial g_2}{\partial \xi} + 8\frac{\partial g_3}{\partial \xi} - \frac{\partial g_4}{\partial \xi}\right],$$

where $g_1 = f_m\omega_{m-1}$.

So

$$\frac{\partial g_1}{\partial \xi} = \frac{1}{6\Delta\xi}[f_{m-1}\omega_{m-2} - 8f_{m-1/2}\omega_{m-3/2} + 8f_{m+1/2}\omega_{m-1/2} - f_{m+1}\omega_m].$$

$$g_2 = f_m\omega_{m-1/2},$$

so

$$\frac{\partial g_2}{\partial \xi} = \frac{1}{6\Delta\xi}[f_{m-1}\omega_{m-3/2} - 8f_{m-1/2}\omega_{m-1} + 8f_{m+1/2}\omega_m - f_{m+1}\omega_{m+1/2}].$$

$$g_3 = f_m\omega_{m+1/2},$$

so

$$\frac{\partial g_3}{\partial \xi} = \frac{1}{6\Delta\xi}[f_{m-1}\omega_{m-1/2} - 8f_{m-1/2}\omega_m + 8f_{m+1/2}\omega_{m+1} - f_{m+1}\omega_{m+3/2}].$$

$$g_4 = f_m\omega_{m+1}$$

so

$$\frac{\partial g_4}{\partial \xi} = \frac{1}{6\Delta\xi}[f_{m-1}\omega_m - 8f_{m-1/2}\omega_{m+1/2} + 8f_{m+1/2}\omega_{m+3/2} - f_{m+1}\omega_{m+2}].$$

Thus,

$$\begin{aligned}\frac{\partial}{\partial \xi}(f\frac{\partial \omega}{\partial \xi})_m = \frac{1}{36\Delta\xi^2}[&f_{m-1}(\omega_{m-2} - 8\omega_{m-3/2} + 8\omega_{m-1/2} - \omega_m)\\ &- 8f_{m-1/2}(\omega_{m-3/2} - 8\omega_{m-1} + 8\omega_m - \omega_{m+1/2})\\ &+ 8f_{m+1/2}(\omega_{m-1/2} - 8\omega_m + 8\omega_{m+1} - \omega_{m+\frac{3}{2}})\\ &- f_{m+1}(\omega_m - 8\omega_{m+1/2} + 8\omega_{m+3/2} - \omega_{m+2})].\end{aligned}$$

We will simplify this, by denoting $\Delta\xi = x$ and using the spectral representation. For example,

$$\begin{aligned}&\omega_{m-2} - 8\omega_{m-3/2} + 8\omega_{m-1/2} - \omega_m\\ &= \int \Omega(k)e^{ik\xi}[e^{-2ikx} - 8e^{\frac{-3ikx}{2}} + 8e^{-ikx} - 1]dk\\ &= \int \Omega(k)e^{ik\xi}[e^{-ikx}(e^{-ikx} - e^{ikx}) - 8e^{-ikx}(e^{-ikx/2} - e^{ikx/2})]dk\\ &= \int \Omega(k)e^{ik\xi}\left[e^{-ikx}\left(-2i\sin kx + 16i\sin\frac{kx}{2}\right)\right]dk\\ &= \int \Omega(k)e^{ik\xi}\left[4i\left(4 - \cos\frac{kx}{2}\right)\sin\frac{kx}{2}e^{-ikx}\right]dk.\end{aligned}$$

Similarly,

$$\omega_{m-3/2} - 8\omega_{m-1} + 8\omega_m - \omega_{m+1/2} = \int \Omega(k)e^{ik\xi}\left[4i\left(4 - \cos\frac{kx}{2}\right)\sin\frac{kx}{2}e^{-ikx/2}\right]dk$$

$$\omega_{m-1/2} + 8\omega_m + 8\omega_{m+1} - \omega_{m+3/2} = \int \Omega(k)e^{ik\xi}\left[4i\left(4 - \cos\frac{kx}{2}\right)\sin\frac{kx}{2}e^{ikx/2}\right]dk$$

$$\omega_m - 8\omega_{m+1/2} + 8\omega_{m+3/2} - \omega_{m+2} = \int \Omega(k)e^{ik\xi}\left[4i\left(4 - \cos\frac{kx}{2}\right)\sin\frac{kx}{2}e^{ikx/2}\right]dk.$$

Thus

$$\frac{\partial}{\partial \xi}\left(f\frac{\partial \omega}{\partial \xi}\right) = \int\int F(k')\Omega(k)\left[k_{\rm eq}^{\rm CD_4}\right]^2 e^{i(k+k')\xi}dkdk',$$

where

$$\begin{aligned}(k_{\rm eq}^{\rm CD_4})^2 &= \frac{1}{36x^2}\left[e^{-ik'x}\left\{e^{-ikx}4i(4-\cos\frac{kx}{2})\sin\frac{kx}{2}\right\}\right.\\ &\quad - 8e^{-\frac{ik'x}{2}}\left\{e^{-\frac{ikx}{2}}4i(4-\cos\frac{kx}{2})\sin\frac{kx}{2}\right\}\\ &\quad + 8e^{\frac{ik'x}{2}}\left\{e^{\frac{ikx}{2}}4i(4-\cos\frac{kx}{2})\sin\frac{kx}{2}\right\}\\ &\quad \left.-e^{ik'x}\left\{e^{ikx}4i(4-\cos\frac{kx}{2})\sin\frac{kx}{2}\right\}\right]\\ &= \frac{4i(4-\cos\frac{kx}{2})\sin\frac{kx}{2}\{e^{-i(k+k')x}-8^{-i(k+k')\frac{x}{2}}+8e^{i(k+k')\frac{x}{2}}-e^{i(k+k')x}\}}{36x^2}\\ &= \frac{i(4-\cos\frac{kx}{2})\sin\frac{kx}{2}}{9x^2}\left\{-2i\sin(k+k')x+16i\sin(k+k')\frac{x}{2}\right\}.\end{aligned}$$

Thus

$$(k_{\rm eq}^{\rm CD_4})^2 = -\frac{4}{9x^2}\left[\left(4-\cos\frac{kx}{2}\right)\left(4-\cos\frac{k+k'}{2}x\right)\sin\frac{kx}{2}\sin\frac{k+k'}{2}x\right].$$

In this case also, the aliased component coming from Region 1 (of Fig 10.8) can be estimated from

$$d_{\rm CD_4}^a(\xi) = \int_0^{k_m}\int_{k_m-k'}^{k_m}(k_{eq}^{\rm CD_4})^2\Omega(k)F(k')e^{i(k+k')\xi}dkdk'.$$

Exercise

1. For the CD_6 scheme for the first derivative given in Equation (5.32), obtain

$$\left(\frac{k_{\rm eq}}{k}\right)_{\rm CD_6} = \frac{\sin 3k\Delta x - 9\sin 2k\Delta x + 45\sin k\Delta x}{30k\Delta x}.$$

2. For the fifth order upwind scheme show that

$$\text{Imag}\left\{\left(\frac{k_{\rm eq}}{k}\right)_{\rm UD_5}\right\} = \frac{12\cos(2k\Delta x) - 30\cos k\Delta x - 2\cos 3k\Delta x + 20}{60k\Delta x}$$

3. Plot $|G_{\rm UD_1}|$ as a function of $k\Delta x$ for different values $N_c = 0.01, 0.1, 1$ and 2. What conclusions you can draw from your plots?

4. Obtain the expression for G_{UD_5} and repeat the above exercise.
5. The 1D wave equation is solved by first order upwind scheme (as given in Equation (8.4)) and by leap frog method (as given in Equation (10.40)). Identify the CFL numbers for which the calculations will be "correct" for each method.

References

[1] ARAKAWA, A., Computational design of long term numerical integration of the equations of fluids motion: Two-dimensional incompressible flow, *J. Comp. Phys.*, **1**, 119–143, (1966).

[2] BASHFORTH, F. & ADAMS, J.C., *An Attempt to Test the Theories of Capillary Action by Computing the Theoretical and Measured Forms of Drops of Fluids,* Cambridge Press, (1883).

[3] CANUTO, C., HUSSAINI, M.Y., QUARTERONI, A. & ZANG, T.A., *Spectral Methods in Fluid Dynamics*, Springer Verlag, New York, (1988).

[4] CEBECI, T. & BRADSHAW, P., *Momentum Transfer in Boundary Layers*, Hemisphere, Washington D.C., (1977).

[5] DURRAN, D.R., *Numerical Methods for Wave equations in Geophysical Fluid Dnamics.* Springer-Verlag, New York, (1999).

[6] HALTINER, G.J. & WILLIAMS, R.T., *Numerical Prediction And Dynamic Meterology.* John Wiley & Sons (2nd ed.), New York, (1980).

[7] HARLOW, F.H. & WELCH, J.E., Numerical calculation of time-dependent viscous incompressible flow of fluid with free surface, *Phys. Fluids*, **8(12)**, 2182–2189, (1965).

[8] HARTEN, A., High Resolution schemes for hyperbolic conservation laws, *J. Comp. Phys.*, **49**, 357–393, (1983).

[9] HEYWOOD, J.G., RANNACHER, R. & TUREK, S., Artifical boundries and flux and pressure conditions for the incompressible Navier-Stokes equations, *Int. J. Num. Methods Fluids*, **22**, 325–352, (1996).

[10] KAWAMURA, T, TAKAMI, H. & KUWAHARA, K., A new higher order upwind scheme for incompressible Navier-Stokes equations, *Fluid Dynamic Res.*, **1**, 145–162, (1985).

[11] KIM, J. & MOIN, P., Application of a fractional-step method to incompressible Navier-Stokes equation, *J. Comp. Phys.*, **58**, 308–322, (1985).

[12] KRAVCHENKO, A.G. & MOIN, P., On the effect of numerical errors in large eddy simulations of turbulent flows, *J. Comp. Phys.*, **131**, 310–322, (1997).

[13] LE, H. & MOIN, P., An improvement of fractional step methods for the incompressible Navier-Stokes equation, *J. of Comp. Phys.*, **92**, 369–379 (1991).

[14] LELE, S.K., Compact finite difference schemes with spectral-like resolution, *J. Comp. Phys.*, **103**, 16–42 (1992).

[15] LEONARD, B.P., A stable accurate convective modeling procedure based on quadratic upstream interpolaration, *Comp. Methods Appl. Mech.*, **19**, 59–98, (1978).

[16] LESIEUR, M & METAIS, O., New trends in large-eddy simulation of turbulence, *Ann. Rev. of Fluid Mech.*, **28**, 45–82, (1996).

[17] LILLY, D.K., On the computational stability of numerical solution of time-dependent non-lnear geophysical fluid dynamics problems U.S. Weather Bureau, *Monthly Weather Review*, **93(1)**, 11–26 (1965).

[18] NAIR, M.T., SENGUPTA, T.K. & CHAUHAN, U.S., Flow past rotating cylinders at high Reynolds number using higher order upwind schemes, *Computers & Fluids*, **27(1)**, 47–70, (1998).

[19] ORSZAG, S.A., On the elimination of aliasing in finite difference schemes by filtering high-wave number components, *J. Atmos. Sc.*, **28**, 1074, (1971).

[20] PHILIPS, N.A., *An Example of Non-linear Computational Instability: The Atmosphere and Sea in Motion*, Rockfeller Institute Press, New York, (1959).

[21] PRESS, W.H., FLANNERY, B.P., TEUKOLSKY, S.A. & VETTERING, W.T., *Numerical Recepies: The Art of Scientific Computing*, Cambridge University Press, (1989).

[22] ROACHE, P.J., *Computational Fluid Dynamics*, Hermosa Publishers, Albuquerque, New Mexico, (1982).

[23] SCHUMANN, U., Direct and large eddy simulation of turbulence: Introduction to the modeling of Turbulence, *Von Karman Institute for Fluid Dynamics*, Lecture Notes, 1987–06 (1987).

[24] SENGUPTA, T.K., CHATTOPADHYAY, M., WANG, Z.Y. & YEO, K.S., By-pass mechanism of transition to turbulence, *J. Fluids & Structures*, **16(1)**, 15–29,(2002).

[25] SENGUPTA, T.K. & NAIR, M.T., Upwind schemes and large eddy simulation, *Int. J. Num. Meth. in Fluids*, **31**, 879–889 (1999).

[26] SENGUPTA, T.K. & PIN, L.Y., Discretization of Laplacian operator and aliasing error, *Proc. of 9th Asian Congress of Fluid Mech.*, (Eds. E. Shirani & A. Pishevar), Isfahan, Iran (2002)

[27] SENGUPTA, T.K. & SENGUPTA, R., Flow past an impulsively started circular cylinder at high Reynolds number, *Comput. Mech.*, **14(4)**, 298–310, (1994).

[28] SENGUPTA, T.K., DE, S. & SARKAR, S., Vortex-induced instability of incompressible wall-bounded shear layer, *J. Fluid Mech.*, **493**, 277-286, (2003).

[29] SHU, C.W. & OSHER, S., Efficient implementation of essentially non-oscillatory shock capturing schemes, *J. Comp. Phys.*, **77**, 439–471, (1988).

[30] SHU, C.W. & OSHER, S., Efficient implementation of essentially non-oscillatory shock capturing schemes II, *J. Comp. Phys.*, **83**, 32–78, (1989).

[31] SMAGORINSKY, J., Some historical remarks on the use of nonlinear viscosities, in *Large Eddy Simulation of Complex Engineering and Geophysical Flows* (Eds. B. Galperin & S.A. Orszag), Cambridge Univ. Press U.K (1993).

[32] TEMAM, R., *Navier-Stokes Equations*, North Holland, Amsterdam (3rd edition), (1984).

11 Higher Order Methods

From our discussions in the previous chapter it is now apparent that it is routinely possible to solve sets of differential equations numerically for practical parameter ranges. This includes the solution of partial differential equations with stringent requirements of resolving wide ranges of spatial and temporal scales. The role of such scales in CFD applications has been discussed in Chapter 4 for the most general cases of unsteady viscous flows. In the literature one comes across a large number of efforts that solve the governing Navier-Stokes equation, resolving all the scales of a turbulent flow in a DNS for moderate Reynolds numbers. At the same time, large RAM machines are available even at desk-top, allowing one to solve problems with larger number of grid points. This necessitates a search for faster methods—faster than the explicit methods discussed in the previous chapters. The higher order methods are expected to fulfill this role, because this will allow one to solve the same problem with fewer grid points. Conversely, higher order methods allow solving problems at higher Reynolds number with the same grid. Thus higher order methods allow solving those problems which were hitherto impossible.

The effectiveness of CFD schemes is generally ranked by the leading order of the Taylor series expanded term which is used to define the truncation error of the scheme. It is expected that higher the order, better is the quality of the scheme. While this may appear as a correct expectation, the discussion in Chapter 10 will make the reader aware that classification of schemes simply based on order of truncation error is not complete. In Chapter 10, the schemes were assessed based on their behaviour in the spectral plane.

For example, in Fig. 10.2 the various central discretization methods for representing a first derivative were compared. While the spectral method provides the maximum accuracy, it is seen that higher the order of the scheme, better is the representation. If one compares these central schemes with the upwind schemes, it is apparent that upwinding is equivalent to introducing numerical dissipation to the central scheme. Thus one can take a very high order central scheme and then add numerical dissipation term proportional to ∇^{2n} of the physical variable. Classifying the numerical methods based on truncation error will render this method to $(2n-1)^{th}$ order, irrespective of the spectral accuracy of the associated central scheme. Similar classification conundrum will make some of the finite volume and finite element methods with $n = 1$ only first order accurate, despite their excellent spectral resolution. It is altogether a different matter that these methods sometimes produce poor results due to the fact that the introduced numerical dissipation interferes with the physical dissipation in the energy-carrying wave number range. In CFD methods for time independent problems, if these are consistent and numerically stable, then by Lax's equivalence theorem the convergence to steady state solution is assured.

For time dependent problems, especially acoustics and electromagnetics problems, the above requirement of equivalence theorem does not provide correct solutions for these wave phenomena. The essential properties of wave phenomena were discussed in Chapter 4, and it was shown that satisfying the *dispersion relation* is equivalent to satisfying the governing equations and boundary conditions in the spectral plane. And for a dynamical system it was pointed out that the group velocity is an important parameter that transports energy of the system. It is in this context that dynamical systems were classified as *dispersive* or *non-dispersive*, depending on the relationship between group velocity and phase speed. For general inviscid compressible flows, via linearizion of Euler equation, Tam & Web (1993) have demonstrated the existence of acoustic, entropic and vortical waves. The acoustic waves are isotropic, non-dispersive, almost non-dissipative.

Since acoustic waves are non-dispersive, the signal-energy propagates with the speed of sound. For inviscid flows the entropic and vortical waves are also seen to be non-dispersive. Irrespective of the fact whether the dynamical system is dispersive or not, the adopted numerical method for solving the time dependent governing differential equations must satisfy the physical dispersion relation.

Many popular CFD methods can produce spuriously dispersive, anisotropic and highly dissipative results. To numerically stabilize codes, large amount of dissipation is added explicitly or implicitly; such methods will not be *dispersion relation preserving*, in general.

In the previous chapter we discussed some of the higher order explicit methods (given by Equations (10.16) and (10.18) and in Exercise 10.1). In this chapter we introduce compact schemes for which the solution and its various order derivatives are treated as unknowns. The derivatives at the grid points are solved from an implicit equation involving the implicit derivatives and the functions. There are two essential features of compact schemes: high spectral accuracy and relatively compact stencils which relate the derivatives with the function. Although these methods do not have the resolution of spectral method, they provide robustness and can solve problem involving complex geometries. Historically, these methods owe their origin to the Padé schemes for ODE, described in Kopal (1962). A general description and application for CFD is given in Lele (1992).

These methods were conceived of as an alternative to the spectral method and are based on Padé approximation. Thus, the derivatives are obtained by solving a set of linear algebraic equations. As the representations are compact, the resulting matrices are band-limited and hence very easily solved. Some of the early works in this field is to be found in Hirsch (1975), Adams (1977) and Lele (1992). Such methods have been optimized to solve the time-domain Maxwell equation in Gaitonde & Shang (1997) and for problems relating to acoustics in Tam & Web (1993) and Fung et al (1996).

11.1 The General Compact Schemes

Consider a function u, whose derivatives are to be evaluated by the application of the compact schemes. For the nth derivative of the said function defined as $u_j^{(n)}$ at the jth node, the general implicit equation is given by

$$\sum_{k=-N_l}^{N_r} a_k u_{j+k}^{(n)} = \frac{1}{h^n} \sum_{k=-M_l}^{M_r} b_k u_{j+k} \tag{11.1}$$

on a uniform grid of spacing h. This implies that such schemes discussed here are only for uniform grids or they are applied to equations in the transformed planes, as defined in Chapter 9. There are recorded attempts of developing compact schemes for non-uniform grids, but they will not be discussed here.

In this implicit relation, the derivatives have a bandwidth spanning N_l and N_r points to the left and right about the node under consideration. The derivatives are related to the function values involving M_l points to the left of the jth node and M_r points to the right of the node. If one puts a restriction that a_k's are symmetric about a_0, then the above relation will result in a central compact scheme only if $N_l = N_r$ and $M_l = M_r$. Assuming that $u(x)$ has a valid Taylor expansion up to order $O(h^{r+n})$ at each x, then Equation (11.1) gives $u_j^{(n)}$ with order $O(h^{r+1})$ accuracy.

The coefficients a_k and b_k are determined such that the order of the scheme is fixed by the leading truncation error term. Let us demonstrate the same for the first and second derivatives of a function, which are mostly required for CFD applications.

11.2 Approximating First Derivatives by Central Scheme

For the generalized Padé scheme, the first derivative is related to the functions by using Equation (11.1) with the following conditions.

$$n = 1,$$
$$N_l = N_r = 2 \quad \text{and}$$
$$M_l = M_r = 3.$$

Also, $a_{-2} = a_2 = \beta, a_{-1} = a_1 = \alpha, a_0 = 1$

$b_3 = -b_{-3} = \frac{c}{6}, b_2 = -b_{-2} = \frac{b}{4}, b_1 = -b_{-1} = \frac{a}{2}, b_0 = 0$

Thus, Equation (11.1) reduces to

$$\beta u'_{j-2} + \alpha u'_{j-1} + u'_j + \alpha u'_{j+1} + \beta u'_{j+2}$$
$$= c\frac{u_{j+3} - u_{j-3}}{6h} + b\frac{u_{j+2} - u_{j-2}}{4h} + a\frac{u_{j+1} - u_{j-1}}{2h}. \tag{11.2}$$

The coefficients can be related by Taylor expansion of terms on the right and left hand side of Equation (11.2) about the jth node. Equating the Taylor series coefficients of various orders, we get the constraints on these coefficients in Equation (11.2). The first unmatched coefficients determine the formal order accuracy of the representation.

Thus one can generate the following sequence of relations among the coefficients of Equation (11.2) when we develop schemes up to

2nd order:

$$a + b + c = 1 + 2\alpha + 2\beta \tag{11.3}$$

4th order:

$$a + 4b + 9c = 6(\alpha + 4\beta) \tag{11.4}$$

6th order:

$$a + 16b + 81c = 10(\alpha + 16\beta) \tag{11.5}$$

8th order:

$$a + 64b + 729c = 14(\alpha + 64\beta) \tag{11.6}$$

10th order:

$$a + 256b + 6561c = 18(\alpha + 256\beta) \tag{11.7}$$

and so on. That is, if we are looking for a second order accurate scheme, then Equation (11.3) has to be satisfied. For fourth order accuracy one needs to satisfy Equations (11.3) and (11.4).

When none of these coefficients are assigned a zero value, then assembling all the equations for different node values obtained from Equation (11.2) will constitute a linear algebraic equation given by

$$[A]\{u_j^{'}\} = [B]\{u_j\}. \tag{11.8}$$

In this equation $[A]$ is a symmetric pentadiagonal matrix with non-zero elements contiguous to the diagonal. In this equation the right hand side is treated as known. Of course this representation is not complete, as one cannot use Equation (11.2) at immediate neighbourhood points of the boundary. However, if one is solving a Cauchy problem (i.e. a pure initial value problem), then one can use Equation (11.8) directly. If one is instead solving a periodic problem, i.e., u as a function of x is periodic over a given interval, then the A matrix will be a periodic pentadiagonal matrix.

In Equation (11.2) if we choose $\beta = 0$ then the corresponding A matrix in Equation (11.8) would be a tridiagonal matrix or periodic tridiagonal matrix. The Thomas algorithm for solving a tridiagonal matrix equation has already been given at the end of Chapter 6. We provide here an algorithm for solving a periodic tridiagonal matrix equation.

11.3 Method for Solving Periodic Tridiagonal Matrix Equation

Consider the matrix equation

$$[A]\{X\} = \{b\} \tag{11.9}$$

that needs to be solved for the unknown vector $\{X\}_{n\times 1}$, where the matrix $[A]$ has a periodic tridiagonal structure given by

$$[A]_{n\times n} = \begin{bmatrix} b_1 & c_1 & 0 & \cdots & \cdots & \cdots & 0 & a_1 \\ a_2 & b_2 & c_2 & 0 & \cdots & \cdots & \cdots & 0 \\ 0 & a_3 & b_3 & c_3 & 0 & \cdots & \cdots & 0 \\ \cdots & \cdots & \cdots & \cdots & \cdots & \cdots & \cdots & \cdots \\ \cdots & \cdots & \cdots & \cdots & \cdots & \cdots & \cdots & \cdots \\ \cdots & \cdots & \cdots & \cdots & \cdots & \cdots & \cdots & \cdots \\ 0 & \cdots & \cdots & \cdots & 0 & a_{n-1} & b_{n-1} & c_{n-1} \\ c_n & 0 & \cdots & \cdots & \cdots & 0 & a_n & b_n \end{bmatrix}. \tag{11.10}$$

As we had seen in Thomas algorithm, here also we would split the $[A]$ matrix into a lower and upper triangular matrices:

$$[A] = [L][U], \tag{11.11}$$

where the matrices $[L]$ and $[U]$ have the structure

$$[L] = \begin{bmatrix} q_1 & 0 & 0 & \cdots & \cdots & \cdots & 0 & 0 \\ p_2 & q_2 & 0 & 0 & \cdots & \cdots & \cdots & 0 \\ 0 & p_3 & q_3 & \cdots & 0 & \cdots & \cdots & 0 \\ \cdots & \cdots & \cdots & \cdots & \cdots & \cdots & \cdots & \cdots \\ \cdots & \cdots & \cdots & \cdots & \cdots & \cdots & \cdots & \cdots \\ \cdots & \cdots & \cdots & \cdots & \cdots & \cdots & \cdots & \cdots \\ \cdots & \cdots & \cdots & \cdots & \cdots & \cdots & \cdots & 0 \\ r_1 & r_2 & r_3 & \cdots & \cdots & r_{n-2} & p_n & q_n \end{bmatrix}, \tag{11.12}$$

and

$$[U] = \begin{bmatrix} 1 & u_1 & 0 & \cdots & \cdots & \cdots & 0 & w_1 \\ 0 & 1 & u_2 & 0 & \cdots & \cdots & 0 & w_2 \\ \cdots & \cdots & \cdots & \cdots & \cdots & \cdots & \cdots & \cdots \\ \cdots & \cdots & \cdots & \cdots & \cdots & \cdots & \cdots & \cdots \\ \cdots & \cdots & \cdots & \cdots & \cdots & \cdots & 0 & \cdots \\ \cdots & \cdots & \cdots & \cdots & \cdots & \cdots & \cdots & w_{n-2} \\ \cdots & \cdots & \cdots & \cdots & \cdots & \cdots & \cdots & u_{n-1} \\ 0 & \cdots & \cdots & \cdots & \cdots & \cdots & 0 & 1 \end{bmatrix}. \tag{11.13}$$

A term-by-term multiplication of $[L]$ and $[U]$ matrix and then equating it with $[A]$ matrix yields the algorithm for the evaluation of p_i, q_i, r_i, u_i and w_i in Equations (11.12) and (11.13).

For example, the first row of $[A]$ provides

$$a_1 = q_1 w_1, \quad b_1 = q_1, \quad c_1 = q_1 u_1.$$

Hence

$$q_1 = b_1; \quad w_1 = a_1/b_1 \quad \text{and} \quad u_1 = c_1/b_1 \tag{11.14}$$

Similarly, for the second row of $[A]$ matrix

$$a_2 = p_2, \quad b_2 = p_2 u_1 + q_2, \quad c_2 = q_2 u_2,$$

which provides $p_2 = a_2$; $q_2 = b_2 - p_2 u_1$ and $u_2 = c_2/q_2$. This can be generalized to obtain p_j, q_j and u_j from

$$\left.\begin{aligned} p_j &= a_j \\ q_j &= b_j - p_j u_{j-1} \\ u_j &= c_j/q_j \end{aligned}\right\} \text{for} \quad j = 2 \quad \text{to} \quad n-2 \tag{11.15}$$

$$p_{n-1} = a_{n-1}; \quad q_{n-1} = b_{n-1} - p_{n-1} u_{n-2}; \quad u_j = (c_j - p_j . w_{j-1})/q_j$$

for $j = (n-1)$ and with the starting value provided as $u_1 = c_1/b_1$. From the last column of $[A]$ matrix, one can get the following recursion relation.

$$p_j w_{j-1} + w_j q_j = 0 \quad \text{for} \quad j = 2 \quad \text{to} \quad n-2. \tag{11.16}$$

This can be used to evaluate w_j, using the starting value of $w_1 = a_1/b_1$.

From the last row of $[A]$ matrix one gets the following equation:

$$r_1 = c_n,$$

$$r_j + r_{j-1} u_{j-1} = 0 \quad \text{for} \quad j = 2 \quad \text{to} \quad n-2, \tag{11.17}$$

$$r_{n-2} u_{n-2} + p_n = a_n, \tag{11.18}$$

and

$$\sum_{j-1}^{n-2} r_j w_j + p_n u_{n-1} + q_n = b_n. \tag{11.19}$$

From Equation (11.18) one can obtain p_n and this can be used in Equation (11.19) to obtain q_n. Having thus obtained the L and U matrices, it is easy to solve for the vector $\{X\}$ by defining an auxiliary variable

$$[U]\{X\} = \{V\}. \tag{11.20}$$

Thus in a sequence, one first solves

$$[L]\{V\} = \{b\}, \tag{11.21}$$

and then solves Equation (11.20) for $\{X\}$. This is easy to perform and is left as an exercise.

11.4 An Example of a Sixth Order Scheme

Let us consider first a case for which $\beta = c = 0$ so that Equation (11.2) reduces to

$$\alpha u'_{j-1} + u'_j + \alpha u'_{j+1} = \frac{a}{2h}(u_{j+1} - u_{j-1}) + \frac{b}{4h}(u_{j+2} - u_{j-2}). \tag{11.22}$$

One can fix the coefficient values by equating the resultant Taylor series on the right and left hand side of the above equation term by term, expressed about the jth node.

$$\text{Coefficient of } u'_j : 1 + 2\alpha = a + b \tag{11.23}$$

$$\text{Coefficient of } u'''_j : \alpha = \frac{a + 4b}{6} \tag{11.24}$$

$$\text{Coefficient of } u^{(v)}_j : \alpha = \frac{a + 16b}{10} \tag{11.25}$$

Different ordered accurate schemes can be obtained for evaluation of the first derivative, by equating successively the coefficient of the Taylor series expansion for Equation (11.22). In all such schemes, satisfaction of relation (11.23) is a must and this can be viewed as the *consistency* requirement of Equation (11.22).

Solving these three equations yields

$$\alpha = 1/3, \quad a = 14/9 \quad \text{and} \quad b = 1/9. \tag{11.26}$$

This is a sixth order scheme, because the leading truncation error term is given by $\frac{4}{7!}h^6 u^{(7)}$. However, if we decide to satisfy only Equations (11.23) and (11.24), the resultant scheme would be only fourth order accurate. For example, Lele (1992) shows that this constitutes a one-parameter family schemes for which $a = \frac{2}{3}(\alpha + 2)$ and $b = \frac{1}{3}(4\alpha - 1)$. One can repeat the exercise for the case of $\beta = 0$ and $c \neq 0$ to come out with different schemes, which may be of same order as the above mentioned schemes. Having obtained the same ordered scheme by choosing different coefficients in Equation (11.2), it is natural to ask—which is best of them all? Also we would like to inquire

(*a*) Is there a physical basis for choosing any one of the compact schemes?

(*b*) Is the departure of the discrete derivatives a discontinuous function of these coefficients?

That is, instead of choosing α, a and b from Equation (11.26), if we perturb the value of these coefficients while satisfying Equation (11.23) the consistency condition, will the representation be poor or will the departure of the discrete representation also depart by a small amount from the exact derivative?

The answer to the first question regarding the uniqueness of representation cannot be given unambiguously in the absence of further research. However, one can point out that the compact schemes are necessarily an auxiliary set of relations to be used for discretizing the governing differential equation. Each term in the discretization must be a consistent representation in the continuum limit of $h \to 0$. While discussing the consistency of Dufort-Frankel method for one-dimensional diffusion equation, it was noted that each derivative term was consistently discretized. But a full discretized equation of this parabolic PDE is equivalant to solving a hyperbolic PDE, if the time step is not an order magnitude smaller than the spatial grid size. While using compact schemes we must keep this in mind. For simple model equations, it is possible to show unambiguously the connection between the discretization of individual derivatives and the discrete equation for consistency with the continuum equation. However, for complex nonlinear PDEs the establishment of consistency is not straightforward. We also note that this problem is avoided in many finite volume and finite elements formulations, where with the appropriate choice of basis functions one solves the PDE by a *weak formulation* and the resultant discrete equation is then consistent. While this takes away the freedom to choose from large possibilities, it provides uniqueness to the discrete formulation, having chosen a particular basis function.

The answer to the second question is very important and has often been overlooked in the debate over the formal accuracy of the leading order truncation error term. It is the understanding (reason provided in this chapter) that the discrete approximation of any operator is a continuous function of the parameters that define the approximation. In fact, Haras & Ta'asan (1994), extended the work of Lele (1992) to obtain optimal values of the coefficients of Equation (11.22) to minimize the departure of the discrete approximation of the compact scheme from Fourier spectral method. This departure was measured over the full range of wave numbers that a fixed grid allow. In doing so, Haras & Ta'asan (1994) focused only on a second order accurate scheme. Lele (1992) also performed an optimization process, but the departure of the compact scheme from the spectral method was minimized at three large wave numbers. In doing so, Lele observed that the optimized fourth order accurate scheme provided better resolution than a tenth order explicit central scheme. All these results indicate that the formal accuracy of the truncation error cannot be the sole criteria. A better depiction would be in the spectral plane—as was shown in the previous chapter.

In analysing any schemes, we will follow a general framework by analysing Equation (11.8). Different discrete methods use different $[A]$ and $[B]$ matrices. For example, for explicit schemes, one can simply set $[A] = [I]$, the identity matrix.

As mentioned before, to solve non-periodic problems, special stencils have to be created for the boundary and near-boundary points. In the following, we describe some of the schemes for first derivatives that are in use. We identify boundary nodes by $j = 1$ and N and the near-boundary nodes by $j = 2$ and $j = N - 1$. The interior nodes are from $j = 3$ and $N - 2$. Adams (1993) has suggested using the following stencils:

$$j = 1: \quad 2u_1' + 4u_2' = \frac{(-5u_1 + 4u_2 + u_3)}{h}, \tag{11.27}$$

$$j = 2: \quad u_1' + 4u_2' + u_3' = 3\frac{(u_3 - u_1)}{h}, \tag{11.28}$$

and

$$3 \leq j \leq N - 2: \quad u_{j-1}' + 3u_j' + u_{j+1}' = \frac{(-u_{j-2} - 28u_{j-1} + 28u_{j+1} + u_{j+2})}{12h}. \tag{11.29}$$

One can write the stencils for $j = N - 1$ and N similar to that given by Equations (11.27) and (11.28). In the above, the near-boundary point stencil ($j = 2$ and $N - 1$) has fourth order accuracy while the interior points have sixth order formal accuracy.

The following fourth order accurate stencils were used in Carpenter et al. (1993).

$$j = 1: \quad u_1^{'} + 3u_2^{'} = \frac{(-17u_1 + 9u_2 + 9u_3 - u_4)}{(6h)}. \tag{11.30}$$

$$2 \leq j \leq N - 1: \quad u_{j-1}^{'} + 4u_j^{'} + u_{j+1}^{'} = \frac{3(u_{j+1} - u_{j-1})}{h}. \tag{11.31}$$

For the boundary point $j = N$, the following stencil is used:

$$u_N^{'} + 3u_{N-1}^{'} = \frac{(17u_N - 9u_{N-1} - 9u_{N-2} + u_{N-3})}{(6h)}. \tag{11.32}$$

Note that the right hand side expression for Equation (11.32) is anti-symmetric with respect to the expression given for the other boundary stencil given by Equation (11.30).

The above compact scheme is contrasted with the following fourth order accurate explicit scheme, also presented in Carpenter et al. (1993).

$$j = 1: \quad u_1^{'} = \frac{(-25u_1 + 48u_2 - 36u_3 + 16u_4 - 3u_5)}{(12h)}. \tag{11.33}$$

$$j = 2: \quad u_2^{'} = \frac{(-3u_1 - 10u_2 + 18u_3 - 6u_4 + u_5)}{(12h)}. \tag{11.34}$$

$$3 \leq j \leq N - 2: \quad u_j^{'} = \frac{(u_{j-2} - 8u_{j-1} + 8u_{j+1} - u_{j+2})}{(12h)}. \tag{11.35}$$

Although the formal order of approximation of both the schemes are the same, they do not provide the same spectral resolution. This is discussed next.

11.5 Order of Approximation vs. Resolution

The order of approximation usually refers to a local Taylor series expansion and is measured by the leading truncation error term. However, resolution refers to the ability of numerical schemes to resolve various length scales involved in a problem, and a best view of it is provided by expressing the unknowns by Fourier-Laplace transform. This has already been introduced in Chapters 4 and 10. Thus the resolution of a scheme is measured by the largest wave number that can be represented. This should not be confused (as often found in the literature on DNS and LES) with the maximum wave number fixed by the Nyquist criterion. The estimate of the latter criterion is $k_m = \pi/h$ for a uniform grid.

To perform Fourier analysis, the unknown $u(x_j)$ is represented in terms of its transform $U(k)$ using the relation,

$$u_j = u(x_j) = \int_{-k_m}^{k_m} U(k)e^{ikx_j}dk. \tag{11.36}$$

To analyse any scheme, like the one given by Equation (11.8), we rewrite the general scheme by

$$\{u'\} = [C]\{u\},$$

where $[C] = [A]^{-1}[B]$. This underscores the difference between explicit and compact schemes. In explicit schemes $[A]$ is an identity matrix and therefore the bandwidth of the stencil is dictated by the bandwidth of the B matrix. As the stencil has to be at least one point wider than the approximation order, one therefore encounters problems of finding suitable and stable, boundary and near-boundary stencils that also ought to be of higher order.

For compact schemes, $[A]$ matrix is usually a band-limited matrix and so is the $[B]$ matrix. However, $[C]$ matrix will not be band limited in general, for the representation given by Equation (11.37). Hence, the compact schemes, despite their *local* appearance actually produce a *global* approximation. Thus for each compact scheme, there is an equivalent explicit scheme that can be constructed—as given by Equation (11.37). One can furthermore manipulate Equation (11.37) to bring it to the spectral representation as given in Equation (11.36). If one uses Equation (11.36) then a derivative of the function can be written as

$$u'(x_j) = \int ikU(k)e^{ikx_j}dk. \tag{11.38}$$

Thus the derivative at the point x_j is evaluated by the quantities evaluated at x_j on the right hand side. In Equation (11.37), on the right hand side the function values at all nodes are involved. However, using Equation (11.36) one can relate the functions at jth and lth node by

$$u_j = \int ikU(k)e^{ikx_j}P_{lj}dk, \tag{11.39}$$

where $x_j = jh$ and

$$P_{lj} = e^{i(l-j)kh} = R_{lj} + iI_{lj} \tag{11.40}$$

is split into a real and imaginary part.

One can, therefore, represent the derivative at the jth node by the variable at the jth node from the equation

$$u'_j = \left\{\sum_{i=1}^{N} C_{jl}P_{lj}\right\} u_j, \tag{11.41}$$

where C_{jl} are the entries of the $[C]$ matrix $(= [A]^{-1}[B])$ and P_{lj} is defined by Equation (11.40). Thus, if one wants to rewrite the right hand side of Equation (11.41), then using Equation (11.36),

$$u_j' = \int_{-k_m}^{k_m} ik_{eq} U(k) e^{ikx_j} dk, \tag{11.42}$$

where, for the discrete scheme, at the jth node

$$ik_{eq}(x_j) = \sum_{l=1}^{N} C_{jl} P_{lj}. \tag{11.43}$$

Thus, having chosen N and constructed $[A]$ and $[B]$ matrices, for a given scheme, one simply constructs the $[C]$ matrix. Thereafter using Equation (11.43) one can estimate k_{eq} at all the nodes. This type of analysis helps one to evaluate the scheme globally over the whole domain. This can be termed as the spectral matrix stability analysis. This has been introduced by Sengupta et al. (2003).

It is to be realized that in general k_{eq} is a complex number, with the real part (k_{eq}^{real}) indicating the spectral resolution of the scheme and the imaginary part (k_{eq}^{imag}) indicating the added numerical dissipation because,

$$i\int k_{eq} U(k) e^{ikx_j} dk = i\int k_{eq}^{real} U(k) e^{ikx_j} dk - \int k_{eq}^{imag} U(k) e^{ikx_j} dk. \tag{11.44}$$

In representing results for various schemes, it is better to plot (k_{eq}/k) versus kh, because, among all possible representations of an unknown, spectral method can be shown as the least-error representation. The above ratio, thus, reflects the efficiency of a method with respect to the best theoretical estimate. For any grid the abscissa will range between zero and π. It is to be noted that for many applications one chooses symmetric stencil for the interior nodes. Theoretically such schemes are non-dissipative. However, one is forced to take asymmetric stencils for non-periodic problems, for boundary and near-boundary points. Also, it has already been pointed out that the compact schemes are global in nature. Hence the asymmetric boundary and near-boundary stencils have their effects on many interior nodes as well.

Since the present analysis is to estimate the spectral properties of derivatives to specifically include near-boundary point treatment, it is important to do so in a manner that the results are not sensitive to the number of grid points chosen.

For illustrative purposes, one can decide to evaluate second derivative at the discrete nodes by using the Equation (11.37) twice, i.e.,

$$\{u''\} = [C]^2\{u\}. \tag{11.45}$$

If we use the stencils given by Equations (11.27) to (11.29) for $N = 5$, then the second derivative at the center of the domain is given by (can be obtained easily by any symbolic manipulator, like *Mathematica*)

$$u_3'' = \frac{1}{h^2}(u_2 - 2u_3 + u_4). \tag{11.46}$$

This is a second order accurate central difference representation, despite the fact that the stencil applied for $l = 3$ (Equation (11.29)) locally is a sixth order accurate representation, while the boundary stencils are fourth order accurate!

A systematic study revealed that taking $N = 20$ and 30 produces no perceptible difference for evaluating the first and second derivatives. Hence, in all the results that are presented here, $N = 30$ is used to discretize the domain between $x = 0$ and $x = 1$.

In Figs. 11.1(a) and 11.1(b) the real and imaginary parts of the equivalent wave number for the estimation of first derivatives using the compact scheme of Adams (1993) (as given by Equations (11.27) to (11.29)) are shown. In the figures only some representative points are shown, while the points that are not shown follow the neighbouring points' behaviour. It is evident that this scheme retains near-spectral accuracy in representing the correct phase up to a large extent within the Nyquist limit, as seen from the real part. The near-boundary points show deficient spectral resolution and the low-pass filter attribute of the scheme. It is to be emphasized that for problems with Dirichlet boundary conditions, the depicted behaviour for $j = 1$ and N are of no concern as the governing equations will be solved for $j = 2$ to $N - 1$, for any application.

The imaginary part, as shown in Fig. 11.1(b), shows a very disturbing feature of the method. For $j \leq 6$ we note that the imaginary part is positive at higher wave numbers, instead of being negative as can be seen from Equation (11.44). This implies that instead of providing numerical dissipation, one is adding *anti-diffusion* at those points which will lead to numerical instability.

This also establishes the superiority of the spectral matrix stability method over the eigenvalue analysis of semi-discrete equation in the physical plane as given in Carpenter et al. (1993) and in

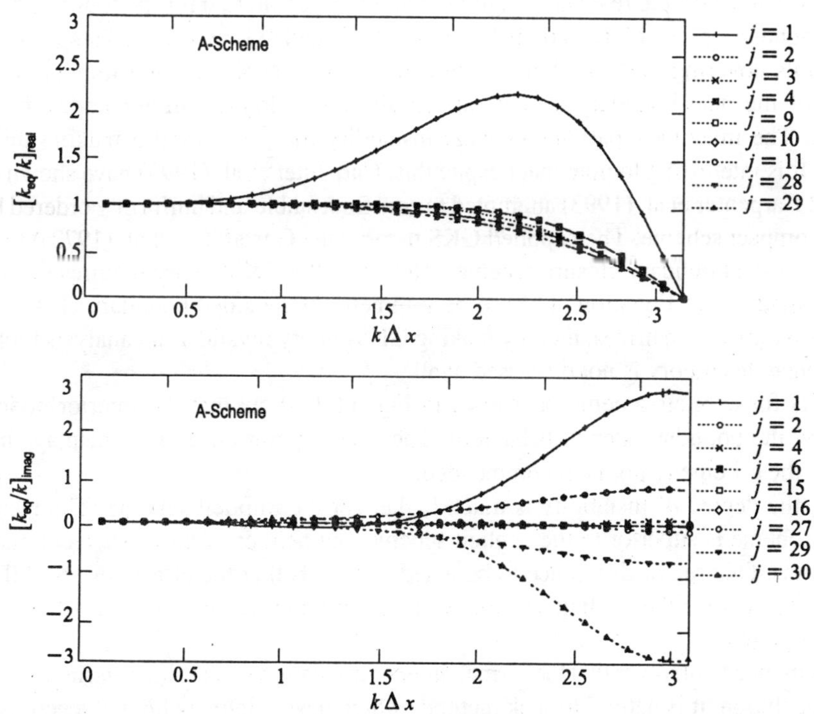

Fig. 11.1 *Spectral resolution and added numerical diffusion of compact scheme proposed by Adams (Equations (11.27 to 11.29)) obtained by the global spectral analysis tool developed in Equation (11.43) and shown as k_{eq}/k vs. $k\Delta x$.*

many other similar efforts. This analysis is capable of not only indicating whether the scheme is stable or unstable globally – it also pinpoints the wave number range over which the scheme is unstable – if the scheme is unstable for some specific points. The second advantage of this method of analysis is that it is independent of the differential equation to be solved. In effect this method comments on the spatial discretization alone unlike the semi-discrete eigenvalue analysis which analyses a particular differential equation. For example, in Fig. 11.1(b) the point-2 is unstable for all wave numbers. For point-1 the instability is more severe than for point-2 at high wave numbers. However, as indicated before, point-1 is of no concern since this is a boundary point. For points-3 to -6, the instability is mild and appears in windows of wave numbers and may not directly lead to a catastrophic breakdown of solution procedure in a computation. Also, if there is a mean convection in the direction of increasing j, then as the flow convects downstream instabilities will decay. As shown in this figure, for the points in the second half of the domain (for $j = 15$ to 30), the scheme is dissipative. The amount of dissipation added for $j = 29$ and 30 are so large that for a convection problem this nature of the scheme will work like a buffer-layer used in many flow-transition problems. The utility of such a layer is to smoothly decay the perturbations to zero at the outflow, so that there is no spurious reflection from the out-flow boundary. A typical use of such a buffer layer can be seen in Sengupta et al. (2002), where explicit method was used to solve Navier-Stokes equation.

In Figs. 11.2 and 11.3 the real and imaginary part of $\frac{k_{eq}}{k}$ are plotted for the fourth order implicit and explicit schemes given by Equations (11.30)–(11.32) and (11.33)–(11.35) respectively. Here, only the points covering half the domain are shown. Once again, for the implicit scheme the asymmetry of the near-boundary stencils affects the resolution at interior nodes as shown in Fig. 11.2(b). The imaginary part shows large instability for $j = 2$ and 3 making this scheme impractical. It is interesting to note that despite this, Carpenter et al. (1993) have shown this to be GKS-stable! Carpenter et al. (1993) attempted to construct stable and uniformly ordered boundary closure for compact scheme. They applied GKS theory (see Gustaffson et al. (1972)) to show the stable nature of the boundary closure schemes. However, the GKS theory requires the associated $[A]$ and $[B]$ matrices in Equation (11.8) to be symmetric. One-sided boundary closure schemes cannot yield symmetric matrices, thereby making GKS-theory invalid as an analysis tool for Padé schemes. Hence this theory is not discussed at all.

The results for explicit scheme, as shown in Fig. 11.3, show that the interior nodes remain unaffected by the boundary stencil behaviour. The $j = 2$ point once again displays large anti-diffusion for the scheme to not be recommended.

However, the degree of instability is much higher for the implicit scheme. The resolution of the implicit scheme is superior to the explicit scheme, but neither of these are recommended for the applications. One major difference to be noted as well is that the effects of instability at any point is strictly local for the explicit scheme, while for implicit scheme methods this also affects neighbouring points.

Having convinced ourselves that the formal order of truncation error is immaterial when it comes to spectral resolution, it is natural to seek methods which have relatively higher spectral accuracy, but are easy to implement numerically, so that the overall computational speed is enhanced. One such scheme reported by Haras & Ta'asan (1994), was developed for periodic problems and did

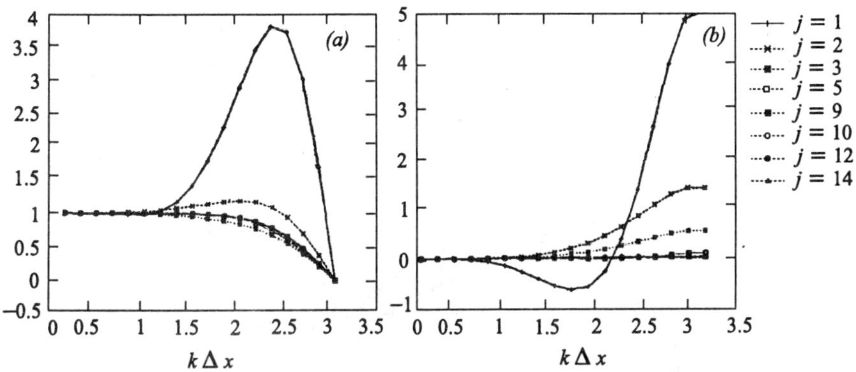

Fig. 11.2 *Spectral resolution and numerical dissipation as given by Equation (11.43) for the implicit scheme due to Carpenter et al., given by Equations (11.30–11.32) and shown as k_{eq}/k vs. $k\Delta x$ for representative points in the domain.*

not address the above mentioned problem of boundary stencil. The starting point was the choice of coefficients for the stencil given by Equation (11.22).

But unlike Lele's sixth order scheme for which the coefficients of Equation (11.22) was fixed by the values given in Equation (11.26), here an optimal search was conducted to minimize the L_2-norm of the departure of the compact scheme from the spectral method over the full Nyquist range, for a class of initial conditions for linear one-dimensional wave equation.

The optimal values for the coefficients in Equation (11.22) were obtained as

$$\left.\begin{aligned} \alpha &= 0.3793894912 \\ a &= 1.57557379 \\ b &= 0.1832051925 \end{aligned}\right\} \tag{11.47}$$

This particular choice of the free parameters makes the scheme only second order accurate (only Equation (11.23) is satisfied, not Equations (11.24) and (11.25)). It is interesting to point out that such a scheme has a much higher spectral resolution than that shown for the sixth order methods of Adams (1993) for the interior nodes. To make a comparison, we show the case where the above inner stencil is used along with boundary and near-boundary stencils of Adams (1993), given by Equations (11.27) and (11.28). The real and imaginary part of $\frac{k_{eq}}{k}$ is shown in Fig. 11.4.

The interior stencils show excellent spectral resolution as shown by the real part in comparison to the case shown in Fig. 11.1. The imaginary part shows large instability for $j = 2$ at intermediate wave numbers and milder instability at intermediate wave number windows for $j = 5$. Although this scheme performs better than the Adams (1993) scheme, it will still not be recommended for practical applications because of the instability caused by the boundary closure. However, this analysis shows the efficacy of optimization process in the search of better schemes for problems with non-periodic boundary condition. The theoretical aspect of the optimization process is discussed next for such general problems.

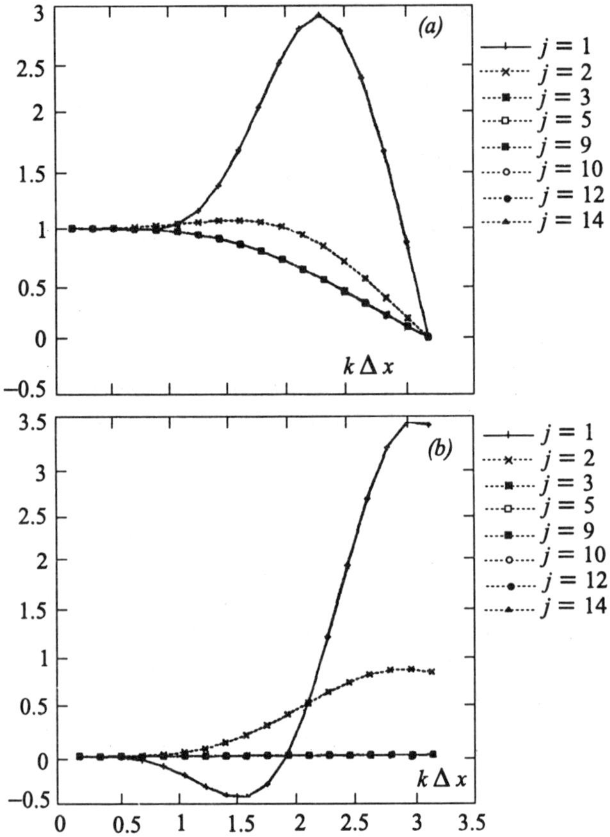

Fig. 11.3 *The spectral analysis of 4th order explicit scheme (Equations 11.33–11.35) due to Carpenter et al.*

11.6 Optimization Problem associated with the Discrete Evaluation of First Derivatives

Let $L(kh)$ be the exact differential operator acting on $U(k)$ to produce the first derivative as given in Equation (11.38). The corresponding numerical estimate of the first derivative is $L_h(kh)$ and is given by Equation (11.42). One can therefore obtain the squared error across all wave numbers given by

$$g_j = \int_{-k_m}^{k_m} |L_h(kh) - L(kh)|_j^2 U^2(k)dk. \tag{11.48}$$

Thus, the global minimization problem will involve minimizing the error over all the computational nodes. Thus, one should minimize

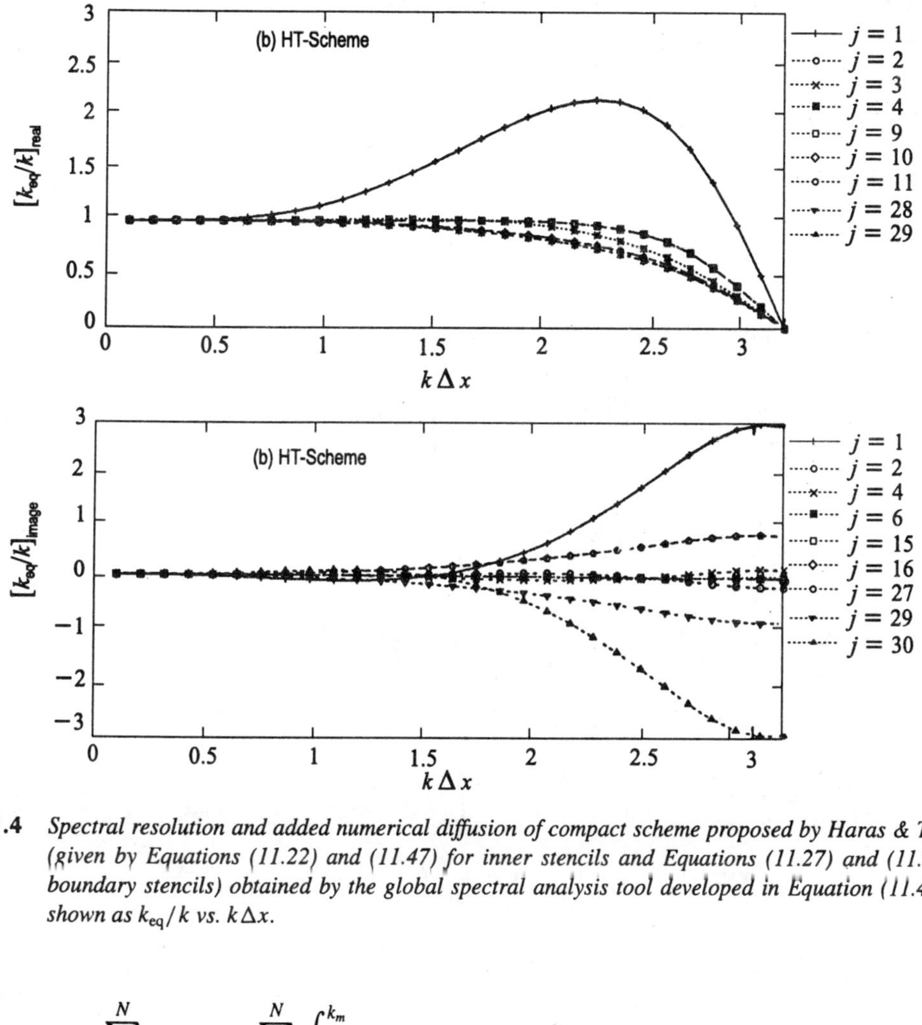

Fig. 11.4 *Spectral resolution and added numerical diffusion of compact scheme proposed by Haras & Ta'asan (given by Equations (11.22) and (11.47) for inner stencils and Equations (11.27) and (11.28) for boundary stencils) obtained by the global spectral analysis tool developed in Equation (11.43) and shown as k_{eq}/k vs. $k\Delta x$.*

$$G(.,.) = \sum_{j=1}^{N} g_j(.,.) = \sum_{j=1}^{N} \int_{-k_m}^{k_m} |L_h(kh) - L(kh)|_j^2 dk \tag{11.49}$$

In the above expressions the argument of the left hand side functions are the parameters over which the problem is to be minimized. The corresponding objective function used in Haras & Ta'asan (1994) is given by

$$\bar{G} = \int_{-k_m}^{k_m} |L_h(kh) - L(kh)|^2 [U(k)]^2 dk, \tag{11.50}$$

where $U(k)$ was taken corresponding to the initial condition of the wave equation which was solved. In the result presented here, one is looking for a more conservative estimate with respect to a white noise process or an excitation given by a Dirac delta function such that spectral weight of the error is same for all wave numbers. Haras & Ta'asan (1994) considered a wave equation with particular type of band-limited spectrum for $U(k)$ corresponding to initial data.

Since for the estimation of first derivative

$$L(kh) = ikh,$$

and from Equation (11.43)

$$L_h(kh) = \sum_{l=1}^{N} C_{jl}(R_{lj} + iI_{lj}), \tag{11.51}$$

therefore from Equation (11.48),

$$\begin{aligned} g_j &= \int_{-k_m}^{k_m} \left| \sum_{l=1}^{N} C_{jl}R_{lj} + i\left(\sum_{j=1}^{N} C_{jl}I_{lj} - kh\right)\right|^2 dk \\ &= \int_{-k_m}^{k_m} \left[\left(\sum_l C_{jl}R_{lj}\right)\left(\sum_m C_{jm}R_{mj}\right)\right. \\ &\quad \left. + \left(\sum_l C_{jl}I_{lj}\right)\left(\sum_m C_{jm}I_{mj}\right) - 2kh\left(\sum_l C_{jl}I_{lj}\right) + k^2h^2\right] dk. \end{aligned} \tag{11.52}$$

The following simplifications can be made for Equation (11.52):

$$I_1 = \int_{-k_m}^{k_m} k^2h^2 dk = 2h^2\frac{k_m^3}{3} = \frac{2}{3}\frac{\pi^3}{h} = \frac{2\pi^3}{3}(N-1) \tag{11.53}$$

$$I_2 = 2h\int_{-k_m}^{k_m}\left(\sum_l C_{jl}I_{lj}\right)k dk = 2h\sum C_{jl}\int k\sin\,[kh(l-j)]dk$$

for $j = l : I_2 \equiv 0$.

For $j \neq l$:

$$\begin{aligned} I_2 &= 2h\sum C_{jl}\left[-k\frac{\cos kh(l-j)}{h(l-j)} + \int \frac{\cos\; kh(l-j)}{h(l-j)}dk\right] \\ &= 2h\sum_l C_{jl}\left[-k\frac{\cos kh(l-j)}{h(l-j)} + \frac{\sin\; kh(l-j)}{h^2(l-j)^2}\right]_{-k_m}^{k_m} \\ &= 2h\sum_l C_{jl}\left[\frac{-2k_m}{h(l-j)}\cos\;\pi(l-j)\right] \\ &= -4k_m\sum_{l=1,l\neq j}^{N}\frac{C_{jl}}{(l-j)}(-1)^{l-j}. \end{aligned}$$

Thus,

$$I_2 = -4\pi(N - 1) \sum_{l=1,1 \neq j}^{N} \frac{C_{jl}}{(l - j)}(-1)^{(l-j)}. \tag{11.54}$$

Let us derive one other term in Equation (11.52).

$$\begin{aligned} I_3 &= \sum_l \sum_m C_{jl}C_{jm} \int_{-k_m}^{k_m} R_{lj} R_{mj} dk \\ &= \sum\sum C_{jl}C_{jm} \int \cos[kh(l-j)] \quad \cos[kh(m-j)] \quad dk \\ &= \sum\sum \frac{C_{jl}C_{jm}}{h} \left\{ \frac{\sin[k_m h(l+m-2j)]}{l+m-2j} + \frac{\sin\,[k_m h(l-m)]}{l-m} \right\}. \end{aligned}$$

For $l = m$:

$$\begin{aligned} I_3 &= \sum\sum C_{jl}C_{jm} \int \cos^2[kh(i-j)]dk \\ &= \sum_{l_{l=m}}\sum_m C_{jl}C_{jm} \left[k_m + \frac{\sin[2k_m h(l-j)]}{h(l-j)} \right] \end{aligned}$$

For $l = j$:

$$I_3 = \sum\sum C_{jl}C_{jm}(2k_m).$$

Similar simplifications can be brought about to show

$$\begin{aligned} I_3 + I_4 &= \int \left[\left(\sum C_{jl}R_{lj}\right)\left(\sum C_{jm}R_{mj}\right) + \left(\sum C_{jl}I_{lj}\right)\left(\sum C_{jm}I_{mj}\right) \right] dk \\ &= 2\pi(N-1)C_{jj}^2 + \pi(N-1) \sum_{l=1,l \neq j}^{N} C_{jl}C_{jl}. \end{aligned} \tag{11.55}$$

Hence,

$$\begin{aligned} g_j = {} & \frac{2\pi^3}{3}(N-1) + 4\pi(N-1) \sum_{l=1,j \neq l}^{N} \frac{C_{jl}}{(l-j)}(-1)^{(l-j)} \\ & + 2\pi(N-1)C_{jj}^2 + \pi(N-1) \sum_{l=1,l \neq j}^{N} C_{jl}C_{jl}. \end{aligned} \tag{11.56}$$

Thus one can write

$$\frac{g_j}{\pi(N-1)} = \frac{2\pi^2}{3} + \sum_{l \neq j} \frac{4C_{jl}}{l-j}(-1)^{(l-j)} + 2C_{jj}^2 + \sum_{l \neq j} C_{jl}^2. \tag{11.57}$$

So far in this analysis no specific choice for scheme has been made. For any scheme, to use the above analysis, one has to simply obtain the corresponding $[C]$ matrix to understand how spectral accuracy can be enhanced by inspecting Equations (11.57) and (11.49).

For a general compact scheme (implicit scheme)

$$[C] = [A]^{-1}[B],$$

and for explicit schemes $[A] = [I]$ and hence

$$[C] = [B].$$

Inspection of Equation (11.57) reveals that the error scales linearly with the number of points. Since the first, third and fourth term on the right hand side of Equation (11.57) are non-negative, the only term that can reduce error is the second term

$$4 \sum_{l \neq j} \frac{C_{jl}}{(l-j)}(-1)^{(l-j)}. \tag{11.58}$$

In Equation (11.58), since C_{jl} is divided by $(l-j)$, the maximum contribution to this term will be for those l's for which $(l-j)$ is minimum, i.e., the immediate neighbouring point of j will produce the maximum effect in changing the error and the terms further from the node in question will produce progressively lesser and lesser effects. This is the motivation to keep any scheme (explicit or implicit) to be compact.

As an example, consider the contribution coming from $l = (j+1)$ and $(j-1)$ in Equation (11.58) as

$$= 4(C_{ll-1} - C_{ll+1}).$$

Thus to minimize g, we must have

$$(a) \quad C_{l,l+1} > 0,$$
and
$$(b) \quad C_{ll+1} > c_{ll-1},$$
or
$$(c) \quad C_{ll-1} < 0.$$

Also, to minimize g_j in Equation (11.57), one can construct a scheme that is as compact as possible so that one has fewer contribution coming from the third and fourth terms.

For explicit schemes, $[C] = [B]$ and one can comment on the least-error schemes by inspecting B matrix itself. It is apprent from Equation (11.57) that all central schemes for first derivative have $C_{jj} = 0$, making the third term zero.

Example 1: Second order central differencing scheme
Here $C_{jj} = 0$ and $C_{jj\pm1} = \pm 1/2$ and therefore from Equation (11.57)

$$\frac{g_j}{\pi(N-1)} = \frac{2\pi^2}{3} - 4.50. \tag{11.59}$$

Example 2: Fourth order central differencing scheme
$C_{jj} = 0$ and $C_{jj\pm1} = \pm 2/3$ and $C_{jj\pm2} = \mp 1/12$

Hence, from Equation (11.57) one gets

$$\frac{g_j}{\pi(N-1)} = \frac{2\pi^2}{3} - 4.7638888. \tag{11.60}$$

As one would expect, the global error norm for the fourth order central scheme is lower as compared to the second order central scheme. The fact that asymmetric stencils give rise to much larger contribution to global error can be seen if one estimates this for the stencil given by Equations (11.33) and (11.34). These are obtained as

$$\frac{g_1}{\pi(N-1)} = \frac{2\pi^2}{3} + 14.733 \tag{11.61}$$

and

$$\frac{g_2}{\pi(N-1)} = \frac{2\pi^2}{3} - 4.15277. \tag{11.62}$$

A comparison among Equations (11.60) – (11.62) reveals that the error is created by the particular choice of stencil (Equations (11.61)) for $j = 1$ and N, while the other points $(2 \leq j \leq (N-1))$ cause a reduction of error norm.

We should take note that not all upwind schemes show a degradation. For example, if one considers a first order upwinding at $j = 1$, then $C_{11} = 1$ and $C_{12} = -1$, so that

$$\frac{g_1}{\pi(N-1)} = \frac{2\pi^2}{3} - 2. \tag{11.63}$$

A comparison of Equation (11.63) with Equations (11.61) and (11.59) shows that the first order upwinding is superior to the fourth order accurate scheme given by Equation (11.33). It is also superior to the second order central difference scheme. It is however, not suggested in any way to the reader that first order upwind scheme is to be used for practical applications in preference to higher order upwind schemes or central schemes. This is because, apart the from spectral resolution, one must not alter the flow physics (as first order upwinding alters physical dissipation!) and also should satisfy DRP property—as discussed in Chapter 10.

11.7 Optimized Compact Scheme for First Derivative

Let us consider the following stencils used in Lele (1992) for evaluation of first derivatives:

$$j = 1 : u_1' + \alpha u_2' = (au_1 + bu_2 + cu_3 + du_4)/h, \tag{11.64}$$

$$j = 2 : u_1' + 4u_2' + u_3' = 3(u_3 - u_1)/h, \tag{11.65}$$

$$3 \leq j \leq N-2 : \alpha_1 u_{j-1}' + u_j' + \alpha_1 u_{j+1}' = \frac{b_1}{4h}(u_{j+2} - u_{j-2}) + \frac{a_1}{2h}(u_{j+1} - u_{j-1}), \tag{11.66}$$

by optimizing the above stencils with α and α_1 as the free parameters. A third order accuracy requirement for relation (11.64) and the fourth order accuracy requirement for Equation (11.66), allows one to relate the rest of the parameters in terms of α and α_1. This gives:

$$a = -\frac{(11+2\alpha)}{6};\ b = \frac{6-\alpha}{2};\ c = \frac{2\alpha - 3}{2};\ d = \frac{2-\alpha}{6};$$

and

$$a_1 = 2\left(\frac{\alpha_1}{3} + 2\right);\ b_1 = \frac{4\alpha_1 - 1}{3}. \tag{11.67}$$

Thus the global optimization requires minimizing the objective function

$$G_1(\alpha, \alpha_1) = \sum_{j=1}^{N} g_j(\alpha, \alpha_1),$$

where g_j is as given in Equation (11.57). In performing the optimization, one has to restrict the value of α_1 to less than or equal to $\frac{1}{2}$, so that the corresponding $[A]$ matrix of Equations (11.64)–(11.66) remains strictly diagonally dominant. In this case direct inversion of $[A]$ is possible without doing any extra work. Although the same logic requires $\alpha \leq 1$, one can still obtain $[C]$ matrix when this condition is mildly violated. Instead of optimizing the objective function, it is evaluated in $(\alpha - \alpha_1)$-plane and plotted as in Fig. 11.5.

It is of particular interest to note that there is a connected region in the middle and right where the L_2 norm is unacceptably high. This is the range where the diagonal dominance is violated and the $[A]$ matrix is singular. Excluding this part, the drawn contour lines identify regions with smoothly varying error norms. The optimum values were found for $\alpha = 1.334$ and $\alpha_1 = 0.41$. In Fig. 11.6 the real and imaginary part of k_{eq}/k are displayed as a function of kh for this optimal parameter. The real part reveals a significant improvement of resolution over the non-optimal cases shown in Figs. 11.1 to 11.3.

The imaginary part of (k_{eq}/k), however, displays instability for $j = 2, 3$ and 5 for high wave numbers. This is due to the fact that the objective function $G_1(\alpha, \alpha_1)$ did not have any constraint on the imaginary part. This will be attempted later.

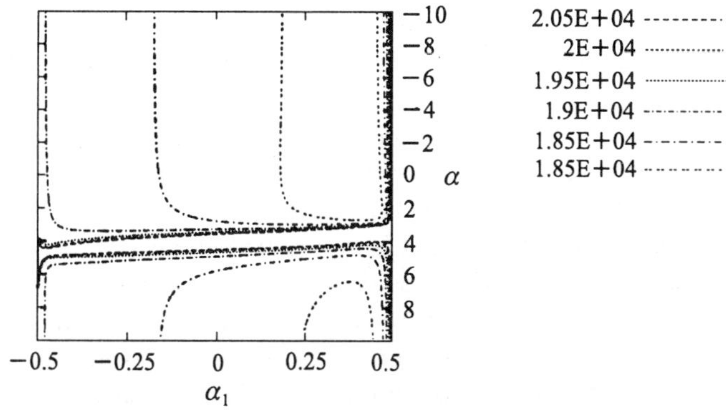

Fig. 11.5 *Contour plots of the L_2-norm of the objective function for the compact scheme defined by Equations (11.64) to (11.66) in the two-parameter $(\alpha-\alpha_1)$-space for evaluating an optimum scheme for the first derivative.*

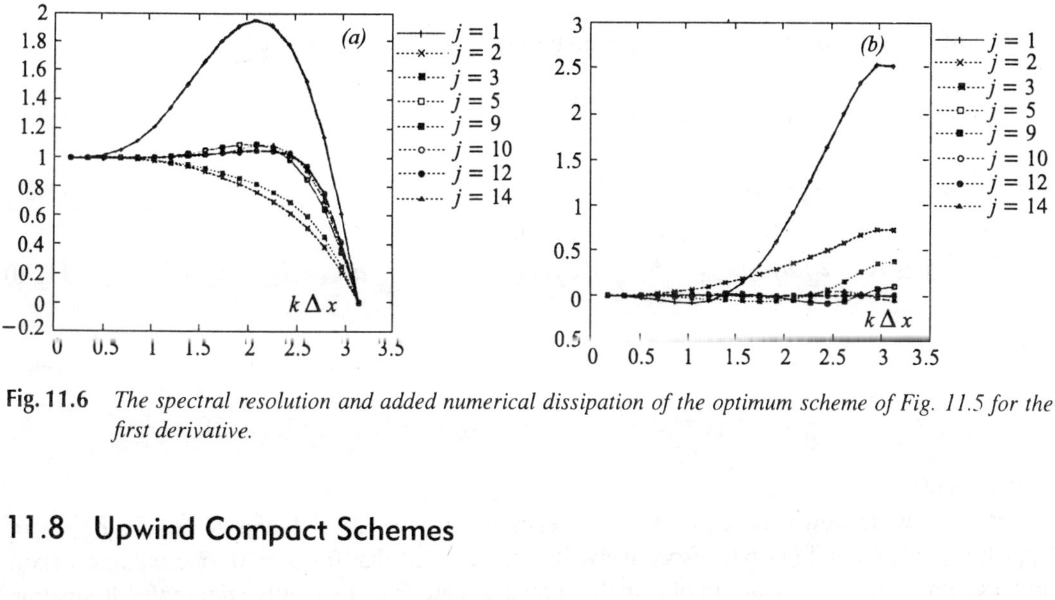

Fig. 11.6 *The spectral resolution and added numerical dissipation of the optimum scheme of Fig. 11.5 for the first derivative.*

11.8 Upwind Compact Schemes

It was stated earlier in support of compact (implicit) scheme over explicit high order scheme, that explicit stencils are wider than the compact stencils and hence more boundary closures are required. At the same time the results of the previous section show that the boundary closures required in compact schemes can be major sources of numerical instability at many points across the domain. From the discussion of the previous chapter we noted that aliasing is also a serious source of error accumulation at high wave numbers and frequencies. A linearly unstable algorithm will deteriorate further in the presence of aliasing. This last aspect is due to the global properties of the compact scheme. Contrast this with upwind-biased higher order explicit schemes which are local, and if boundary stencils are stable then the numerical procedure will not be prone to instabilities. The explicit dissipation of this class of higher order upwind biased scheme helps in controlling aliasing error.

The robustness and the success of higher order upwind-biased schemes in controlling aliasing error makes the case of using upwinded compact schemes. In recent years, many upwind compact schemes have been developed. Tolstykh (1991) proposed a fifth order compact upwind scheme for moisture transport equation in atmosphere. Christie (1985) proposed compact upwind schemes for convection dominated problems. Zingg et al. (1995) have used a fifth order upwind scheme with built-in filter term for solving linear wave equation. Adams & Shariff (1996) proposed fifth order upwind compact scheme using central grid stencils to solve a shock-turbulence interaction problem. Zhong (1998) has presented various upwind compact schemes for the DNS of hypersonic boundary layer transition. We will discuss this scheme to explain the features of upwinded compact schemes.

In these schemes, central stencils are used with adjustable built-in numerical dissipation. The order of accuracy of the upwind schemes are one-order lower than the maximum orders the central stencils can achieve, so that there is an adjustable coefficient in the leading dissipative truncation term.

The fifth order upwind scheme suggested by Zhong (1998) is given by

$$j = 1 : 6u_1' + 18u_2' = \frac{-17u_1 + 9u_2 + 9u_3 - u_4}{h}, \tag{11.68}$$

$$j = 2 : u_1' + 4u_2' + u_3' = 3\frac{(u_3 - u_1)}{h}, \tag{11.69}$$

$$3 \leq j \leq N - 2 : b_{j-1}u_{j-1}' + b_j u_j' + b_{j+1}u_{j+1}' = \frac{1}{h}\sum_{k=-2}^{2} a_{j+k}u_{j+k}, \tag{11.70}$$

where,

$$a_{j\pm2} = \pm\frac{5}{3} + \frac{5\alpha}{6}; \quad a_{j\pm1} = \pm\frac{140}{3} + \frac{20\alpha}{3}; \quad a_j = -15\alpha; \quad b_{j\pm1} = 20 \pm \alpha;$$

and $b_j = 60$.

One can write down the corresponding relation for $j = N - 1$ and N by analogy with Equations (11.69) and (11.68) respectively. It is to be noted that for $\alpha = 0$, one obtains a sixth order central scheme. The origin of α in these relations are from explicitly adding the dissipation term $\frac{\alpha}{6!}h^5\frac{\partial^6 u}{\partial x^6}$. The interior stencil corresponding to $\alpha = 0$ is the same as given in Adams (1977).

Zhong (1998) investigated the stability of the scheme for linear wave equation by matrix stability analysis for $\alpha = -2, -1$ and 0. This scheme can be investigated by looking at k_{eq}/k as obtained from Equation (11.43). The real and imaginary parts of k_{eq}/k are shown in Fig. 11.7 for $\alpha = -1$. Despite the observations in Zhong (1998) to the contrary, one notices large instabilities for the near boundary points.

This unstable behaviour of the near-boundary points faced by the central schemes due to Adams (1993), Haras & Ta'asan (1994) and the upwind scheme of Zhong (1998) is essentially due to the chosen implicit boundary closure. Thus, one way of solving this instability problem is to use explicit stencils for boundary closures. The following boundary stencil is proposed in Sengupta et al. (2003),

$$u_1' = \frac{1}{2h}[-3u_1 + 4u_2 - u_3]. \tag{11.71}$$

For the near-boundary point ($j = 2$) a combination of the following second order accurate explicit stencils are used.

$$u_2' = \frac{1}{2h}[-3u_2 + 4u_3 - u_4] \tag{11.72}$$

and

$$u_2' = \frac{u_3 - u_1}{2h} - \frac{12\beta}{h}\left(\frac{h^4}{4!}u_2^{iv}\right). \tag{11.73}$$

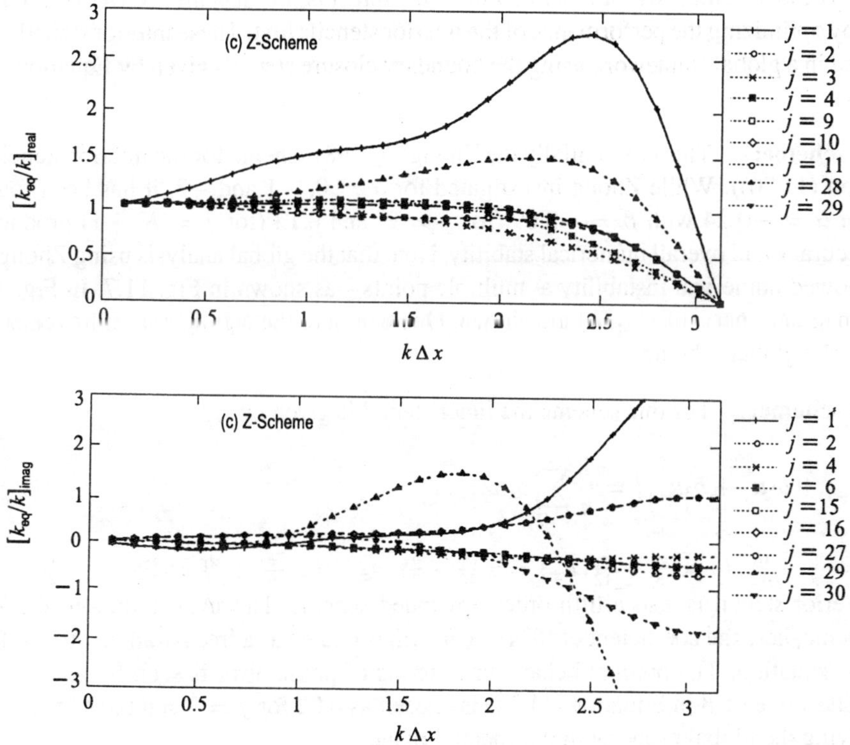

Fig. 11.7 *Spectral resolution and added numerical diffusion of compact scheme proposed by Zhong (given by Equations (11.68–11.70)) obtained by using Equation (11.43) and shown as k_{eq}/k vs. $k\Delta x$ for the full domain.*

The representation of Equation (11.72) has a dispersion term as a leading truncation error $(-\frac{4}{h} \times \frac{h^3}{3!} u_2^{'''})$, and Equation (11.73) has both dispersion and dissipation terms implicit in the stencil.

It has been noted that a 2:1 blend of Equations (11.73) and (11.72) performs the best in reducing numerical instability at the interior nodes. Thus, the following near-boundary stencil is recommended for compact schemes as given below.

$$u_2' = \frac{1}{h}\left[\left(\frac{2\beta}{3} - \frac{1}{3}\right)u_1 - \left(\frac{8\beta}{3} + \frac{1}{2}\right)u_2\right] + \frac{1}{h}\left[+(4\beta + 1)u_3 - \left(\frac{8\beta}{3} + \frac{1}{6}\right)u_4 + \frac{2\beta}{3}u_5\right], \tag{11.74}$$

with β as a floating constant, which fixes the built-in fourth order dissipation terms.

11.9 Some New Compact Schemes (Sengupta et al. (2003))

These schemes are proposed as a significant improvement over already discussed schemes. As these are upwind schemes, we call them *optimal upwind compact schemes* (OUCS) and these are arrived at by optimizing the performance of the interior stencils first. These interior stencils are then investigated in a global framework using the boundary closure stencils given by Equations (11.71) and (11.74).

OUCS - 1 scheme: This is essentially the Zhong's (1998) scheme for the interior stencil (given by Equation (11.70)). While Zhong investigated for $\alpha = 0, -1$ and -2, it has been found that a choice of $\alpha = -0.24$ with $\beta = -0.09$ (for $j = 2$) and 0.12 (for $j = N - 1$) produced best spectral accuracy and overall numerical stability. Note that the global analysis using Zhong (1998) scheme showed numerical instability at multiple points—as shown in Fig. 11.7. In Fig. 11.8 the real and imaginary parts of (k_{eq}/k) are shown. One can note the significant improvement in the stability of the global scheme.

OUCS - 2 scheme: For this scheme the inner stencil is given by

$$b_1 u_{j-1}' + b_2 u_j' + b_3 u_{j+1}' = \frac{1}{h}\sum_{k=2}^{2} a_k u_{j+k}, \tag{11.75}$$

where $b_1 = \frac{b_2}{3} - \frac{\alpha_1}{12}$; $b_3 = \frac{b_2}{3} + \frac{\alpha_1}{12}$; $a_{\pm 2} = \pm\frac{b_2}{36} + \frac{\alpha_1}{72}$; $a_{\pm 1} = \pm\frac{7b_2}{9} + \frac{\alpha_1}{9}$ and $a_0 = -\frac{\alpha_1}{4}$.

This interior stencil is also a fifth order upwinded scheme. However, unlike in the Zhong's (1998) scheme, here the coefficient of $(\partial^6 u/\partial x^6)$ term is treated as a free parameter to fix the order of the representation. The optimal behaviour in terms of phase error is seen for $\alpha_1 = -40$ and $b_2 = 36$. The value of β in Equation (11.74) is chosen as 0.02 for $j = 2$ and 0.09 for $j = N - 1$ while studying the global property in the spectral plane.

In Fig. 11.9 the real and imaginary parts of k_{eq}/k are shown for all possible wave numbers. One sees further improvements on spectral resolution and numerical stability for this scheme over OUCS-1, which was in itself a bigger improvement over Zhong's (1998) scheme.

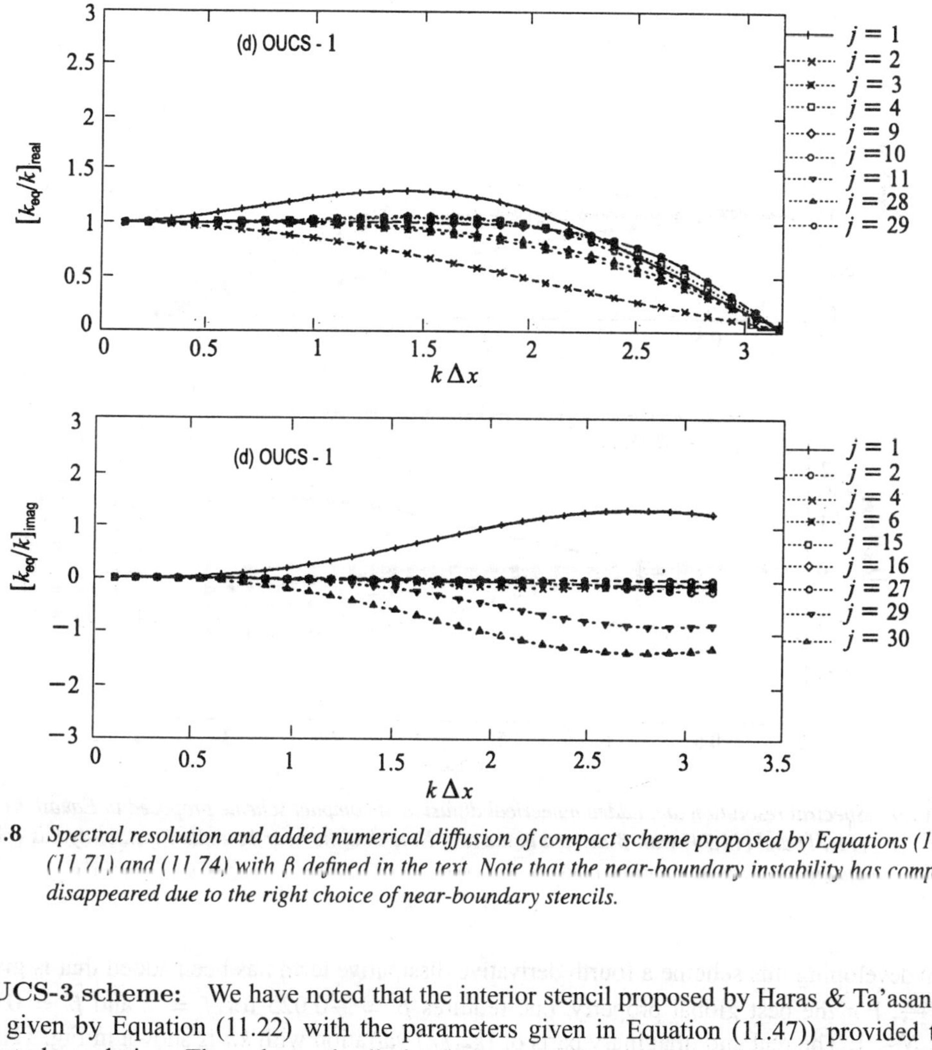

Fig. 11.8 *Spectral resolution and added numerical diffusion of compact scheme proposed by Equations (11.70), (11.71) and (11.74) with β defined in the text. Note that the near-boundary instability has completely disappeared due to the right choice of near-boundary stencils.*

OUCS-3 scheme: We have noted that the interior stencil proposed by Haras & Ta'asan (1994) (as given by Equation (11.22) with the parameters given in Equation (11.47)) provided the best spectral resolution. The scheme is dissipationless, causing poor performance for non-periodic problems (as exhibited in Fig. 11.4). In Fig. 11.4, the global analysis was performed using the boundary closures of Adams (1993) (given by Equations (11.27) and (11.28)). In the proposed scheme which is based on the Haras-Taa'san scheme, the following interior stencils have been used:

$$p_{j-1}u'_{j-1} + u'_j + p_{j+1}u'_{j+1} = \frac{1}{h}\sum_{k=-2}^{2} q_k u_{j+k}, \tag{11.76}$$

where $p_{j\pm1} = D \pm \frac{\eta}{60}$; $q_{\pm2} = \pm\frac{F}{4} + \frac{\eta}{300}$; $q_{\pm1} = \pm\frac{E}{2} + \frac{\eta}{30}$;

and $q_0 = -\frac{11\eta}{150}$ with $D = 0.3793894912$; $E = 1.57557379$; $F = 0.183205192$ and $\eta = -2.0$.

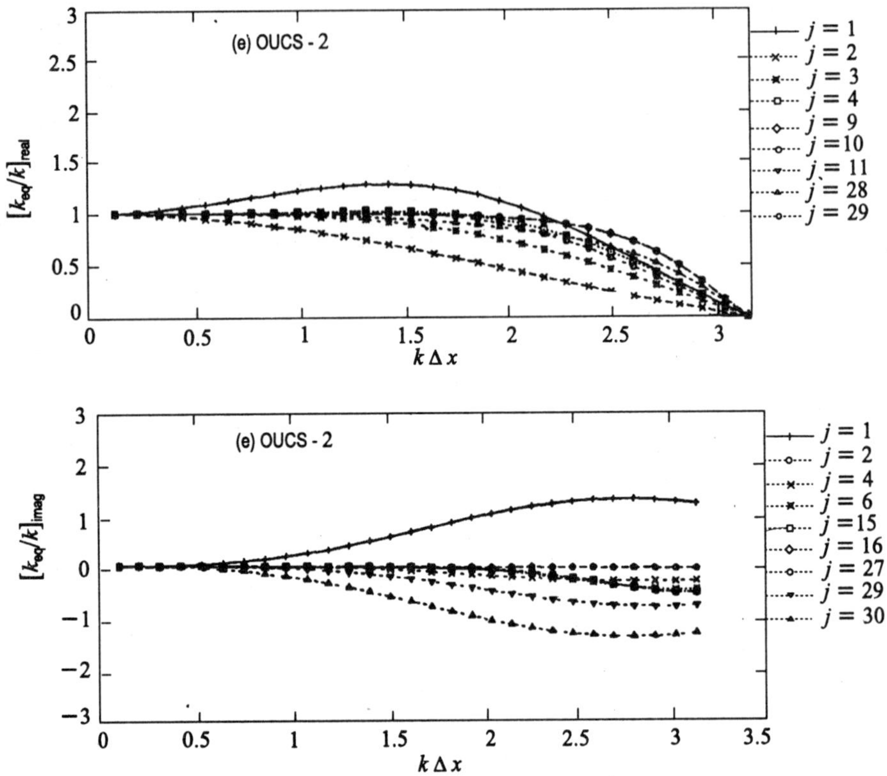

Fig. 11.9 *Spectral resolution and added numerical diffusion of compact scheme proposed in Equations (11.75), (11.71) and (11.74) with β defined in the text. Note that the near-boundary instability has completely disappeared for this scheme also.*

In developing this scheme a fourth derivative dissipative term has been added that is given by $\eta h^3 \frac{\partial^4 u}{\partial x^4}$. For the best global property, one requires $\beta = -0.025$ for $j = 2$ and $\beta = 0.09$ for $j = N - 1$. The real and imaginary parts of (k_{eq}/k) variation with kh is shown in Fig. 11.10.

The spectral resolution of this scheme is comparable to the other two schemes in this class. The numerical stability is also improved for this scheme, as compared to the other two schemes.

11.10 Approximating Second Derivatives

A lot of attention has been devoted to studying the spectral properties of schemes for estimating first derivatives. For CFD this is considered natural because the most important nonlinear convection processes involve the accurate estimation of the first derivative. However, the viscous diffusion terms are equally important in the shear layers. Many important events in turbulent flows are

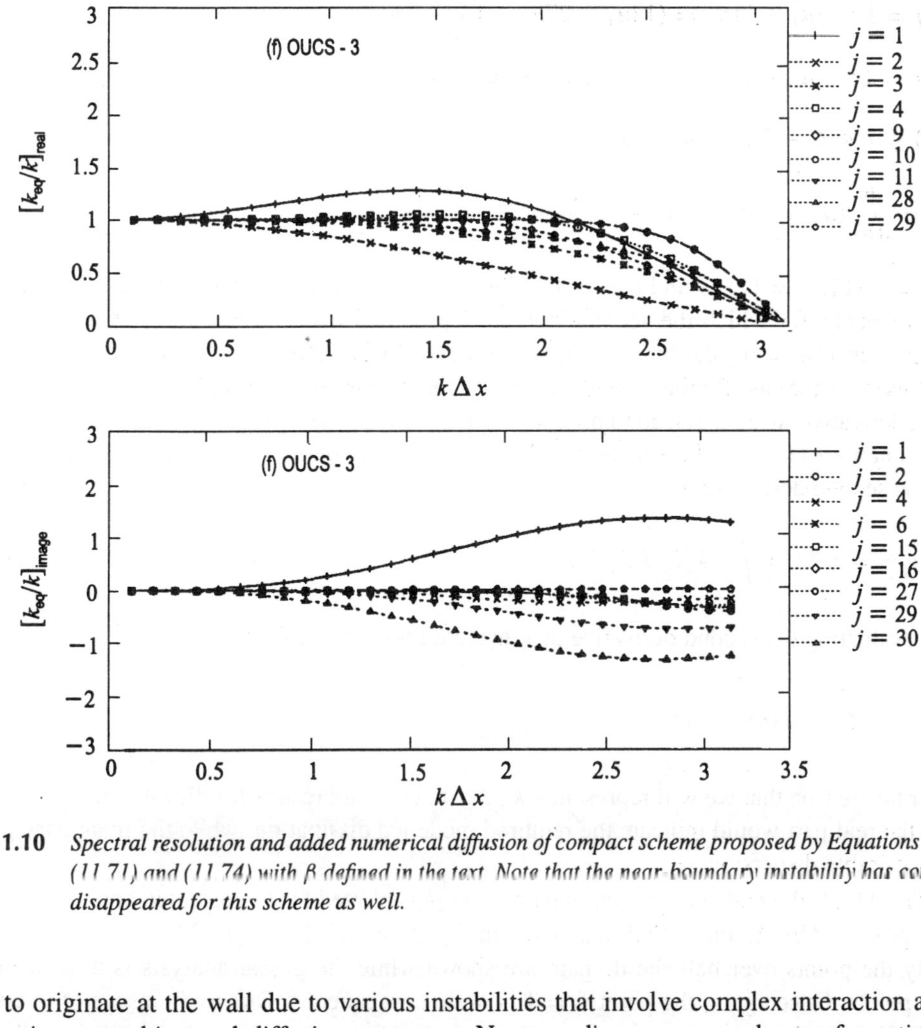

Fig. 11.10 *Spectral resolution and added numerical diffusion of compact scheme proposed by Equations (11.76), (11.71) and (11.74) with β defined in the text. Note that the near-boundary instability has completely disappeared for this scheme as well.*

seen to originate at the wall due to various instabilities that involve complex interaction among convection, stretching and diffusion processes. Next we discuss some schemes for estimating the second derivative using the same global approach for spectral resolution and numerical stability.

To evaluate second derivatives, one can repeat the operation of evaluating first derivative twice, as implied in Equations (11.1) and (11.8) to obtain

$$u'' = [C]u, \tag{11.77}$$

where $[C] = [A]^{-1}[B][A]^{-1}[B]$. If this is used in actual computations, then there would be a necessity to store the first derivatives. There is the other way of evaluating the second derivatives by applying the general principal of compact differencing. For example, Lele (1992) suggested the usage of the following stencils for evaluating second derivative.

$$j = 1: \quad u_1'' + 11u_2'' = (13u_1 - 27u_2 + 15u_3 - u_4)/h^2, \tag{11.78}$$

$$j = 2: \quad u_1'' + 10u_2'' + u_3'' = 12(u_3 - 2u_2 + u_1)/h^2, \tag{11.79}$$

$$3 \leq j \leq N-2: \quad \alpha u_{j-1}'' + u_j'' + \alpha u_{j+1}''$$

$$= \frac{b}{4h^2}(u_{j-2} - 2u_j + u_{j+2}) + \frac{a}{h^2}(u_{j-1} - 2u_j + u_{j+1}), \tag{11.80}$$

with $\alpha = 2/11, a = 12/11$ and $b = 3/11$. This is formally a sixth order accurate scheme. There are two reasons for which the second method is preferred over the first method. Firstly, at the Nyquist limit ($kh = \pi$) the basic operation involved in Equation (11.77) makes $-k_{\text{eq}}/k^2 = 0$, while this is not the case for the second method. Secondly, the repeated application of the algorithm for first derivative twice will lead to poorer spectral resolution at all kh as compared to the method using Equations (11.78)–(11.80). Using the spectral representation of Equation (11.36) one can write the second derivative as

$$u_j'' = u''(x_j) = \int -k^2 U(k) e^{ikx_j} dk, \tag{11.81}$$

while, evaluating the second derivative in a discrete grid is equivalent to

$$u_j'' = \int k_{\text{eq}} U(k) e^{ikx_j} dk. \tag{11.82}$$

It is for this reason that we will represent $-k_{\text{eq}}/k^2$ to compare results for different methods. Once again, the real part would indicate the required modelled dissipation, while the imaginary part is the undesirable dispersion.

In Fig. 11.11, the real and imaginary parts of $-\frac{k_{\text{eq}}}{k^2}$ is plotted for the repeated application of the stencil proposed by Adams (1993) and given in Equations (11.27) to (11.29).

Only the points over half the domain are shown while the global analysis is done using all the 30 points. Once again, the point $j = 1$ is of no consequence. Due to the usage of one-sided stencils for boundary closure there appears dispersion effects. It needs to be pointed out that the alternating signs for different length scales implies that different structures of different length scale will migrate in opposite directions. While solving a multi-dimensional problem, this feature will make the method anisotropic. In Fig. 11.12 the real and imaginary parts of $-k_{\text{eq}}/k^2$ is plotted for the stencil given by Equations (11.78)–(11.80).

It can be seen that the Lele's (1992) scheme is superior to Adams' (1993) scheme applied twice for the evaluation of the second derivative. Also, Lele's scheme has much lesser dispersion error.

In Fig. 11.13 the real part and imaginary part of $-k_{\text{eq}}/k^2$ is plotted for different wave number for OUCS-3 scheme.

The spectral resolution of this method remains good up to $kh \simeq 2.5$, but the dispersion error of this scheme is quite adverse compared to Adams' or Lele's method.

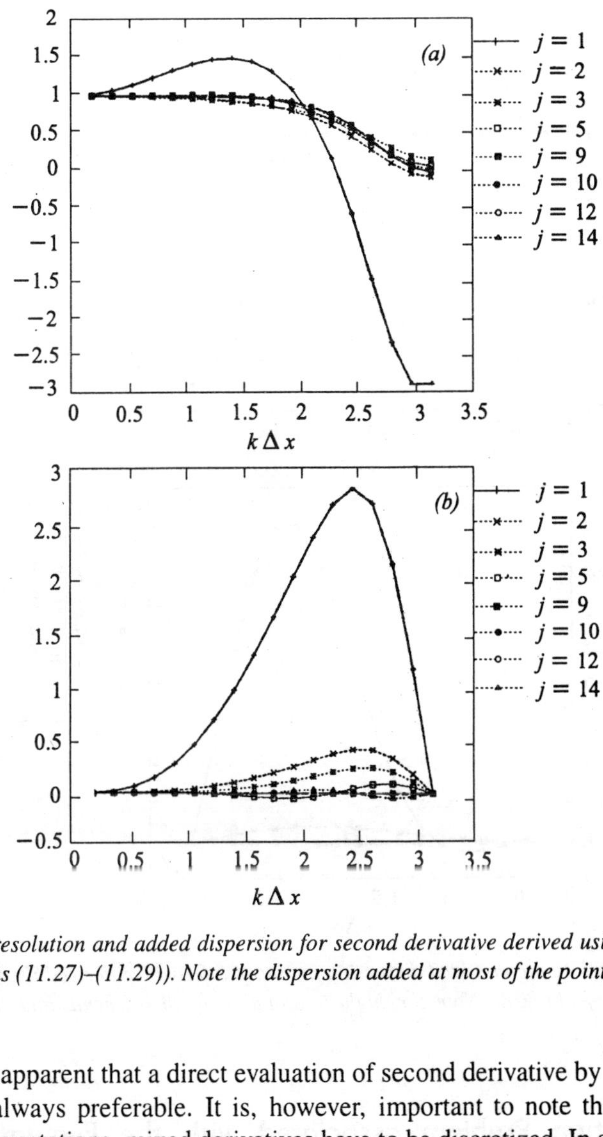

Fig. 11.11 *Spectral resolution and added dispersion for second derivative derived using Adams' scheme twice (Equations (11.27)–(11.29)). Note the dispersion added at most of the points at high wave numbers.*

Thus it becomes apparent that a direct evaluation of second derivative by a method like Lele's method would be always preferable. It is, however, important to note that in many practical transform-plane computations, mixed derivatives have to be discretized. In such cases, they have to be evaluated by using the first derivative methods in multi-directions successively.

It is to be noted that if one replaces Equation (11.78) by the explicit scheme as

$$u_1'' = \frac{u_1 - 2u_2 + u_3}{h^2}, \tag{11.83}$$

the spectral resolution improves while dispersion error reduces for all wave numbers. This is left as an exercise for the reader.

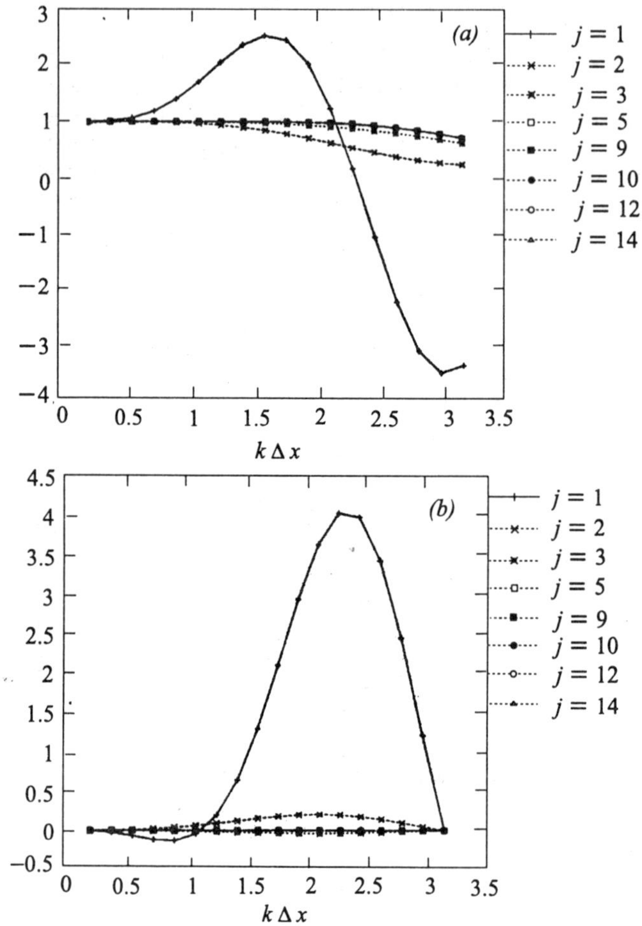

Fig. 11.12 *Spectral resolution and added dispersion for second derivative derived using Lele's scheme (Equations (11.78)–(11.80)). Note the higher resolution of this scheme and lesser dispersion at all points.*

11.11 Optimization Problem associated with the Evaluation of Second Derivatives

Following the method discussed for optimizing the stencil for the first derivative, we proceed for the L_2-norm of error of the discrete method over the spectral method from Equation (11.48) given by

$$g_j = \int_{-k_m}^{k_m} |L_h(kh) - L(kh)|_j^2 dk.$$

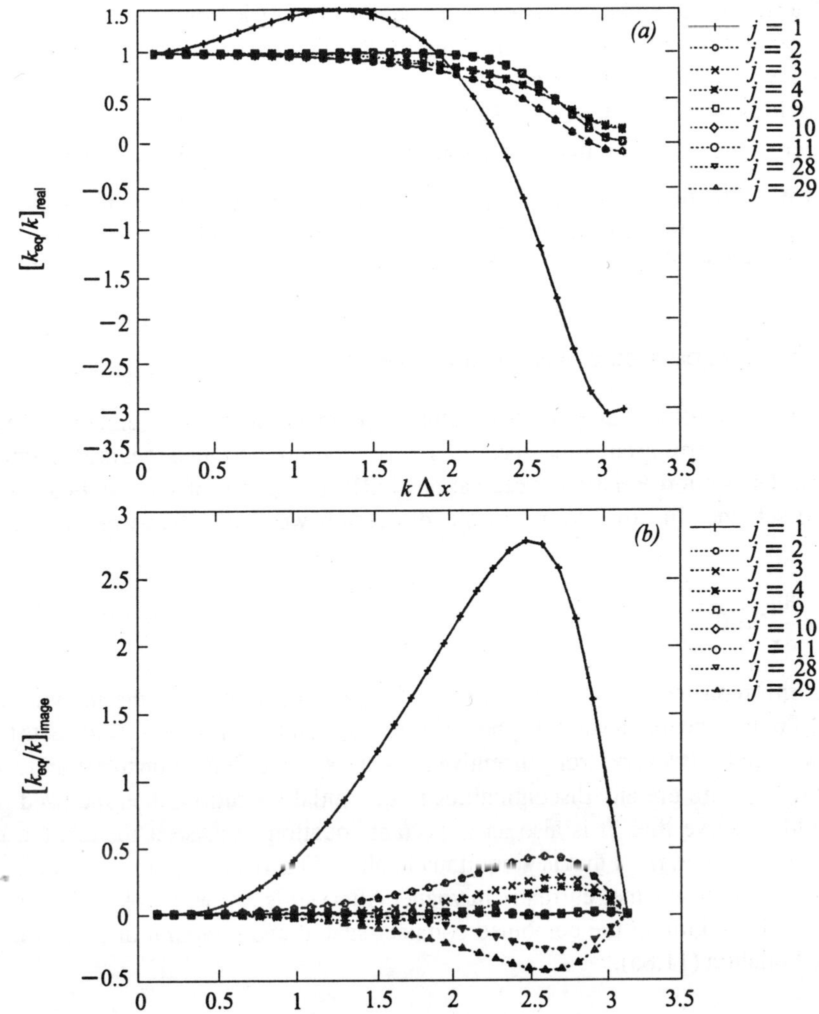

Fig. 11.13 *Spectral resolution and added dispersion for second derivative derived using the scheme given by Equations (11.71), (11.74) and (11.76) twice.*

Here $L(kh) = -k^2h^2$ and $L_h(kh) = \sum_{l-1}^{N} C_{jl}.P_{lj}$

Thus $g_j = \displaystyle\int_{-k_m}^{k_m} \left| \sum_{l=1}^{N} C_{jl} P_{lj} + k^2h^2 \right|^2 dk$ represents the objective function for the minimization problem for the white-noise or shot-noise process. This can be simplified to

$$\frac{g_j}{\pi(N-1)} = \frac{2\pi^4}{5} + 2C_{jj}\left(\frac{2\pi^2}{3} + C_{jj}\right) + \sum_{l \neq j} C_{jl}\left[C_{jl} + \frac{4(-1)^{l-j}}{(l-j)^3}\right]. \tag{11.84}$$

The first term on the right hand side of Equation (11.85) is contributed from the spectral differential operator. From this and the structure of the $[C]$ matrix in Equation (11.77), one can comment on the terms that reduce error.

For the application of Adams scheme (given in Equations (11.27) to (11.29)) twice to evaluate the second derivative, it is seen that all $C'_{jj}s$ are negative except C_{11}. Hence the term $\frac{4\pi^2 C_{jj}}{3}$ will minimize error. Also, the term $\sum_{l \neq j}^{N} \frac{4C_{jl}(-1)^{(l-j)}}{(l-j)^3}$ can reduce the global error if the coefficients in the stencil alternate in sign. The second and fourth order explicit central difference schemes have these properties and are thus low-error schemes.

11.12 Other Properties of Compact Schemes

Based on the discussions in Chapter 10 on the other important properties of numerical schemes, it is noted that they are governing equation-specific. The two most important issues are the amplification factor and the Dispersion Relation Preservation (DRP) properties of the combined space-time discretization scheme. In order to aid in our discussion we would focus on simple 1D-wave equation

$$\frac{\partial u}{\partial t} + c\frac{\partial u}{\partial x} = 0. \tag{11.85}$$

For $c > 0$, this represents a right running wave. The specific property of the analytic solution of Equation (11.86) that the initial solution preserves its shape and amplitude as it convects to the right with the phase speed c, is a very strong incentive to use it as a model equation to calibrate numerical methods. Thus, if there are any discontinuities in the initial condition, then the used numerical scheme should preserve this. It is also an important equation because it models the important physical process of convection that is ubiquitous in all CFD activities. Let us first describe a few temporal discretization schemes currently in vogue. Afterwards, we will look at the amplification factor and the DRP region of the combined compact-spatial and temporal discretization scheme as applied to Equation (11.85).

11.13 Temporal Discretization Schemes

(a) Euler time marching scheme

This is one of the most popular and easy-to-implement scheme. If we indicate time steps by the superscript n, then

$$\left.\frac{\partial u}{\partial t}\right|_j^n \simeq \frac{u_j^{n+1} - u_j^n}{\Delta t}. \tag{11.86}$$

The truncation error of this scheme is $O(\Delta t)$. For Equation (11.85) this term effectively adds dissipation.

(b) Multi-level time integration scheme

There are many multi-level time integration schemes available. They can be analyzed in a single general platform. Here we investigate a class of three time-level schemes.

This is one of the earlier multi-level time integration schemes. If the equation to be solved is written as

$$\left.\frac{\partial u}{\partial t}\right|_j^n = L(u_j^n), \tag{11.87}$$

then

$$u_j^{n+1} = \alpha_1 u_j^n + \alpha_2 u_j^{n-1} + \beta_1 \Delta t L(u_j^n) + \beta_2 \Delta t L(u_j^{n-1}). \tag{11.88}$$

This scheme is designed to obtain higher order accuracy in time discretization and if one demands at least second order accuracy, then one must have

$$\alpha_1 = 1 - \alpha_2; \quad \beta_1 = \frac{1}{2}(\alpha_2 + 3) \quad \text{and} \quad \beta_2 = \frac{1}{2}(\alpha_2 - 1). \tag{11.89}$$

Here α_2 is the free parameter and if one makes the choice

$$\alpha_2 = 1 \quad \text{then} \quad \alpha_1 = 0 = \beta_2; \quad \text{and} \quad \beta_1 = 2 \tag{11.90}$$

then this is the mid-point leap frog method. On the contrary if one chooses

$$\alpha_2 = 0 \quad \text{then} \quad \alpha_1 = 1; \quad \beta_1 = \frac{3}{2} \quad \text{and} \quad \beta_2 = -\frac{1}{2}, \tag{11.91}$$

then this is the second order Adams-Bashforth time integration scheme. It is possible to develop other schemes which have higher time accuracy.

11.14 Higher Order Temporal Scheme

If one uses higher order spatial discretization schemes, then it is natural to seek corresponding higher order temporal schemes. However, higher order temporal schemes like Runge-Kutta schemes have been in use with low order spatial discretization scheme, because of their attractive stability characteristics and relaxed time-step requirement which particularly allows faster approach to steady state. We will discuss a few Runga-Kutta schemes because of the possibility of selecting low-storage methods. These are a class of multi-stage predictor–corrector methods, evaluating and correcting the time derivative at a given time in more than one sub-step.

Two-stage Runge-Kutta scheme

Time marching from $n\Delta t$ to $(n+1)\Delta t$, the integration is achieved in two steps.

Here the predictor or first stage is

$$u^{(1)} = u^{(n)} + \Delta t L(u^{(n)}). \tag{11.92}$$

The second or the corrector stage is

$$u^{(n+1)} = \frac{u^{(n)} + u^{(1)}}{2} + \frac{\Delta t}{2} L(u^{(1)}). \tag{11.93}$$

Three-stage Runge-Kutta scheme

Here the three steps of the scheme for the integration of Equation (11.88) are:

$$\text{First step}: u^{(1)} = u^{(n)} + \Delta t L[u^{(n)}] \tag{11.94}$$

$$\text{Second step}: u^{(2)} = \frac{3}{4} u^{(n)} + \frac{1}{4} u^{(1)} + \frac{\Delta t}{4} L[u^{(1)}] \tag{11.95}$$

$$\text{Third step}: u^{(n+1)} = \frac{u^{(n)}}{3} + \frac{2u^{(2)}}{3} + \frac{2\Delta t}{3} L[u^{(2)}] \tag{11.96}$$

Williamson (1980) examined the subset of all possible three-stage Runga-Kutta schemes for minimal computer storage requirement and recommended the following scheme.

$$u^{(1)} = u^{(n)} + \left\{ \frac{\Delta t}{3} L[u^{(n)}] \right\} \tag{11.97}$$

$$u^{(2)} = u^{(1)} + \frac{15}{16} \left\{ \Delta t L[u^{(1)}] - \frac{5}{9} u^{(1)} \right\} \tag{11.98}$$

$$u^{(n+1)} = u^{(2)} + \frac{8}{15} \left\{ \Delta t L[u^{(2)}] - \frac{153}{128} \left(\Delta t L[u^{(1)}] - \frac{5}{9} u^{(1)} \right) \right\} \tag{11.99}$$

The low storage requirement comes from the observation that $u^{(l)}$ and the quantities within the brackets can be over-written three times during each integration step.

Four-stage Runge Kutta scheme

This scheme is of higher order and is potentially attractive. However, it has been shown by Durran (1999) that its high efficiency factor is misleading. The steps used are:

$$u^{(1)} = u^n + \alpha_1 \Delta t L[u^n],$$
$$u^{(2)} = u^n + \alpha_2 \Delta t L[u^{(1)}],$$
$$u^{(3)} = u^n + \alpha_3 \Delta t L[u^{(2)}],$$

and

$$u^{(n+1)} = u^n + \frac{1}{6} \Delta t L[u^{(3)}], \tag{11.100}$$

where $\alpha_1 = 1/2$, $\alpha_2 = 1/2$ and $\alpha_3 = 1$.

If an amplification factor $\left(G = \frac{u^{n+1}}{u^n}\right)$ is plotted against the various resolved time scales for three- and four stage Runge-Kutta schemes, then one can see that stable time steps for three stage Runge-Kutta scheme can be exceeded by the four stage Runge-Kutta scheme. However, some high frequency events will be severely attenuated by a four-stage RK scheme. Thus, for steady state calculations four-Stage Runge-Kutta scheme will be preferred. However, for time accurate calculations three stage Runge-Kutta time integration scheme will be preferred if instability is avoided.

11.15 Numerical Stability Analysis

In discussing spatial discretization schemes, we have seen their potential unstable nature by the way the schemes add or remove dissipation. However, in solving actual equations it is imperative to discuss the numerical instability aspect of both the spatial and temporal discretization considered together. For this purpose we will consider the model of the one-dimensional wave equation given by Equation (11.85). Once again our aim would be to investigate the numerical stability scale-by-scale, and hence a spectral analysis will be preferred.

If we extended the definition given in Equation (11.36) to include time variation of the dependent variable

$$u(x,t) = \int\int U(k,\omega)e^{i(kx-\omega t)}dkd\omega, \tag{11.101}$$

or alternately

$$u(x,t) = \int U(k,t)e^{ik(x-ct)}dk, \tag{11.102}$$

then we can define an amplification factor of the numerical scheme by

$$G(k) = \frac{U(k,t+\Delta t)}{U(k,t)}. \tag{11.103}$$

This amplification factor will depend on both the spatial discretization (via its dependence on k) and the time discretization scheme. For semi-discrete analysis ($\Delta t \to 0$), this amplification will thus be equal to 1 and this is its theoretical estimate.

The following classification will be used to describe schemes:

The scheme is monotonically stable when: $0 < G < 1$.

The scheme is neutrally stable when: $G = 1$.

The scheme is unstable when: $G > 1$ or $G < -1$.

The scheme is stable, but the solution oscillates in sign when: $-1 < G < 0$.

For the general compact schemes, the spatial derivative is obtained from Equation (11.41) as

$$\frac{\partial u_j^n}{\partial x} = \left\{\sum_{l=1}^{N} C_{jl}P_{lj}\right\} u_j^n. \tag{11.104}$$

The amplification factors for some typical time discretization schemes are given in the following.

(a) Euler time marching scheme

Here Equation (11.85) takes the following discrete form.

$$\frac{u_j^{n+1} - u_j^n}{\Delta t} + c\left\{\sum_{l=1}^{N} C_{jl} P_{lj}\right\} u_j^n = 0.$$

Thus using the definition given by equation (11.104), one gets

$$G = 1 + N_c h \sum_{l=1}^{N} C_{jl} P_{lj}, \tag{11.105}$$

where $N_c = \frac{c\Delta t}{h}$ is the familiar Courant-Friedrich-Lewy (CFL) number. Since G is a complex number, one can talk about in-phase and out-of-phase component of G. One can also describe a method by noting the absolute value of G.

(b) Adams-Bashforth time integration scheme

This scheme is given by equation (11.91) as

$$u^{n+1} = u^n + \frac{\Delta t}{2}\left[3\left(\frac{\partial u}{\partial t}\right)^n - \left(\frac{\partial u}{\partial t}\right)^{n-1}\right], \tag{11.106}$$

and when this is used in Equation (11.85), using equation (11.105) one gets a quadratic equation for G given by

$$G^2 + (3a + 3ib - 1)G - (a + ib) = 0, \tag{11.107}$$

where $a + ib = \frac{N_c h}{2} \sum_{l=1}^{N} C_{jl} P_{lj}$.

As discussed in Chapter 10, the two amplification factors G_1 and G_2, as obtained by solving equations (11.107), represents *physical* and *computational modes*. The exact values of G_1 and G_2 are given by

$$G_1 = \frac{1 - 3a + E}{2} + i\left[\frac{-3b + F}{2}\right], \tag{11.108a}$$

$$G_2 = \frac{1 - 3a - E}{2} + i\left[\frac{-3b - F}{2}\right], \tag{11.108b}$$

where

$$E = \left[\frac{1}{2}(9a^2 - 9b^2 - 2a + 1) + \frac{1}{2}[(9a^2 - 9b^2 - 2a + 1)^2 + 4b^2(9a - 1)^2]^{1/2}\right]^{1/2}$$

and

$$F = \frac{b(9a - 1)}{E}.$$

In general, for small CFL numbers, G_1 is close to 1 while G_2 is close to zero.

(c) Four-stage Runge-Kutta scheme

Once again using Equation (11.104) in

$$L(u) = -c\frac{\partial u}{\partial x}$$

we get

$$\Delta t L(u_j) = -N_c h\left\{\sum_{l=1}^{N} C_{jl}P_{lj}\right\}u_j = -Au_j,$$

where

$$A = N_c h\sum_{l=1}^{N} C_{jl}P_{lj}.$$

The individual steps of Runge-Kutta scheme involves evaluation of the following:
Step 1:

$$u^{(1)} = (1-\alpha_1 A)u^n.$$

Step 2:

$$u^{(2)} = u^n - \alpha_2(1-\alpha_1 A)Au^n = [1-\alpha_2(1-\alpha_1 A)A]u^n.$$

Step 3:

$$u^{(3)} = u^n - \alpha_3[1-\alpha_2(1-\alpha_1 A)A]Au^n = [1-\alpha_3\{1-\alpha_2(1-\alpha_1 A)A\}A]u^n.$$

Step 4:

$$u^{n+1} = u^n - [1 - \alpha_3[1 - \alpha_2(1 - \alpha_1 A)A]A]Au^n.$$

Substituting the values of α_1, α_2 and α_3 and simplifying

$$G = \frac{U^{n+1}}{U^n} = \left[1 - A + \frac{A^2}{2} - \frac{A^3}{6} + \frac{A^4}{24}\right]. \tag{11.109}$$

In Fig. 11.14 the variation of G are shown as a function of kh for different spatial discretization schemes in conjuction with Euler time integration scheme. In the figure the K-scheme is the Kawamura scheme defined by Equation (10.18) with $\beta = 6$; the A-scheme is given by the stencils (Equations (11.27) and (11.29)); The HT- scheme is given by the stencils (Equations (11.27), (11.28) and (11.22)) with the constants given by Equations (11.47); the Z-scheme is given by the stencils (Equations (11.68)–(11.70)), and the OUCS schemes are given by Equations (11.70), (11.75) and (11.76) along with the boundary closure schemes given by Equations (11.71) and (11.74). Two CFL number values are chosen to show their effects on numerical stability. Higher values of CFL number cause significant attenuation of the small scales. The qualitative behaviour of the G-plots are identical for both the cases because the amplification factors are linear functions of the CFL number.

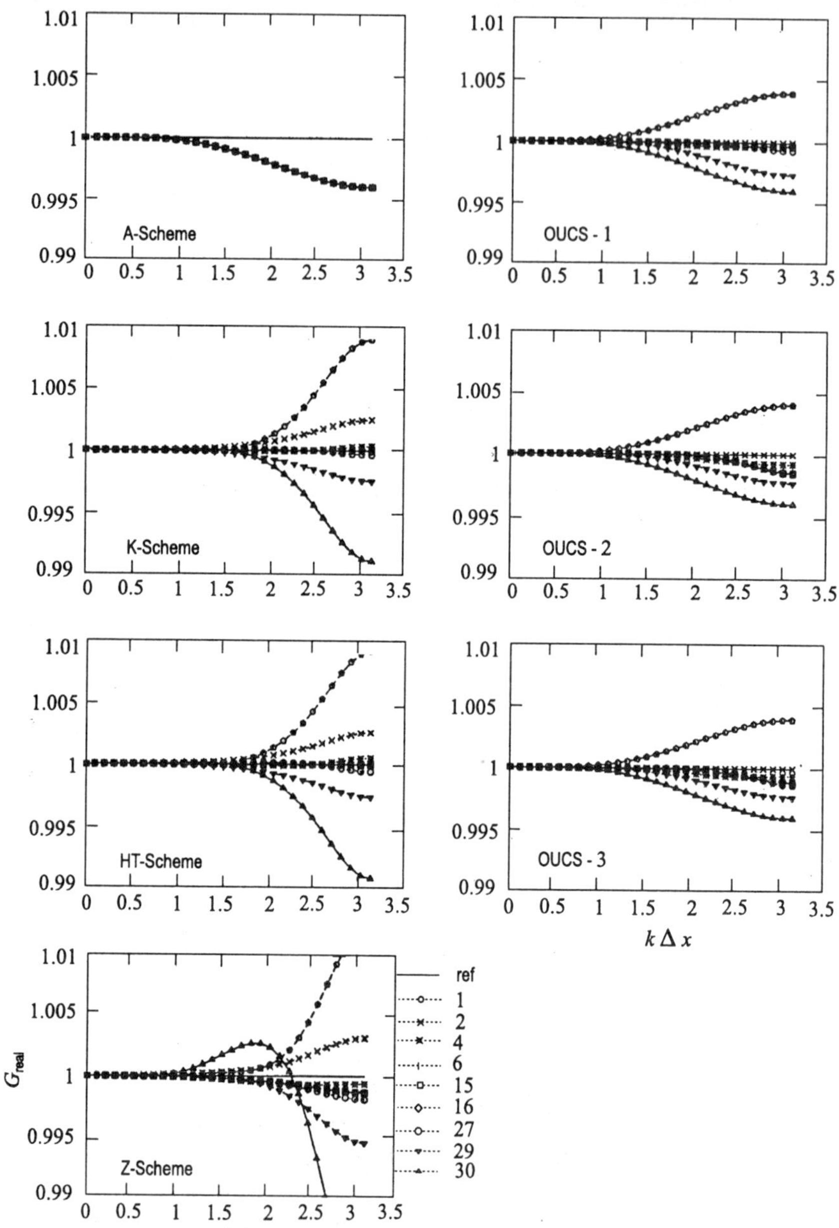

Fig. 11.14 *(a) The amplification factor for 1D wave equation discretized by various spatial schemes and Euler time integration scheme for the CFL number of 0.001.*

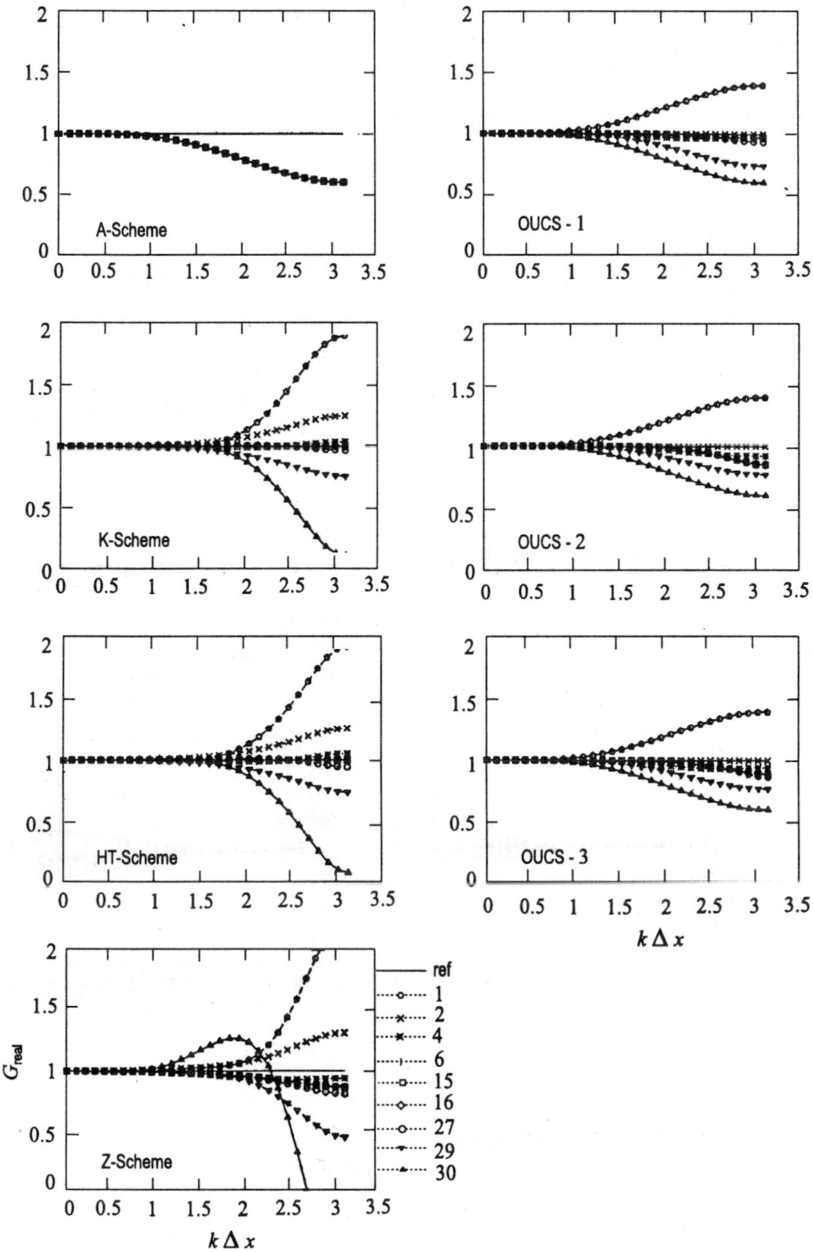

Fig. 11.14 *(b) The amplification factor for 1D wave equation discretized by various spatial schemes and Euler time integration scheme for the CFL number of 0.1.*

In Fig. 11.15, the amplification factors G_1, G_2 for Adams-Bashforth time integration scheme are shown. The first amplification factor G_1, shown in Figure 11.15(a), shows features that are identical to that of Euler time integration scheme. The second amplification factor is very close to zero and is the *computational mode*, shown in Figure 11.15(b).

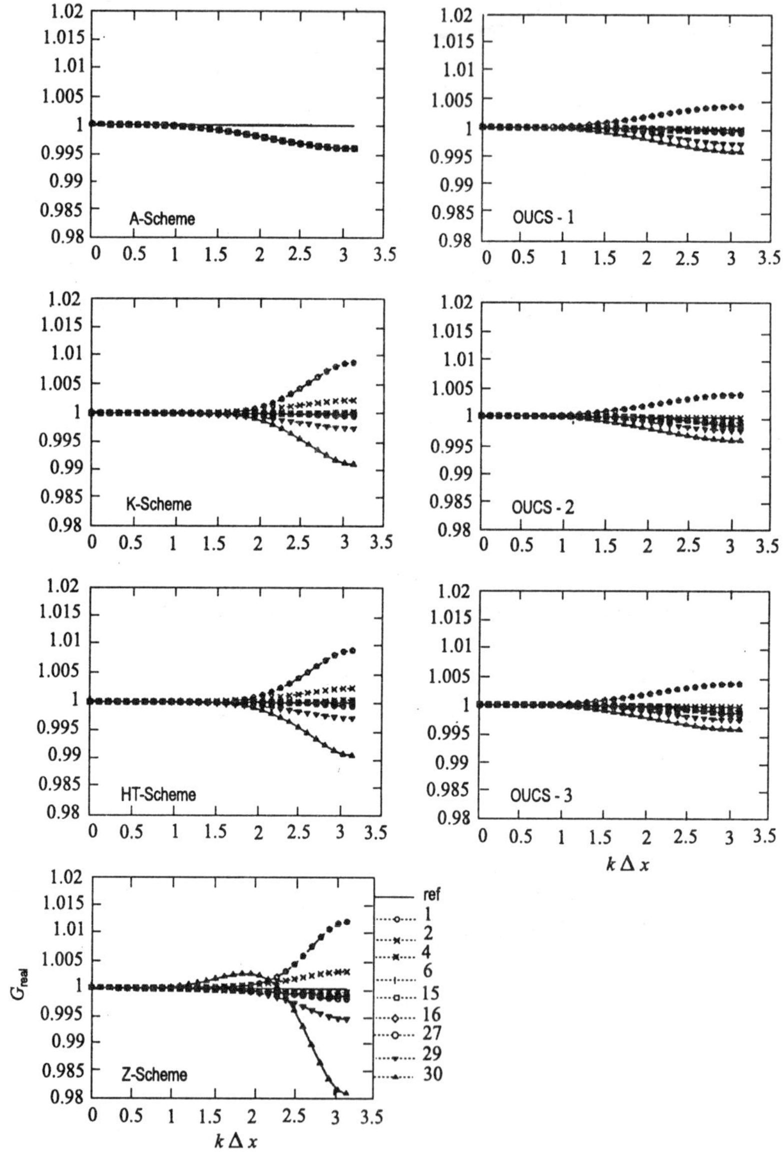

Fig. 11.15 *(a) The physical mode (G_1) of the amplification factor for 1D wave equation discretized by various spatial schemes and Adams-Bashforth time integration scheme for a CFL number of 0.001.*

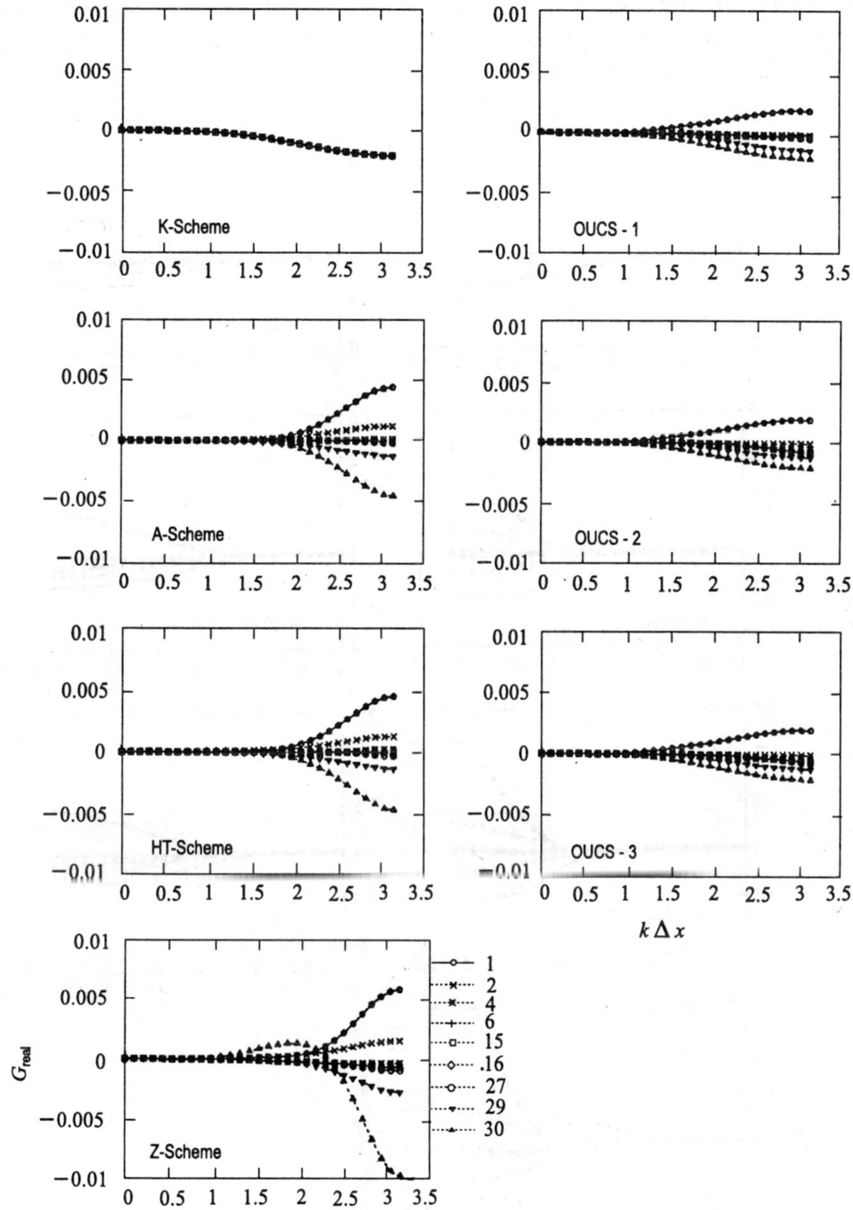

Fig. 11.15 *(b) The computational mode (G_2) of the amplification factor for 1D wave equation discretized by various spatial schemes and Adams-Bashforth time integration scheme for a CFL number of 0.001.*

The higher CFL number case ($N_c = 0.1$) for the Adams-Bashforth time integration scheme is shown in Fig. 11.16. Note here that there are qualitative changes for both G_1 and G_2 for $N_c = 0.1$ as compared to the case for $N_c = 0.001$.

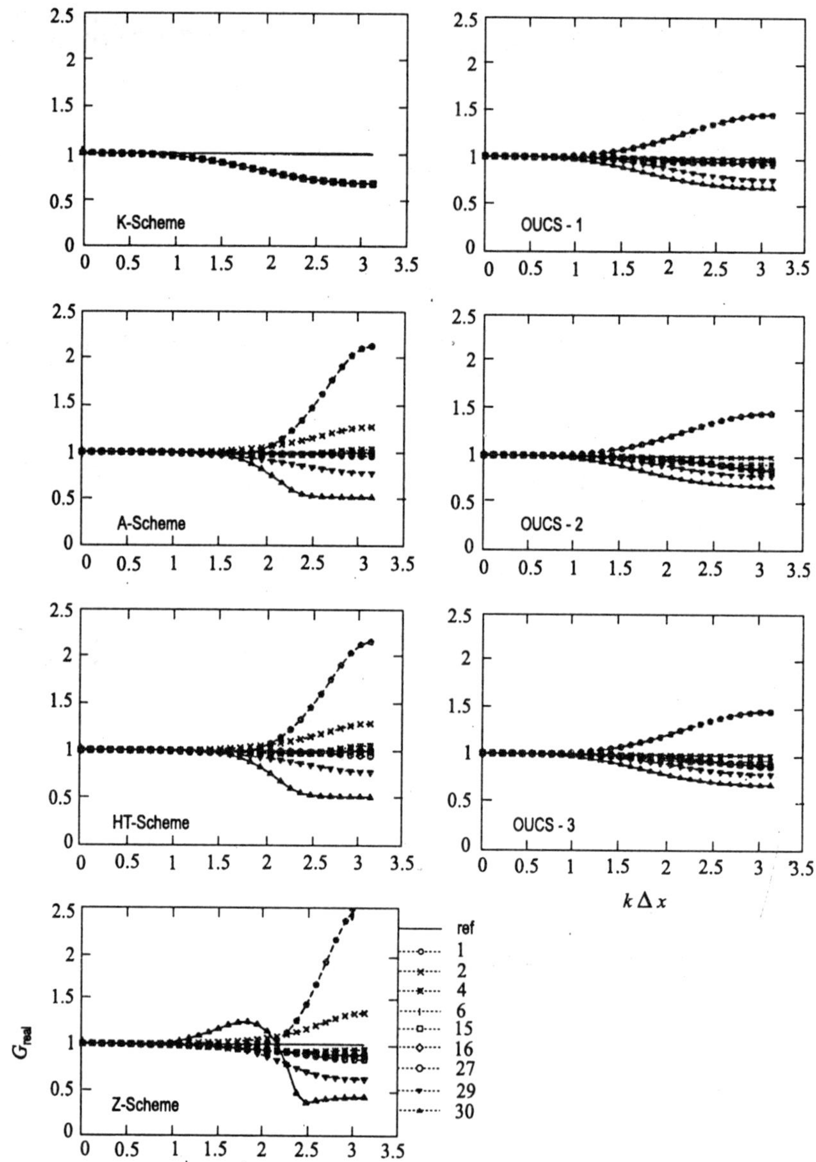

Fig. 11.16 *(a) The physical mode (G_1) of the amplification factor for 1D wave equation discretized by various spatial schemes and Adams-Bashforth time integration scheme for a CFL number of 0.1.*

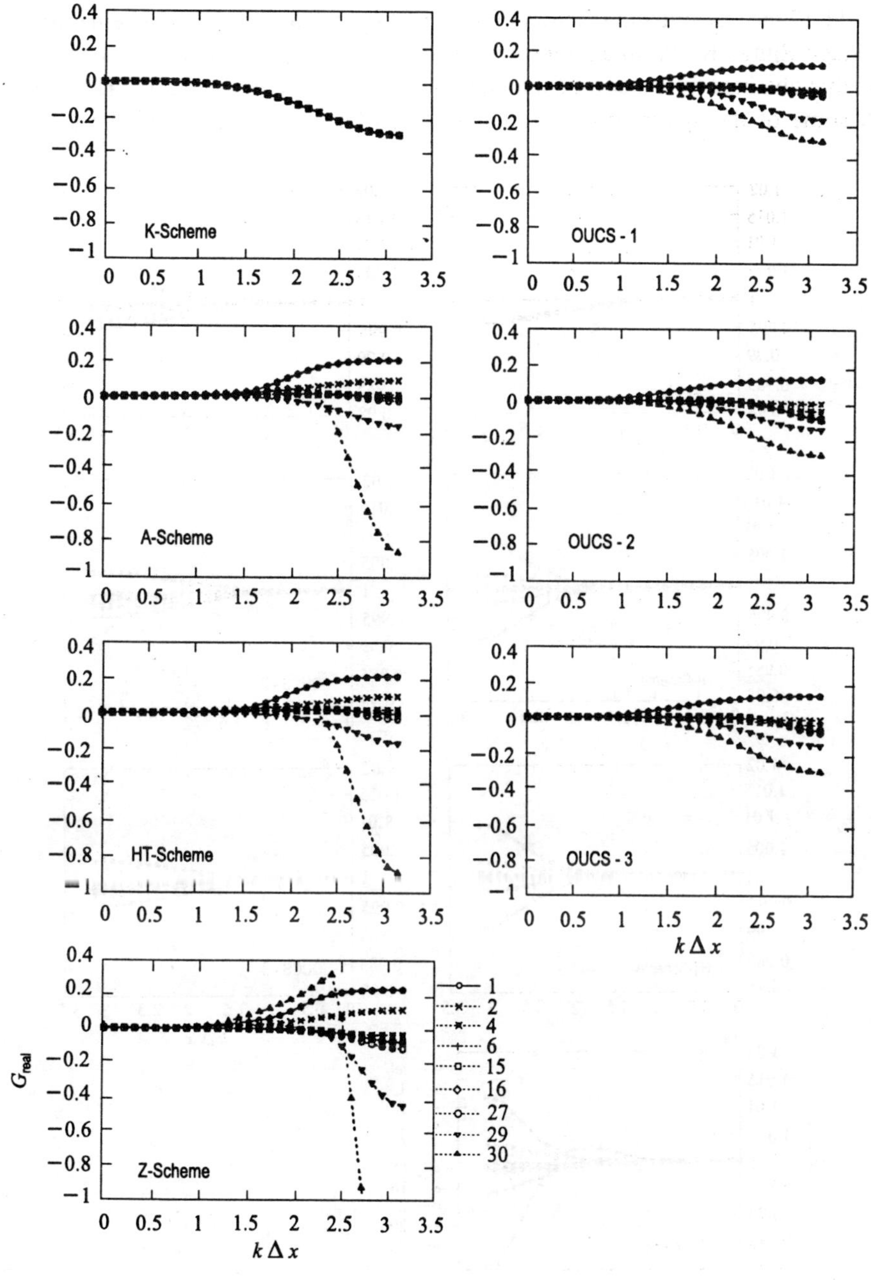

Fig. 11.16 *(b) The computational mode (G_2) of the amplification factor for 1D wave equation discretized by various spatial schemes and Adams-Bashforth time integration scheme for a CFL number of 0.1.*

In Fig. 11.17 the amplification factors for different spatial discretization schemes are shown for the four-stage Runge-Kutta time integration method. The amplification factors are qualitatively similar to that obtained for Euler time integration scheme shown in Fig. 11.14. The wave numbers at which some of the spatial discretization methods become unstable are the same for both

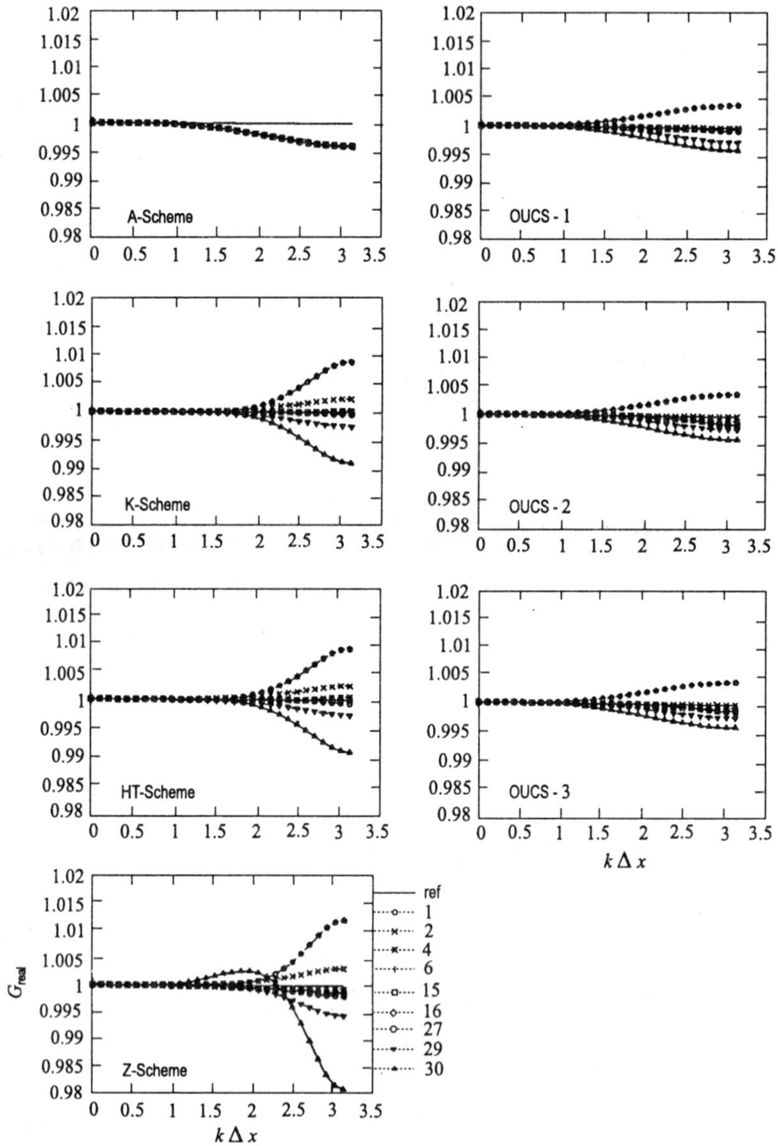

Fig. 11.17 *The amplification factor for 1D wave equation discretized by indicated spatial schemes and four-stage Runge- Kutta time integration scheme for a CFL number of 0.001.*

(a)

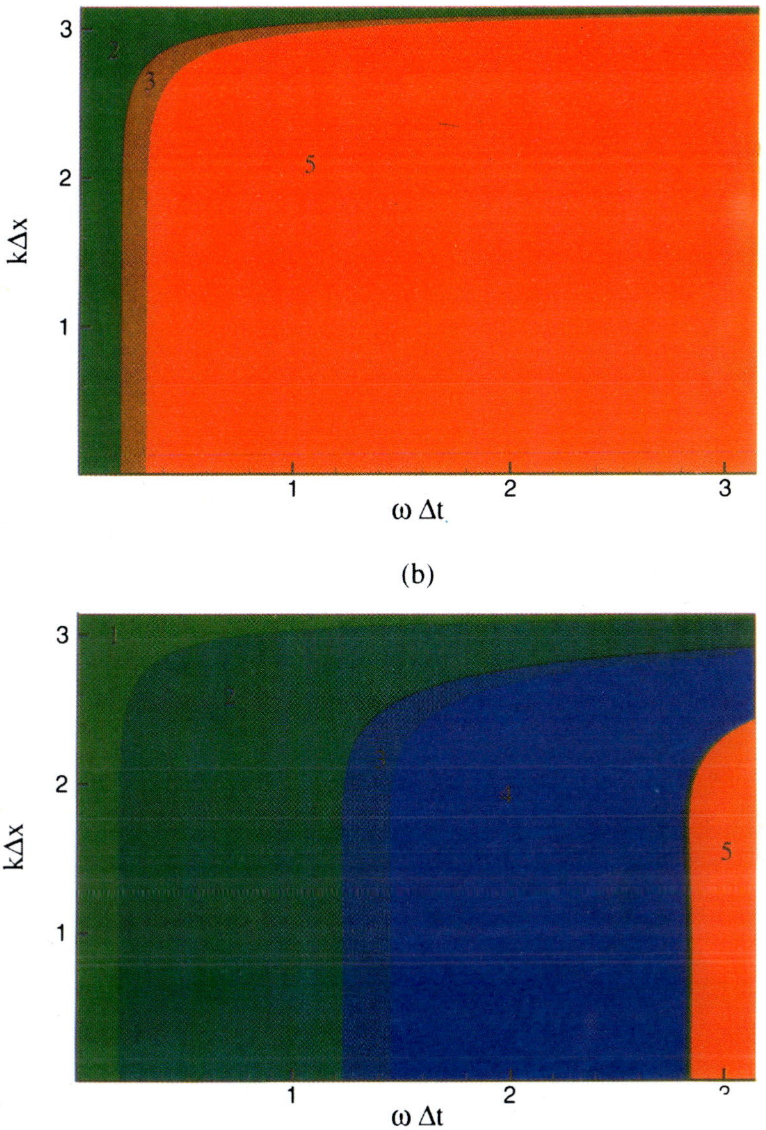

The contours of amplification factor G for solving 1D wave equation by OUCS-3 scheme for spatial discretization. Time integration is by Euler and RK4 methods. In the figure, Region $1 : |G| = 1$*; Region* $(1+2) : 1 < |G| < 0.98$*; Region* $(1+2+3) : 1 < |G| < 0.95$*; Region* $4 : 0.95 < |G| < 1$ *and Region 5:* $|G| > 1$. RK_4 *time integration is much more desirable because of larger neutrally stable range of* $(k\Delta x - \omega\Delta t)$ *- plane available to it.*

these time advancement schemes, only the degree of instability is higher for R-K scheme at higher N_c.

In concluding this subsection, we make the following observations. A global analysis that includes the effect of boundary closure, shows that the central compact schemes are unstable near one boundary. The presented results show that this is the case at the left boundary whose effect percolates in the domain interior. Contrarily, explicit schemes (like third order upwinding scheme) do not have this problem. Upwind schemes like Zhong's (1998), try to overcome this problem in the interior by the implicit dissipation built-in the stencil. However, these are still not enough to overcome the instability. Increasing the CFL number increases the dissipation; the instability, if present at high wave numbers, also increases. The role of *computational modes* as in Adams-Bashforth is still not analyzed properly. Many DNS results presented recently have used some of these numerically unstable schemes. It should be a matter of deep concern if the computed turbulence of these DNS studies are artifact of numerical instability!

11.16 Dispersion Relation Preservation

It has been observed often that the nonlinear numerical instability experienced at a high Reynolds number cannot be suppressed by reducing the CFL number, i.e. by reducing the time step value. One of the reasons is due to pile-up of energy at high wave numbers due to aliasing, and the numerical dissipation added is not sufficient enough at these high wave numbers. There are other occasions, even when the numerical instabilities are avoided, when one is faced with difficulties associated with solutions that seem to match well for some time with experiments but de-correlate as time progresses. This brings us to the second criteria for DNS that relates to the dispersion relation which the governing time dependent PDE has in the continuum sense, while the numerical solution does not satisfy it.

For many wave propagation problems the group velocity is a more meaningful physical quantity than the phase speed as the disturbance energy propagates with it (Lighthill (1978)). For example, the one-dimensional wave equation has the dispersion relation-as given in chapter 4: $\omega = kc$. This implies that the group velocity is equal to the phase speed (c). The system is thus non-dispersive, i.e it is shape preserving (also amplitude preserving-since this is a non-dissipative system). In general, for a nonlinear system, it is a priori not known whether it is dispersive or non-dispersive. However, to test the effectiveness of a space-time discretization scheme, one can numerically solve the one-dimensional wave equation. One can obtain the numerical group velocity and compare it with the physical group velocity (phase speed c, in this case) to comment on the dispersion relation preservation (DRP) property of the numerical scheme.

A new class of methods has been identified by Tam and Web (1993) as DRP method for which the numerical and physical dispersion relation matches over a large range of wave numbers and circular frequency. This property will be assessed here for the space-time discretization schemes by comparing the numerical and physical group velocity. For linear 1D wave equation the physical group velocity is

$$V_g = \frac{d\omega}{dk} = c. \tag{11.110}$$

For the numerical scheme used to discretize Equation (11.85), the numerical dispersion relation is

$$\omega_{\text{eq}} = ck_{\text{eq}}, \tag{11.111}$$

which gives the numerical group velocity as

$$(V_g)_{\text{num}} = c\frac{(dk_{\text{eq}}/dk)}{(d\omega_{\text{eq}}/d\omega)}. \tag{11.112}$$

For the spatial discretization scheme, let

$$\frac{dk_{\text{eq}}}{dk} = p + iq,$$

and for the temporal discretization scheme,

$$\frac{d\omega_{\text{eq}}}{d\omega} = r + is.$$

Then from Equation (11.112),

$$(V_g)_{\text{num}} = c\frac{pr + qs}{r^2 + s^2}. \tag{11.113}$$

Thus

$$\frac{(V_g)_{\text{num}}}{V_g} = \frac{pr + qs}{r^2 + s^2}. \tag{11.114}$$

This ratio on the right hand side can be calculated for a chosen numerical scheme and a region in the $(kh-\omega\Delta t)$- plane be marked where this departs from one (the ideal ratio) by a small tolerance.

From Equation (11.43),

$$\frac{dk_{\text{eq}}}{dk} = \sum_{l=1}^{N} c_{jl}(x_l - x_j)e^{ikh(l-j)}. \tag{11.115}$$

For the representative time iteration schemes, one can similarly obtain:

(a) Euler time-stepping

$$r + is = e^{-i\omega\Delta t} \tag{11.116}$$

(b) Mid-point leap frog method

$$r + is = \cos\omega\Delta t \tag{11.117}$$

(c) Second order Adams-Bashforth method

$$\left.\begin{aligned} r &= [a_3c_3 + b_3d_3]/[a_3^2 + b_3^2] \\ s &= [a_3d_3 - b_3c_3]/[a_3^2 + b_3^2] \end{aligned}\right\} \tag{11.118}$$

where

$$a_3 = 9 + \cos(2\omega\Delta t) - 6\cos(\omega\Delta t),$$

$$b_3 = \sin(2\omega\Delta t) - 6\sin(\omega\Delta t).$$

$$c_3 = 8\cos(\omega\Delta t) - 4,$$

$$d_4 = -4\sin(\omega\Delta t).$$

(d) Runge-Kutta schemes

Here instead of calculating $(V_g)_{\text{num}}$ from Equation (11.113), one estimates it directly from the wave Equation (11.85).

If we define

$$a_4 + ib_4 = h\sum_{l=1}^{N} c_{jl}e^{ik(x_l - x_j)} \tag{11.119}$$

and

$$a_5 + ib_5 = \sum_{l=1}^{N} c_{jl}(x_l - x_j)e^{ik(x_l - x_j)}, \tag{11.120}$$

then $(V_g)_{\text{num}}$ is obtained for the three RK schemes as

RK_2 scheme:

$$(V_g)_{\text{num}} = \frac{c}{r_1}[N_c b_4 b_5 - N_c a_4 a_5 + a_5]; \tag{11.121}$$

RK_3 scheme:

$$(V_g)_{\text{num}} = \frac{c}{r_1}\left[a_5 + N_c(b_5b_4 - a_4a_5) + \frac{N_c}{2}a_5a_4^2 - N_ca_4b_4b_5 - \frac{N_c}{2}a_5b_4^2\right]; \tag{11.122}$$

RK_4 scheme:

$$(V_g)_{\text{num}} = \frac{c}{r_1}\left[a_5 + \frac{2}{3}N_c(b_4b_5 - a_4a_5) + \frac{a_5N_c^2}{6}(3a_4^2 - b_4^2 - N_ca_4^3 + N_ca_4b_4^2)\right]$$

$$+ b_4 \frac{N_c^2}{6}(-6a_4b_5 - 2b_4a_5 - 3N_ca_4^2b_5 - N_cb_5b_4^2 + 2N_ca_4a_5b_4]; \tag{11.123}$$

where $r_1 = \cos(\omega\Delta t)$.

The ratio of equation (11.114) is plotted next in the admissible range of kh and $\omega\Delta t$. In Fig. 11.18 the DRP zone is marked for the CD_2 scheme, where the right hand side of equation (11.114) is (1 ± 0.05).

In panel-(a), the Euler time integration scheme is considered. Panel (b) depicts the mid-point leap frog time advancement scheme which seems to be an improvement over Euler time marching

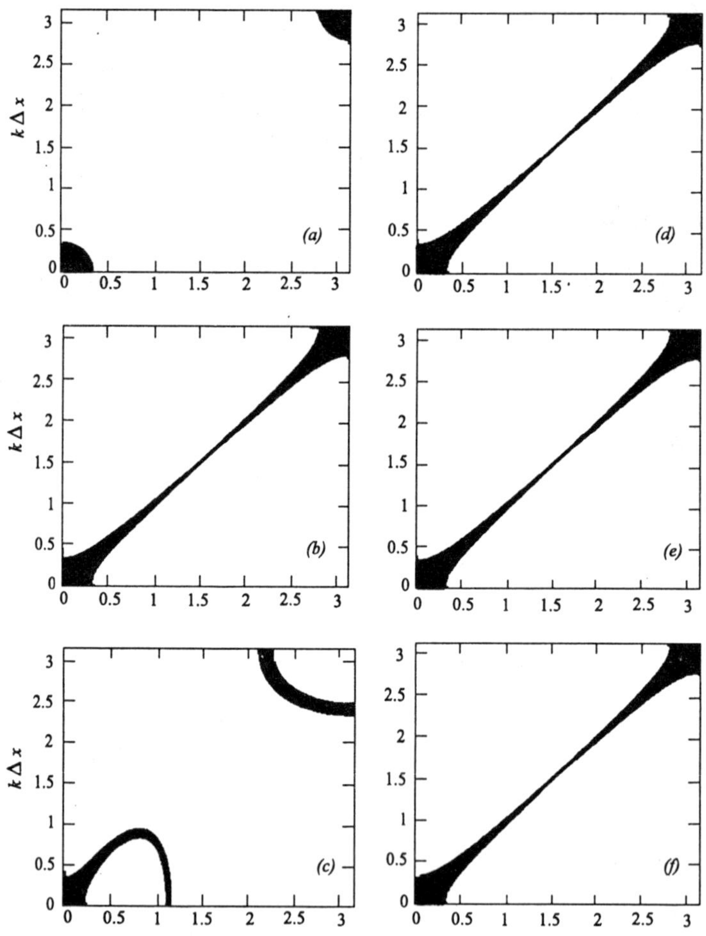

Fig. 11.18 *The DRP region in $(k\Delta x - \omega\Delta t)$-plane for 1D wave equation obtained by matching numerical group velocity and physical group velocity with 5% tolerance. Spatial discretization is by second order central differencing and temporal discretization by (a) Euler scheme, (b) mid- point leap frog scheme, (c) Adams-Bashforth scheme, (d) two-stage Runge-Kutta scheme, (e) three-stage Runge-Kutta scheme and (f) four-stage Runge-Kutta scheme for a CFL number 0.001.*

scheme. The fact that the amplification factor G for this scheme is identically equal to 1 for all wave number for the inviscid wave equation makes it an attractive numerical method, and in many weather prediction codes this is often adopted. Comparatively, the second order Adams-Bashforth scheme has poorer DRP property. Due to the presence of the *computational mode*, this scheme is not recommended despite the fact that Haltiner & Williams (1980) have recommended this method and Kim & Moin (1985) have used it for DNS. All the Runge-Kutta schemes of time integration have a DRP region in no way different from the mid-point leap frog method. It is a moot point as to whether the extra memory and computational effort is justified with no additional proportional returns. This spatial discretization scheme has an in-built symmetry of the DRP zone for all the time advancement strategies. This type of spatial discretization scheme would therefore be ideal if the underlying wave phenomenon is such that $c\Delta t = h$ due to this symmetry. It is to be noted that the DRP- patch near the highest wave number and frequency is not practical. Here the numerical group velocity is close to the physical group velocity, but phase errors are very large.

In contrast the DRP region shown in Fig. 11.19 for the fourth order central differencing scheme for spatial derivative does not have the above mentioned symmetry property for all the time advancement schemes that were used for Fig. 11.18. For all the time-advancement strategies in this case, the DRP property improves with respect to the patch near the origin. Allowable grid spacing improves by more than hundred percent for this spatial discretization. However, the second order Adams-Bashforth scheme continues to display poor phase and dispersion errors. All the Runge-Kutta schemes have identical range of wave numbers and frequencies where the DRP property is maintained, as is observed for the leap frog method.

The sixth order spatial difference scheme results are shown in Fig. 11.20 for the same time-advancement schemes. The range of k over which DRP property is valid increases further as compared to CD_2 and CD_4 schemes. As before, the second order Adams-Bashforth scheme has a smaller range of circular frequencies over which the DRP is approximately valid. This is somewhat unexpected because Adams-Bashforth scheme being second order, one would have expected an improvement over Euler scheme at least!

The fact that the first order upwinding seriously compromises spectral accuracy is also reflected in its very poor DRP property. This is worse than the CD_2 scheme. Contrarily, the third order upwind scheme results shown in Fig. 11.21 have regions of validity of DRP property in the $(kh - \omega\Delta t)$ plane that is similar to CD_4 scheme.

The scheme that is considered next, is given by Equation (10.18) with $\beta = 1$. However, the most popular third order upwind scheme is the one for which $\beta = 6$ and is variously used in Kawamura et al. (1985) and Sengupta & Sengupta (1994). The DRP region of this case—which we term as Kawamura scheme—is shown in Fig. 11.22. A direct comparison between these two sets of results is discussed for the Euler time integration scheme. The Kawamura scheme ($\beta = 6$) shows a larger region of validity for DRP property for higher values of k and ω. This is one of the reasons why third order upwind scheme with $\beta = 6$ produces accurate results with optimum effort. Thus the role of numerical dissipation is more than numerical stablization—it also enhances the scheme's ability to satisfy the dispersion relation provided it is chosen with care.

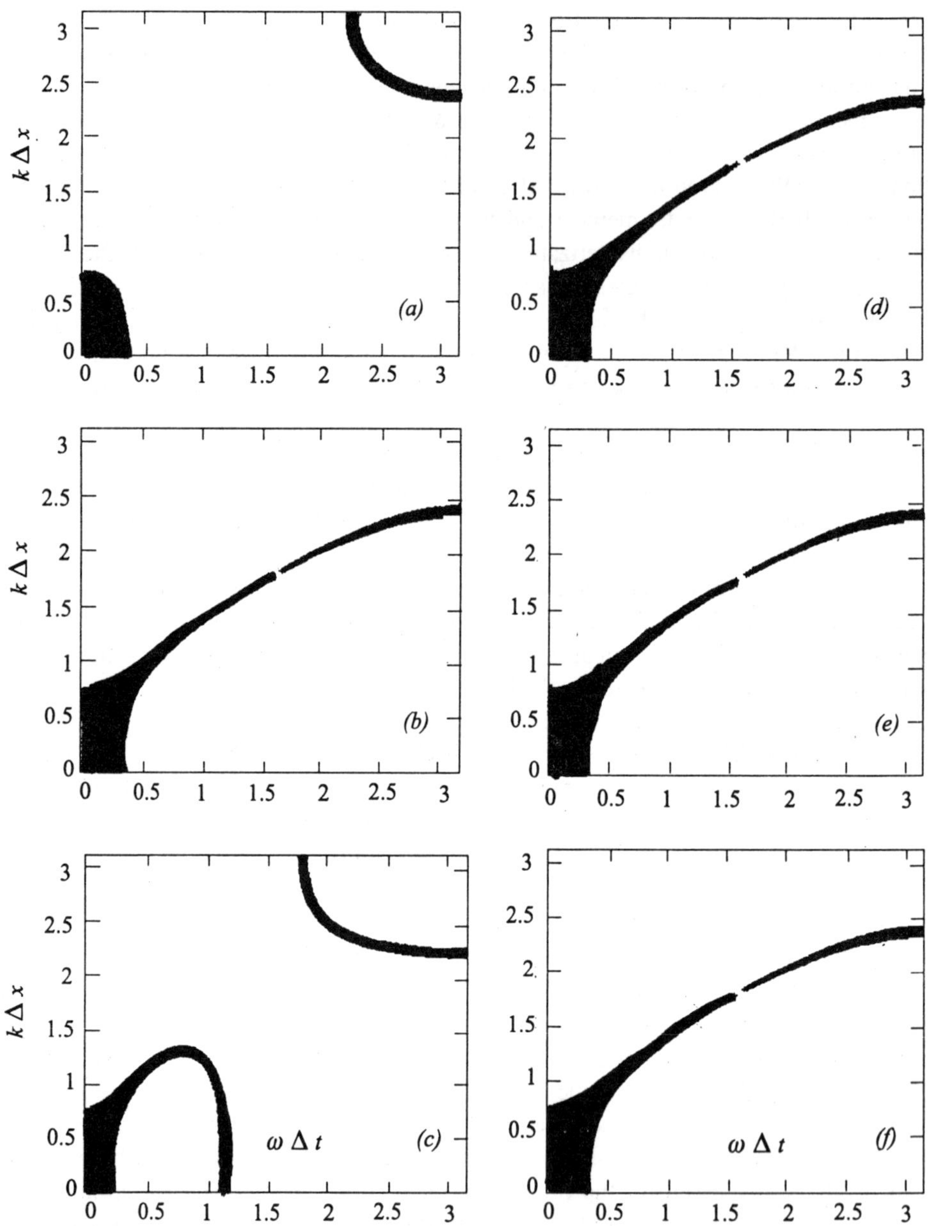

Fig. 11.19 *The DRP region in $(k\Delta x - \omega\Delta t)$-plane for 1D wave equation obtained by matching numerical group velocity and physical group velocity with 5% tolerance. Spatial discretization is by fourth order central differencing and temporal discretization by schemes of Fig. 11.18.*

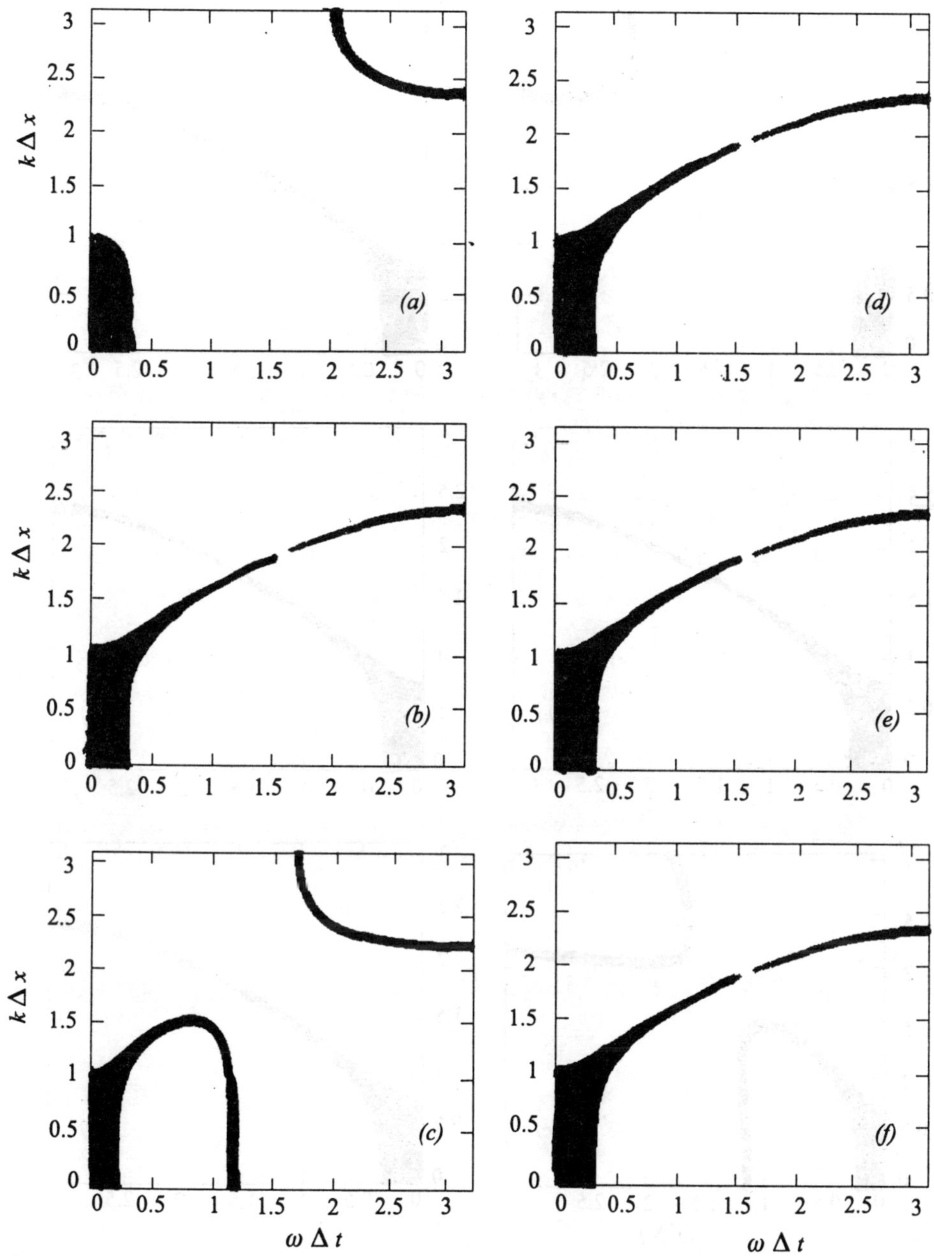

Fig. 11.20 *The DRP region in $(k\Delta x - \omega\Delta t)$-plane for 1D wave equation. Spatial discretization is by sixth order central differencing and temporal discretization as in Fig. 11.18 for a CFL number 0.001.*

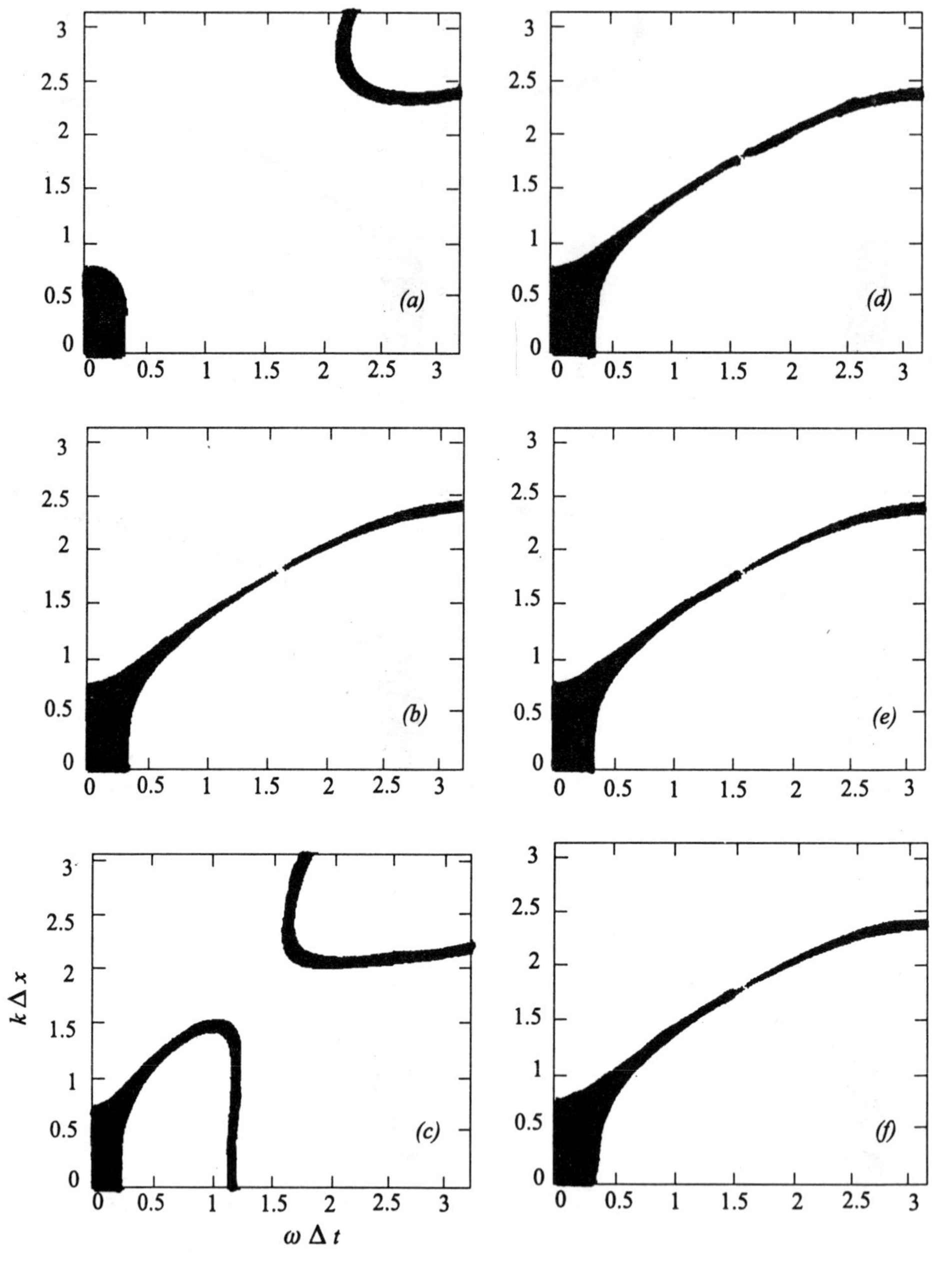

Fig. 11.21 *The DRP region in $(k\Delta x - \omega\Delta t)$-plane for third order upwind scheme given by Equation (10.18) with $\beta = 1$ and temporal discretization as in Fig. 11.18 for a CFL number of 0.001.*

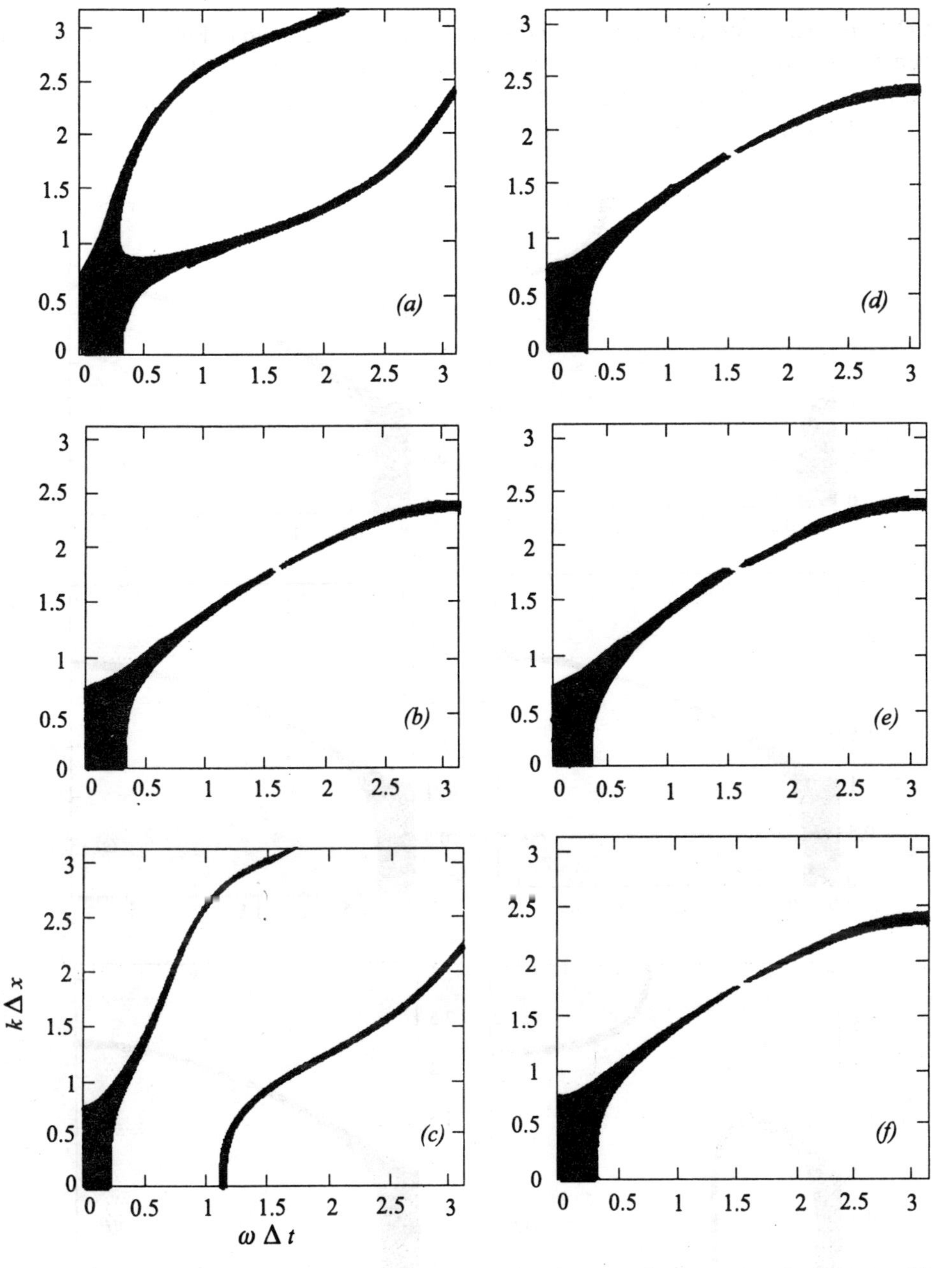

Fig. 11.22 *The DRP region in $(k\Delta x - \omega\Delta t)$-plane for third order upwind scheme given by Equation (10.18) with $\beta = 6$ and temporal discretization as in Fig. 11.18 for a CFL number of 0.001.*

The DRP properties of fifth order upwind spatial scheme are shown in Fig. 11.23 which shows improvement in k-range over third order upwinding schemes. Ranges of kh and $\omega\Delta t$ for this is identical to Fig. 11.20 for CD_6 scheme. This can be used provided appropriate stable boundary closures are derived for this seven point stencil.

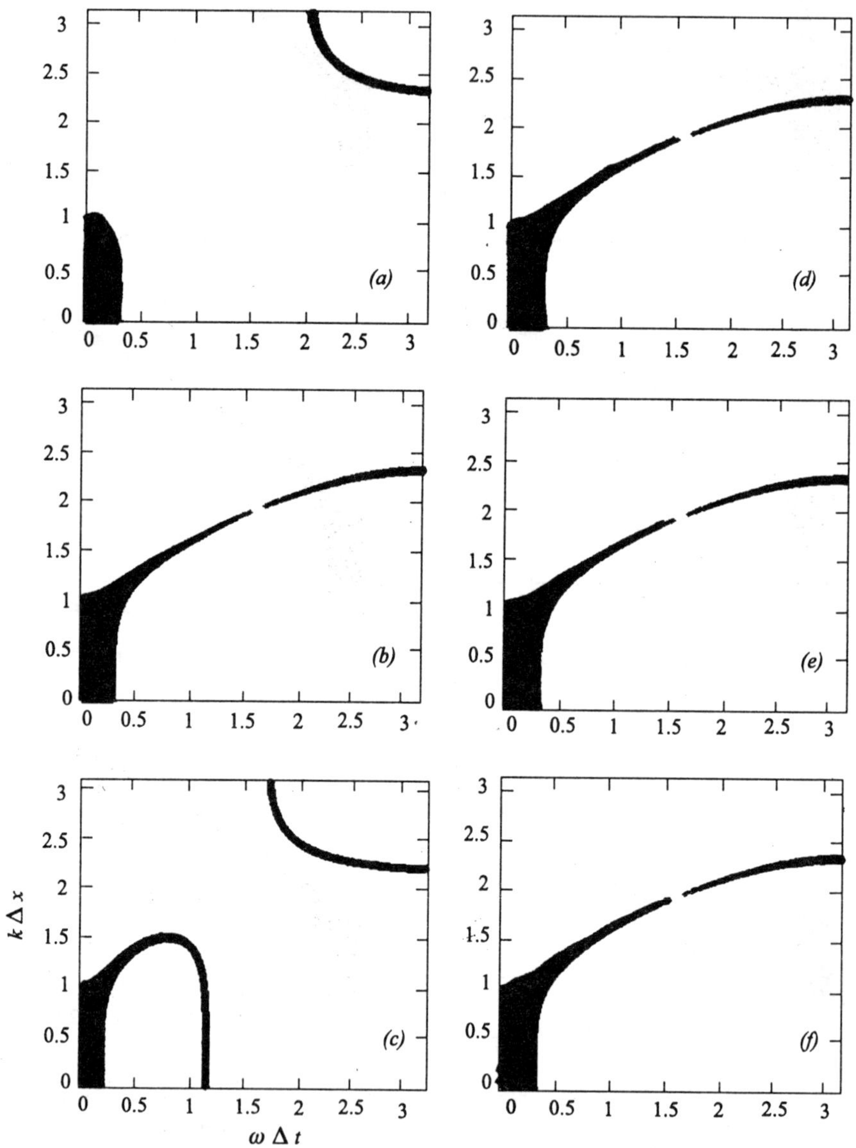

Fig. 11.23 *The DRP region in $(k\Delta x - \omega\Delta t)$-plane for fifth order upwind scheme with temporal discretization as in Fig. 11.18 for a CFL number of 0.001.*

Compared to all these explicit schemes, compact schemes provide larger k-range of validity with respect to DRP property. In Fig. 11.24 results for the Adams' scheme (stencils given by Equations (11.27) to (11.29)) are shown. The range of wave numbers upto which the DRP is valid is given by $kh \simeq 1.5$. This is one and half times higher than that achieved by CD_6 or fifth order upwind scheme.

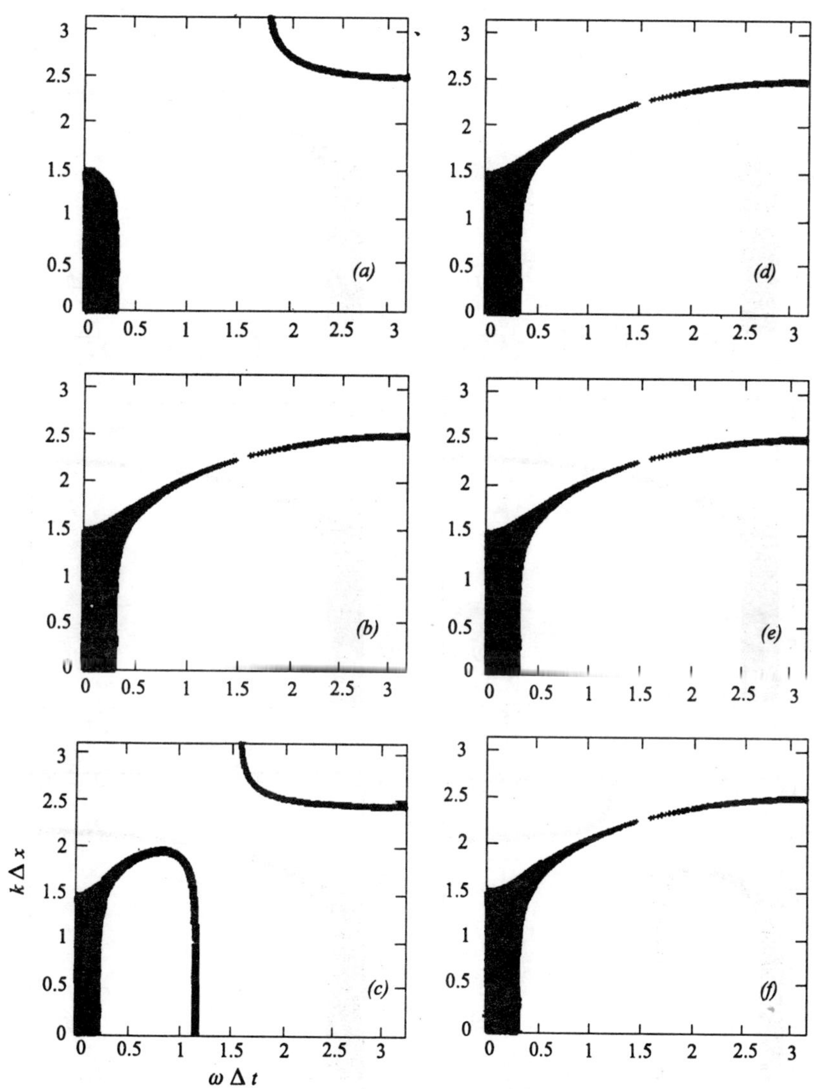

Fig. 11.24 *The DRP region in $(k\Delta x - \omega\Delta t)$-plane for Adams' spatial scheme with temporal discretization and other conditions same as in Fig. 11.18 for a CFL number of 0.001.*

The frequency ranges do not change at all, except that the second order Adams-Bashforth scheme has poorer temporal limits. Figure 11.25 shows results for the Haras-Ta'asan scheme (stencil given by Equations (11.22), (11.47), (11.27) and (11.28)), where one can see a further improvement for values of kh. For this spatial discretization scheme, Euler time integration seems to perform the best for the required computational effort and memory.

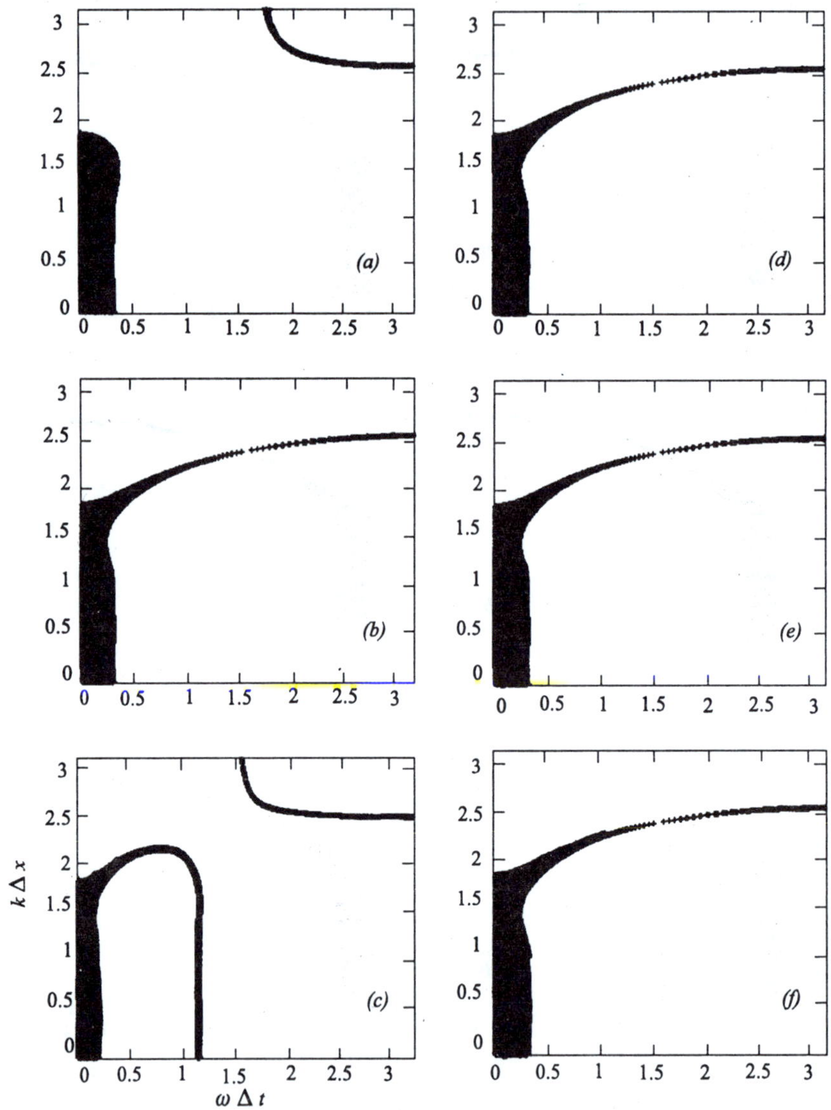

Fig. 11.25 *The DRP region in $(k\Delta x - \omega\Delta t)$-plane for Haras & Ta'asan's spatial scheme with temporal discretization and other conditions same as in Fig. 11.18 for a CFL number of 0.001.*

Figures 11.26, 11.27 and 11.28 display the DRP region of validity for OUCS-1, OUCS-2 and OUCS-3 schemes. Among these three schemes, OUCS-3 has the best DRP property.

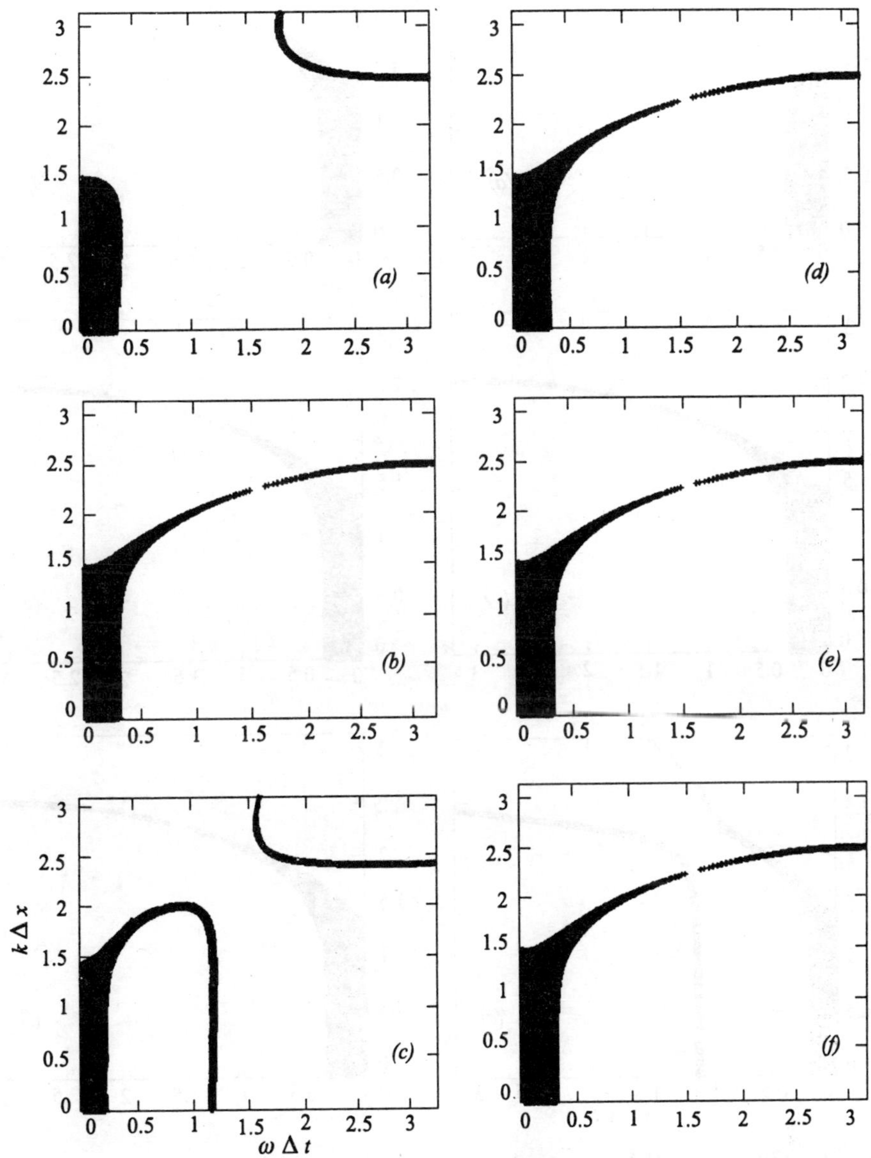

Fig. 11.26 *The DRP region in $(k\Delta x-\omega\Delta t)$ plane for OUCS-1 spatial scheme with temporal discretization and other conditions same as in Fig. 11.18 for a CFL number of 0.001.*

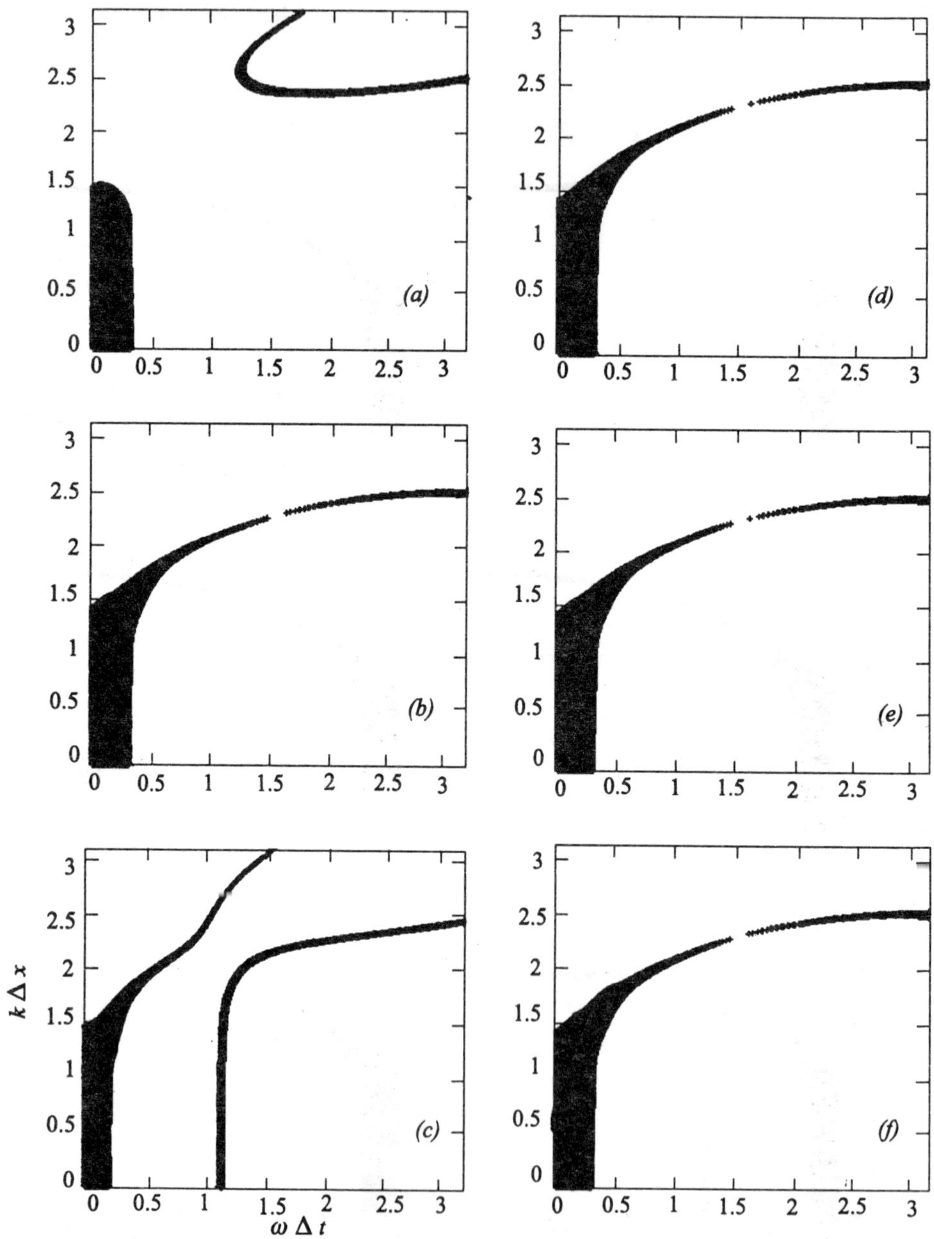

Fig. 11.27 *The DRP region in ($k\Delta x - \omega\Delta t$)-plane for OUCS-2 spatial scheme with temporal discretization and other conditions same as in Fig. 11.18 for a CFL number of 0.001.*

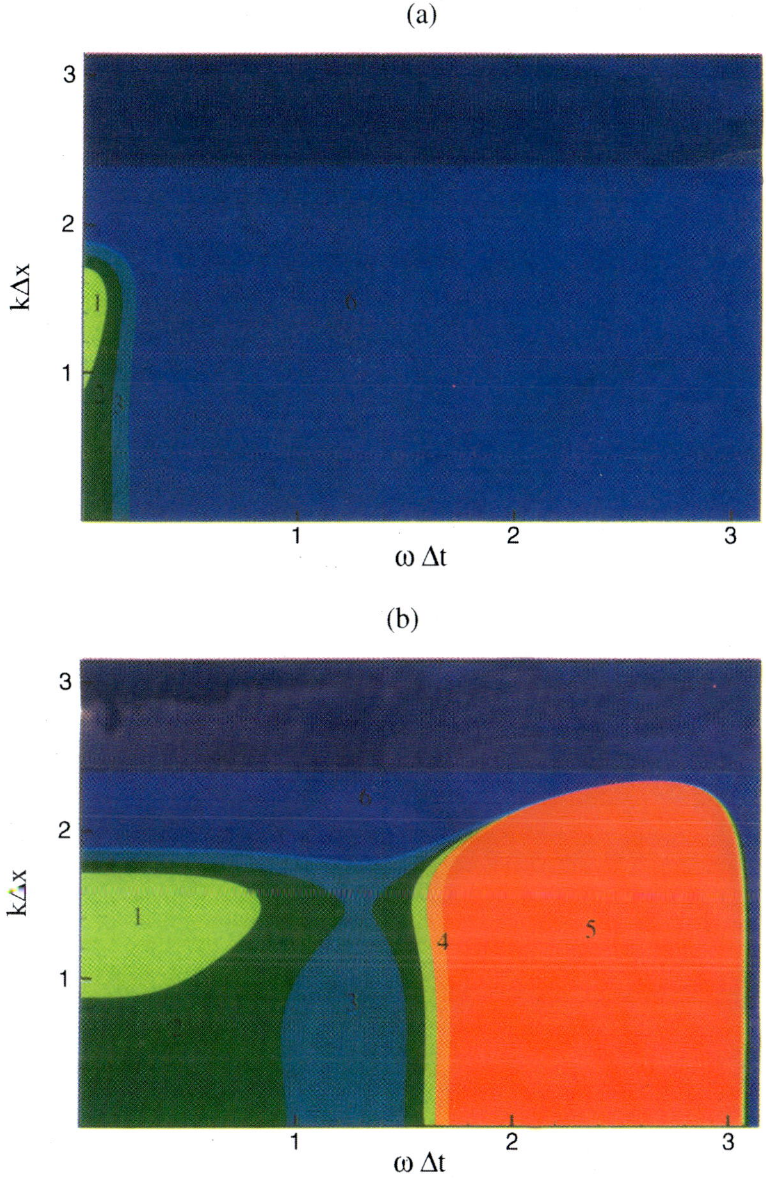

The contours of DRP region in $(k\Delta x - \omega\Delta t)$-plane for solving 1D wave equation by OUCS-3 spatial-scheme and Euler/RK4 time-integration schemes. In the figure: Region $(1+2)$: $0.98 < V_g/c < 1.02$; Region 3: $0.95 < V_g/c < 0.98$; Region 4: $1.02 < V_g/c < 1.05$; Region 5: $V_g/c > 1.05$; Region 6: $0 < V_g/c < 0.95$ and Region 7: $V_g/c < 0$. Note that both the time integration schemes show that for very small length scales there are spurious waves that actually moves upstream!

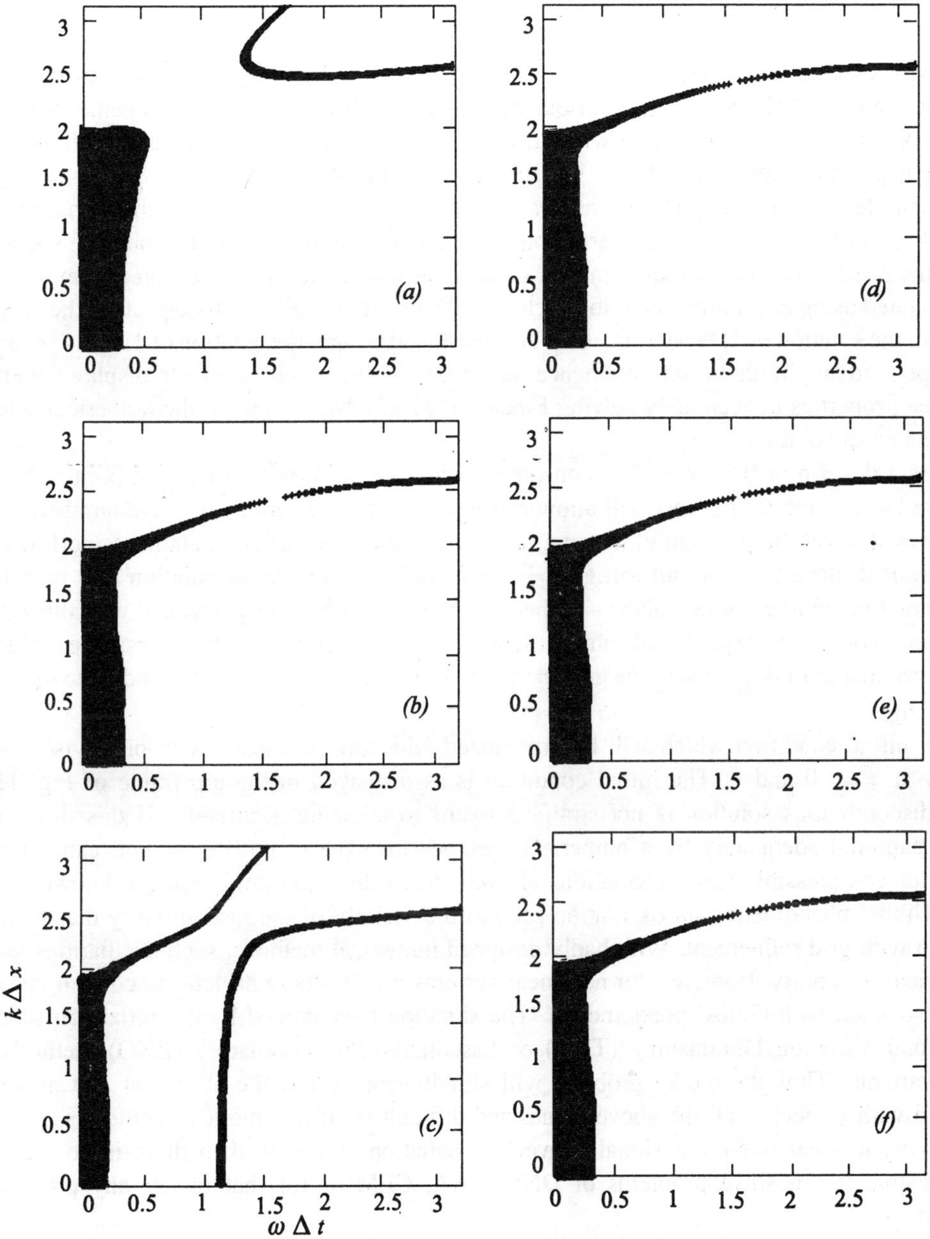

Fig. 11.28 *The DRP region in $(k\Delta x - \omega\Delta t)$-plane for OUCS-3 spatial scheme with temporal discretization and other conditions same as in Fig. 11.18 for a CFL number of 0.001.*

Solution of Linear Wave Equation

In previous sections, analyses have shown the important properties of spectral accuracy, numerical stability and the DRP properties of various explicit and higher order compact schemes for spatial discretization used in conjunction with different time integration strategies for the linear wave equation given by Equation (11.86). We arrived at certain conclusions based on these analyses. For example, for time integration schemes, it was found that explicit Euler scheme has the best amplification factor and DRP property considering computational effort (number of operations per time steps) and storage requirements. Hence, various spatial discretization schemes will be investigated using Euler time integration scheme. We would see the various aspects of the analyses through the solution of 1D wave equation. As mentioned before, the solution of 1D wave equation is shape- and amplitude-preserving, hence this model problem will be able to display the effects of these properties in accurately solving Equation (11.85), by comparing the numerical solution with the exact solution.

Here a domain of $0 \leq x \leq 1$ is considered with 200 uniformly distributed points. The use of very large number of points will allow resolving a very large range of wave numbers which in turn will avoid the problem of aliasing. For the same reason the computation is performed in the physical plane using a uniform grid. For nonuniform grids the calculation will have to be performed in the transformed plane — where the points will be equispaced and will allow direct application of various explicit and compact schemes. However, this will also necessitate evaluation of the product term $\frac{c}{x_\xi}\frac{\partial u}{\partial \xi}$, where the grid transformation is given by $x = x(\xi)$, and that would lead to aliasing .

The initial condition which will be propagated with time is chosen with piecewise discontinuity at $x = 0$ and 1. The initial condition is as displayed in the top frame of Fig. 11.29. This discontinuous solution is not straightforward to compute accurately. If this discontinuity is captured adequately by a numerical method, then the candidate method can form the basis for compressible flow calculations as well. Such discontinuities also are known to produce Gibbs' phenomena—an oscillation seen in the vicinity of the discontinuity that seems to worsen with grid refinement. With badly designed numerical methods, such oscillations lead to numerical instability. However, for nonlinear systems it is far more difficult to contain instabilities associated with Gibbs' phenomenon. The situation is contained by discretization schemes like Total Variation Diminishing (TVD) or Essentially Non-Oscillatory (ENO) methods and their variants. Thus the model problem will simultaneously test the ability of any numerical method with respect to all the above mentioned difficulties. If a numerical method is incapable of solving a linear one-dimensional convection equation correctly, then there is no reason to believe that it can solve problems of DNS, CAA, CEM or for that matter any problem of physics.

Furthermore, it should also be emphasised that the above observation is simply based on the performance of schemes in the physical plane. The moment one has to compute using nonuniform grid in the transformed plane, the problems do get compounded. None the less, the present analysis performed in the physical plane using a uniform grid should alert the readers of the usefulness of different numerical schemes.

For the presented results, the phase speed is taken as $c = 0.05$ so that the signal given by the initial condition leaves the computational domain at $t = 20$. As the initial condition is localized between 0 and 1, the solution should be identically equal to zero for $t \geq 20$ in the computational domain. Thus, a long time integration would allow one to talk about asymptotic stability of the methods, especially to investigate the effects of the boundary closures of the compact and higher order upwind schemes.

The numerical and exact solutions of Equation (11.85) are shown in Fig. 11.29 at different time instants when the Adams' compact scheme (1993) is used for spatial discretization and Euler integration scheme is used for time advancement. The reason that no other time integration method is considered is due to the following: The multi-level time integration like Adams-Bashforth scheme is not desirable due the presence of *computational mode*. Of course one can use mid-point leap frog method which has excellent phase and dispersion error property for inviscid equations. However, if one wants to use it for viscous flows the method is unstable (in fact it is exactly like the Richardson's method that was discussed in Chapter 6), and hence it is not considered here. The multi-step Runga-Kutta methods are also equally good from the point of view of phase and dispersion error. However, they do not give any greater improvement over Euler time integration scheme, at the same time these methods require more storage and computations. Of course, with Euler time advancement strategy one can estimate the effects of various truncation error terms.

The piecewise discontinuity of the initial condition at the boundary which is purposely introduced here, propagates inside the computing domain with passage of time. One can clearly identify the Gibbs' phenomenon only to the left of the discontinuity. This is not difficult to explain by looking at the effective equation solved by considering the leading truncation error terms.

The interior stencil for the Adams' (1993) scheme is given by

$$u'_{j-1} + 3u'_j + u'_{j+1} = \frac{1}{12h}[-u_{j-2} + 28(u_{j+1} - u_{j-1}) + u_{j+2}].$$

The left and right hand side of this expression can be expanded by Taylor series to give

$$\text{L.H.S.} \equiv 5u'_j + h^2 u'''_j + \frac{h^4}{12}u_j^{(5)} + \frac{h^6}{360}u_j^{(7)}$$

and

$$\text{R.H.S.} \equiv 5u'_j + h^2 u'''_j + \frac{h^4}{12}u_j^{(5)} + \frac{13}{7}\left(\frac{h^6}{360}\right)u_j^{(7)}.$$

Thus, using the stencil (11.29) into Equation (11.85) is equivalent to solving the following equation

$$\frac{\partial u}{\partial t} + c\frac{\partial u}{\partial x} - \frac{ch^6}{420}\frac{\partial^7 u}{\partial x^7} = 0. \tag{11.124}$$

This leading truncation error term – the last term on the left hand side represents a leftward dispersion that can be seen in Fig. 11.29. Note that this dispersion term is present at all points

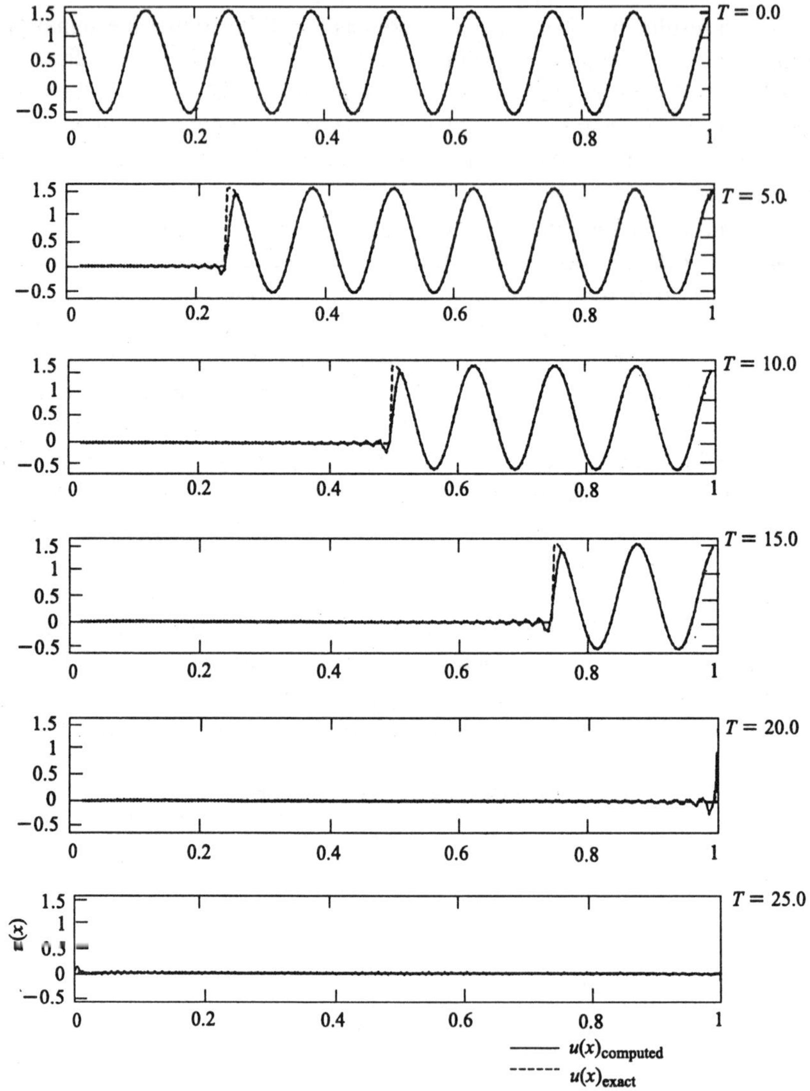

Fig. 11.29(a) *The exact and computed solution of one-dimensional wave equation shown for different time instants. The computations are performed using Adams' spatial scheme and Euler time stepping. The initial condition is depicted in the top frame and the information propagates to the right with the phase speed of* $c = 0.05$. *The computational domain* $0 \leq x \leq 1$ *is discretized using 201 equispaced points and the time step for integration is 0.0001.*

in the domain. However, at those points where the solution is smooth the seventh derivative is negligibly small and its effect is not seen there. But near the discontinuity this leading truncation term dominates and hence its effect is seen as Gibbs' phenomenon.

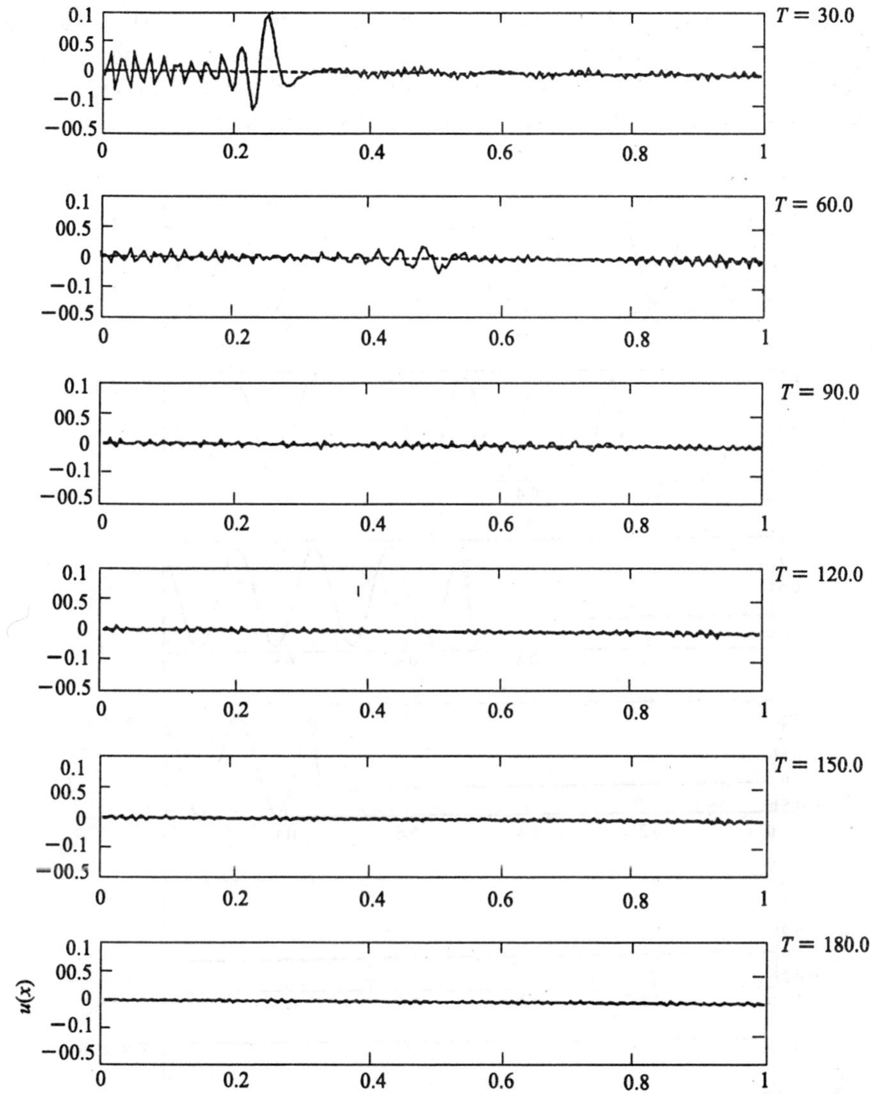

Fig. 11.29(b) *The exact and computed solution of one-dimensional wave equation shown for different time instants. The computations are performed using Adams' spatial scheme and Euler time stepping. The initial condition is depicted in the top frame and the information propagates to the right with the phase speed of* $c = 0.05$. *The computational domain* $0 \le x \le 1$ *is discretized using 201 equispaced points and the time step for integration is 0.0001.*

Apart from the Gibbs' phenomenon one can also note the effect of numerical instability brought about by the boundary closure problem shown in Fig. 11.1(b). It is clearly visible at $t = 25$ in the form of high wave number fluctuations including a large error near the left boundary. Ideally, at this time the solution should have been zero for the whole computed domain. The growing error is governed by Equation (11.85) and one can see the convection of unstable disturbances at $t = 30$. It was seen in Fig. 11.29(b) that as one moves in-board, the disturbances decay and this sequence of error amplification and subsequent decay are seen in other time frames too.

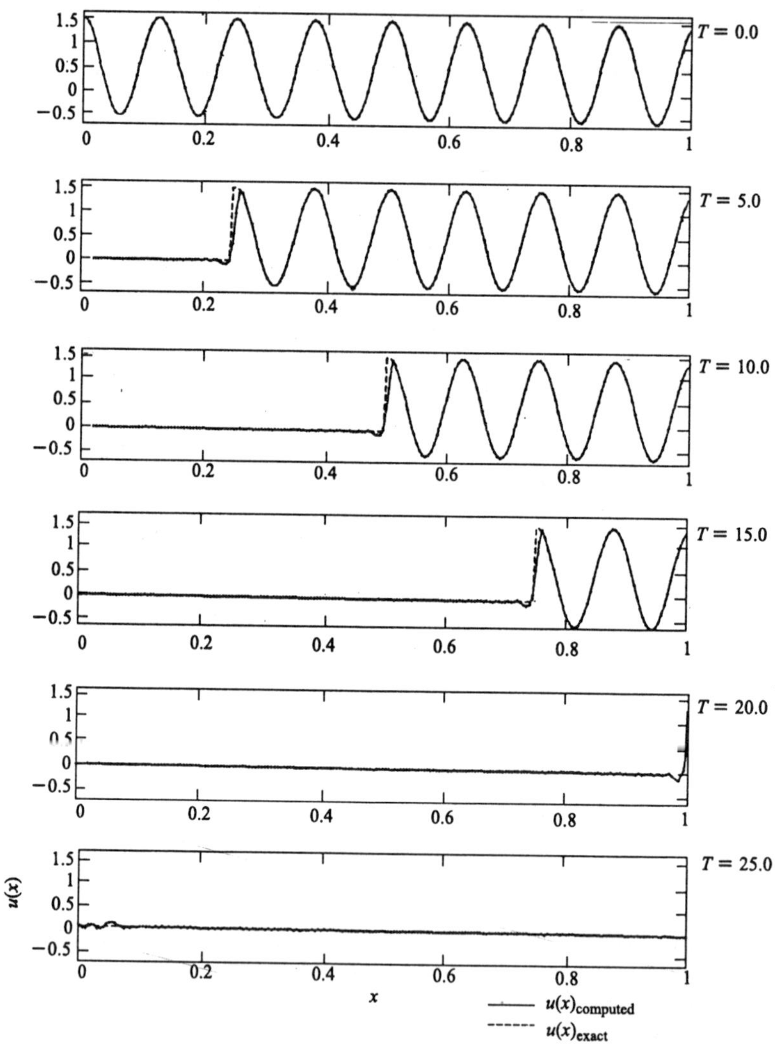

Fig. 11.30(a) *The exact and computed solution of one-dimensional wave equation shown for different time instants. The computations are performed using Haras-Ta'asan spatial scheme and Euler time stepping. Other conditions are as in Figure 11.29.*

In Figs. 11.30 the results are shown for the linear wave equation when one uses Haras-Ta'asan (1994)'s inner stencil with Adams (1993) boundary closure scheme. The results show identical qualitative features that were also seen in Figure 11.29.

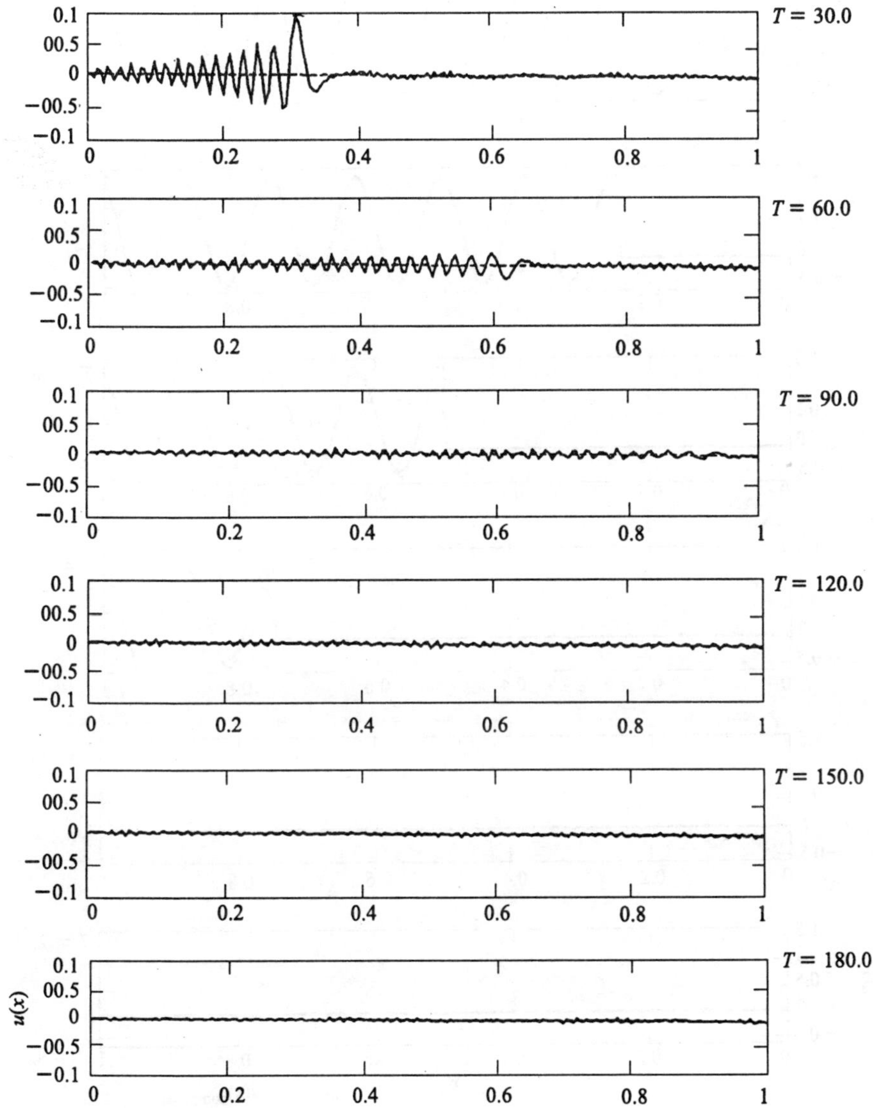

Fig. 11.30(b) *The exact and computed solution of one-dimensional wave equation shown for different time instants. The computations are performed using Haras-Ta'asan spatial scheme and Euler time stepping. Other conditions are as in Figure 11.29.*

In Figures 11.31 to 11.33 the results for the three OUCS schemes are shown. OUCS-1 has the same inner stencil as the one suggested by Zhong (1998), with the dissipation parameter

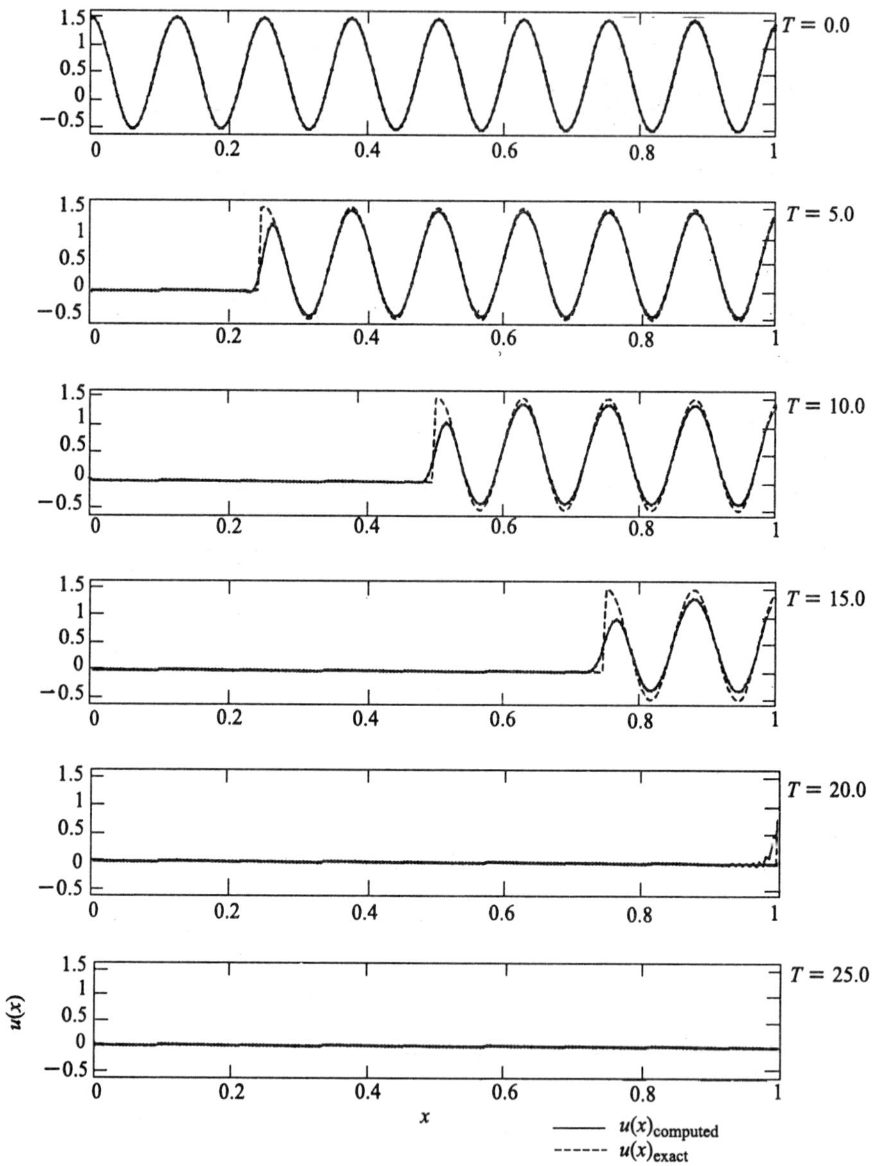

Fig. 11.31 *The exact and computed solution of one-dimensional wave equation shown for different time instants. The computations are performed using OUCS-1 spatial scheme and Euler time stepping. Other conditions are as in Fig. 11.29.*

$\alpha = -0.24$ in Equation (11.70). This stencil has large numerical dissipation and hence cannot capture the discontinuity at the left with sufficient accuracy. Even in smoother part of the flow the wave amplitude is under-predicted due to attenuation – as can be seen for result shown at $t = 10$ and 15. The same excessive dissipation also suppresses the Gibbs' phenomenon. The solution beyond $t = 25$ shows the desired zero solution in the domain.

In comparison with OUCS-1 scheme, the OUCS-2 schemes predicts better quality solution in terms of capturing the continuous and discontinuous part of the solution. The trivial solution beyond $t = 20$ in the computational domain is computed correctly.

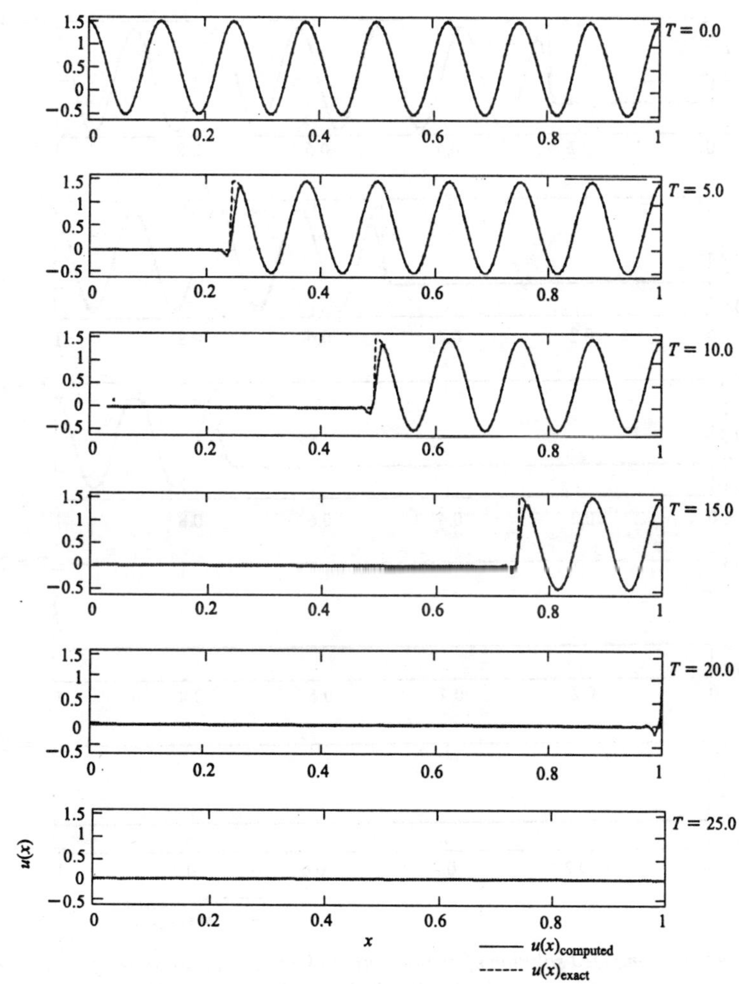

Fig. 11.32 *The exact and computed solution of one-dimensional wave equation shown for different time instants. The computations are performed using OUCS-2 spatial scheme and Euler time stepping. Other conditions are as in Fig. 11.29.*

The upwinded scheme based on Haras and Ta'asan (1994) stencil is the OUCS-3 scheme. The upwinding implied in Equation (11.76) by a value of $\eta = -2$ is excessive, as the regular and discontinuous part of the solutions are under-predicted due to excessive numerical dissipation.

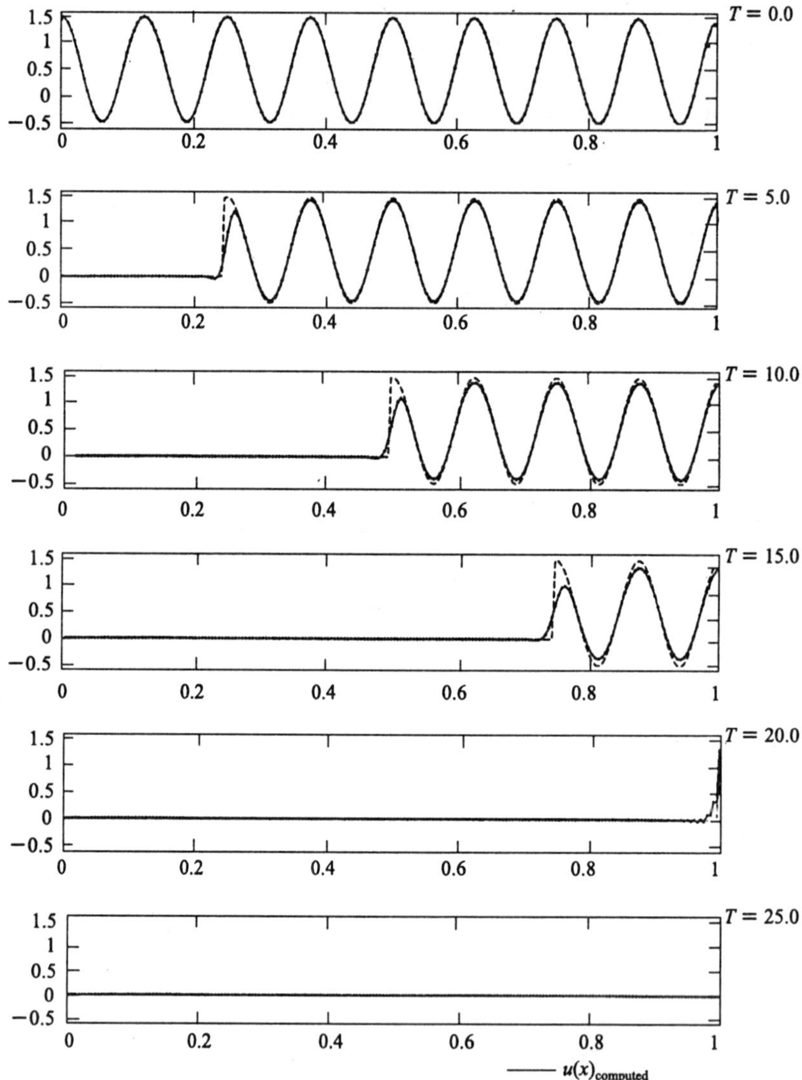

Fig. 11.33 *The exact and computed solution of one-dimensional wave equation shown for different time instants. The computations are performed using OUCS-3 spatial scheme and Euler time stepping. Other conditions are as in Fig. 11.29.*

Finally in Fig. 11.34, the results for the third order upwind scheme as used in Kawamura et al. (1985) and Sengupta & Sengupta (1994), is presented. In this case the discontinuity is smeared over a large region and the solution is over-damped.

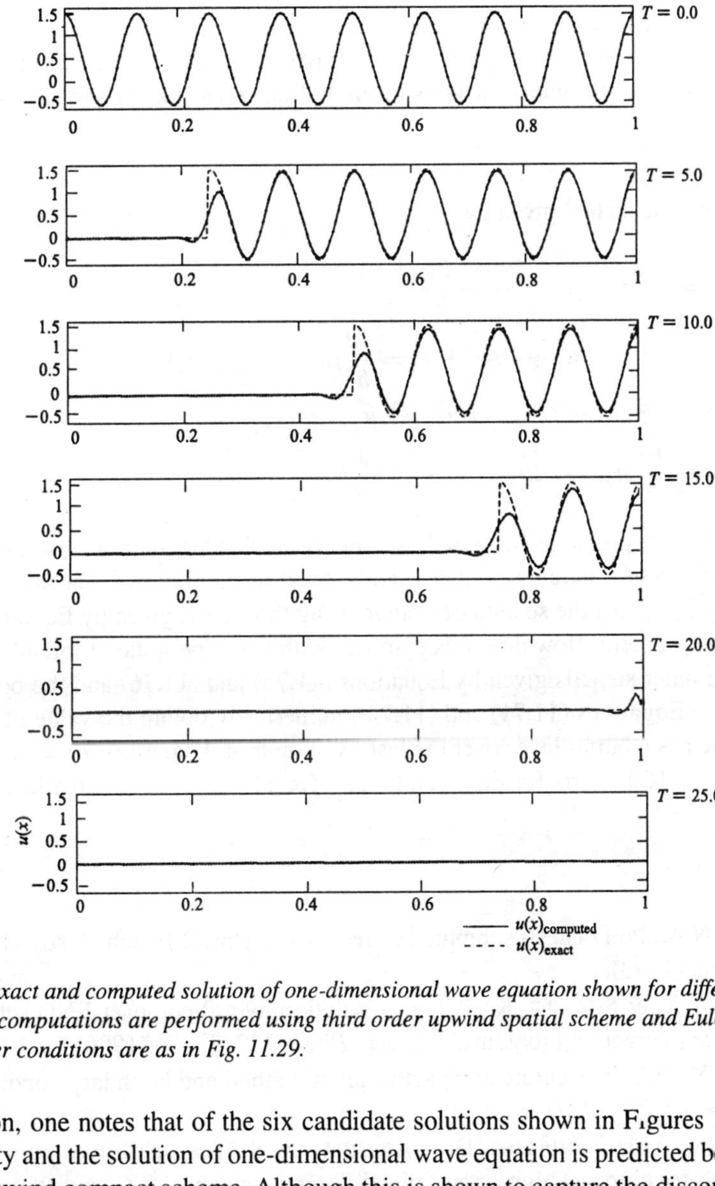

Fig. 11.34 *The exact and computed solution of one-dimensional wave equation shown for different time instants. The computations are performed using third order upwind spatial scheme and Euler time stepping. Other conditions are as in Fig. 11.29.*

In conclusion, one notes that of the six candidate solutions shown in Figures 11.29 to 11.34, the discontinuity and the solution of one-dimensional wave equation is predicted best by OUCS-2 scheme—an upwind compact scheme. Although this is shown to capture the discontinuous better, there is no reason to believe that one can solve compressible flow problems directly using this type of compact schemes. For this problem it is seen that for all these schemes the Gibbs' phenomenon is due to the leading truncation error term that indicates generation of spuriously dispersing waves from the discontinuity in one direction only. If one decides to work in a transformed plane instead, then there is a possibility that Gibbs' phenomenon will cause creation of spurious waves in both directions from the discontinuity. This cannot be simply solved by the use of compact schemes alone. However, such an issue is beyond the scope of the present discussions.

It is noted that some of the compact schemes performed better, solely due to the improvement brought about by better boundary closures given by Equations (11.71) and (11.74).

Exercise

1. Optimize the stencil given by

$$j = 1: \qquad u_1'' = \frac{u_1 - 2u_2 + u_3}{h^2},$$

$$j = 2: \qquad u_1'' + 10u_2'' + u_3'' = \frac{12}{h^2}(u_3 - 2u_2 + u_1),$$

$$3 \le j \le N-2: \qquad \alpha u_{j-1}'' + u_j'' + \alpha u_{j+1}''$$

$$= \frac{b}{4h^2}(u_{j-2} - 2u_j + u_{j+2}) + \frac{a}{h^2}(u_{j-1} - 2u_j + u_{j+1}),$$

and similar expressions for $j = N-1$ and N, so that the interior stencil is at least second or fourth order accurate (to relate a and b to α) for minimum L_2-norm.
2. Plot the k_{eq}/k for the second derivative using the stencil given by Equation (11.80) for a periodic problem. How does this compare with a non-periodic problem?
3. For the inner stencils given by Equations (11.75) and (11.76) and the boundary stencils given by Equations (11.71) and (11.74), numerically obtain the value of β at $N-1$, so that one has a buffer layer over the last five points at the outflow ($j = N$).
4. Obtain the $[C]$ matrix for Equation (11.37) for a Fourier spectral method.

References

[1] ADAMS, N.A., Ph.D Thesis submitted to Technische Univ. Munich, Also as DLR-FB 93–29, Göttingen, (1993).

[2] ADAMS, N.A. & SHARIFF, K., A high-resolution hybrid compact-ENO scheme for shock turbulence interaction problems, *J. Comp. Phys.*, **127**, 27–51 (1996).

[3] ADAMS, Y., Highly accurate compact implicit method and boundary conditions, *J. Comp. Phys.*, **24**, 10–22 (1977).

[4] CARPENTER, M.H., GOTTLIEB, D. & ABARBANEL, S., The stability of numerical boundary treatments for compact high-order finite difference schemes, *J. Comp. Phys.*, **108**, 272–295 (1993).

[5] CHRISTIE, I., Upwind compact finite difference schemes, *J. Comp. Phys.*, **59**, 353–368 (1985).

[6] DURRAN, D.R., *Numerical Methods for Wave Equations in Geophysical Fluids Dynamics*, Springer Verlag, New York (1999).

[7] FUNG, K.Y., MAN, R.S.O & DAVIS, S., Implicit high-order compact algorithm for computational acoustics, *AIAA J.*, **34**, 2029–2037 (1996).

[8] GAITONDE, D. & SHANG, J.S., Optimized compact-difference based finite volume schemes for linear wave phenomena, *J. Comp. Phys.*, **138**, 617–643 (1997).

[9] GUSTAFSSON, B., KREISS, H.A., & SUNDSTRÖM A., Stability theory for difference approximations of mixed initial boundary value problems II, *Math. Comp.*, **26** 649–686 (1972).

[10] HALTINER, G.J. & WILLIAMS, R.T., *Numerical Prediction and Dynamic Meteorology*, Second Edition, John Wiley & Sons, New York (1980).

[11] HARAS, Z. & TA'ASAN, S., Finite difference scheme for long time integration, *J. Comp. Phys.*, **114**, pp 265–279 (1994).

[12] HIRSCH, R.S., Higher order accurate difference solution of fluid mechanics problems by compact differencing technique, *J. Comp. Phys.*, **19**, 90–109 (1975).

[13] KAWAMURA, T., TAKAMI, H. & KUWAHARA, K., New higher-order upwind scheme for incompressible Navier-Stokes equations, *Fluid Dynamic Res.*, **1**, 145–162 (1985).

[14] KIM, J. & MOIN, P., Application of a fractional-step method to incompressible Navier-Stokes equation, *J. Comp. Phys.*, **58**, 308–322 (1985).

[15] KOPAL, Z., *Numerical Analysis*, Springer Verlag, New York (1966).

[16] LELE, S.K., Compact finite difference schemes with spectral like resolution, *J. Comp. Phys.*, **103**, 16–42 (1992).

[17] LIGHTHILL, M.J., *Waves in Fluids*, Cambridge Univ. Press, U.K. (1978).

[18] SENGUPTA, T.K., CHATTOPADHYAY, M., WANG, Z. Y. & YEO, K. S., By-pass mechanism of transition to turbulence, *J. Fluids & Struct.*, **16(1)**, 15–29 (2002).

[19] SENGUPTA, T.K. & SENGUPTA, R., Flow past an impulsively started circular cylinder at high Reynolds number, *Comput. Mech.*, **14(4)**, 298–310 (1994).

[20] SENGUPTA, T.K., GANERIWAL, G. & DE, S., Analysis of central and upwind compact schemem, *J. Comput. Phys.*, **192(2)**, 677–694 (2003).

[21] TAM, C.K.W. & WEBB, J.C., Dispersion-relation-preserving finite difference schemes for computational acoustics, *J. Comp. Phys.* **107**, 262–281 (1993).

[22] TOLSTYKH, A.I., On a class of non centered compact difference schemes of fifth order based on Padé approximates, *Soviet Math. Dokl.*, **44(1)**, 69– (1991).

[23] VICHNEVETSKY, R. & BOWLES, J.B., *Fourier Analysis of Numerical Approximation of Hyperbolic Equations*, SIAM Publ., Philadelphia, USA (1982).

[24] WILLIAMSON, J., Low storage Runge-Kutta schemes, *J. Comp. Phys.*, **35**, 48–56 (1980).

[25] ZINGG, D.W., LOMAX, H. & JURGENS, H., High accuarcy finite difference schemes for linear wave propagation, *SIAM. Sc. Comp.*, **17(2)**, 328–346 (1995).

[26] ZHONG, X., High-order finite difference schemes for numerical simulation of hypersonic boundary-layer transition, *J. Comp. Phys.*, **144**, 622–709 (1998).

12 Introduction to Finite Volume and Finite Element Methods

In Chapter 11, we discussed a class of high accuracy methods—the compact finite difference schemes—some of which can be used for DNS of incompressible flows. It was seen via the spectral analysis (as developed in Chapter 4 and 10) that the ability of a numerical method is best judged by the spatial and temporal resolution provided by it rather than the formal order of the truncation error terms. In this respect a few of the compact methods were preferred (because of the compact stencils used) over the explicit higher order methods. Within the class of compact schemes, it was seen that a second order method proposed by Haras & Ta'asan (1994) provided higher resolution than the sixth order methods of Adams (1993) and the fifth order upwind method of Zhong (1998). In Haras & Ta'asan (1994), the high resolution scheme was obtained by seeking an optimum that minimized the L_2-norm of the departure of the compact scheme from spectral method in solving one-dimensional wave equation subject to a specific class of initial conditions. The other compact methods were obtained by investigating the formal order of truncation error.

In none of these cases there is anything unique for the auxiliary relations used given by the Padé schemes for derivative(s) and function values at the nodes. While this can be viewed as providing flexibility in choosing a method, if sufficient care is not taken it may lead to inconsistency, even when care is exercised to achive numerical stability. Can one obtain auxiliary relations which are based on physical principles and consistency while using compact stencil?

Finite volume and finite element methods can be viewed in this respect as a logical development in providing compact discrete equations based on the physical statement of the problem in terms of the governing equation. Both these methods and other integral methods are said to satisfy the conservation principles in a weak sense.

In finite volume methods (FVM), the physical conservation processes are fulfilled in a macroscopic control volume. The physical "densities" of conserved properties in the control volume is related to the fluxes entering and leaving it through the control surfaces. These fluxes are created and governed by the physical convection and diffusion processes.

Thus, in finite volume methods, it is not even necessary to start from the governing differential equations which presupposes the continuity of functions and derivatives. For example, in the presence of a shock wave the flow has mass and momentum conserved across the shock, but physical variables like pressure and velocity need not be continuous across the shock.

In contrast, finite element methods (FEM) like finite difference methods, start with the governing partial differential equations. Over a chosen finite dimensional element these equations are integrated pre-supposing a particular mathematical and physical structure of the solution with an unknown scaling of the solution at the nodes of elements. Obviously, to keep the formulation tractable, the assumed spatial structure of the function (or the footprint of the function) is taken to be as small as possible. These unknown scaling coefficients are in general taken to be a function of time. Thus, in a sense, the finite element method can be viewed as solving PDEs using seperation of variables methods in terms of space and time dependence. However, the final discrete equations bring in the consistent space-time dependence of the problem, quite unlike the finite difference method.

In following the discretization processes in FVM and FEM, the main motivation is to work directly in the physical plane, thereby avoiding the problem of aliasing error discussed in Chapter 10, those arising due to grid mapping, as seen in Eq. (10.57) for a linear operator. These alternative methods have become major tools in analysing flow problems involving complex geometries. But can these methods be more than engineering tools? It is to be noted that compact schemes can be designed for FVM and there are methods such as spectral element methods which can provide high accuracy FEM solutions. In this chapter, FEM and FVM methods are introduced and compared with finite difference methods.

12.1 Finite Volume Method

This method is explained for the solution of Navier-Stokes equation in the Cartesian frame (as given by the Equation (2.36)) in two dimensions by

$$\frac{\partial U}{\partial t}+\frac{\partial F}{\partial x}+\frac{\partial G}{\partial y}=\left(\frac{\partial Q}{\partial x}+\frac{\partial R}{\partial y}\right), \tag{12.1}$$

where the state vector U, the convective flux vectors F and G, and viscous flux vectors Q and R are given by

$$U = \lfloor \rho \quad \rho u \quad \rho v \quad E \rfloor^T \tag{12.2a}$$

$$F = \lfloor \rho u \quad \rho u^2 + p \quad \rho uv \quad u(p+E) \rfloor^T \tag{12.2b}$$

$$G = \lfloor \rho v \quad \rho uv \quad \rho v^2 + p \quad v(p+E) \rfloor^T \tag{12.2c}$$

$$Q = \lfloor 0 \quad \tau_{xx} \quad \tau_{xy} \quad u\tau_{xx} + v\tau_{xy} - q_x \rfloor^T \tag{12.2d}$$

$$R = \lfloor 0 \quad \tau_{xy} \quad \tau_{yy} \quad u\tau_{xy} + v\tau_{yy} - q_y \rfloor^T . \tag{12.2e}$$

In the above equations, viscous stresses are given by

$$\tau_{xx} = (\lambda + 2\mu)\frac{\partial u}{\partial x} + \lambda\frac{\partial v}{\partial y}, \tag{12.3a}$$

$$\tau_{xy} = \mu\left(\frac{\partial u}{\partial y} + \frac{\partial v}{\partial x}\right), \tag{12.3b}$$

$$\tau_{yy} = (\lambda + 2\mu)\frac{\partial v}{\partial y} + \lambda\frac{\partial u}{\partial x}, \tag{12.3c}$$

and the heat conduction terms are given by,

$$q_x = -\kappa \frac{\partial T}{\partial x} \quad \text{and} \quad q_y = -\kappa \frac{\partial T}{\partial y}, \tag{12.4}$$

where μ and λ are the coefficient of viscosity (most of the time related by Stokes relation: $3\lambda + 2\mu = 0$) and κ is the coefficient of thermal conductivity.

If one is interested in simulating flows of gas that can be treated as perfect, then the static pressure p can be related to the total internal energy per unit volume E by

$$p = (\gamma - 1)\left[E - \frac{\rho}{2}(u^2 + v^2)\right]. \tag{12.5}$$

12.2 Finite Volume Discretization

Here Equation (12.1) is integrated over a finite control volume of area S, i.e.,

$$\int\int_S \frac{\partial U}{\partial t} dS + \int\int_S \left(\frac{\partial F}{\partial x} + \frac{\partial G}{\partial y}\right) dS = \int\int_S \left(\frac{\partial Q}{\partial x} + \frac{\partial R}{\partial y}\right) dS. \tag{12.6}$$

The core of the finite volume method is to replace the surface integrals of the flux vector terms in Equation (12.6) to line integrals by using Green's theorem which is given by

$$\int\int_S \left(\frac{\partial B}{\partial x} - \frac{\partial A}{\partial y}\right) dxdy = \oint_{\partial S} (Adx + Bdy), \tag{12.7}$$

where the area integral on left hand side is equivalent to the line integral of the right hand side. Here, for convective flux terms in Equation (12.6),

$$B \equiv F \text{ and } A \equiv -G.$$

And for the viscous flux terms

$$B \equiv Q \text{ and } A \equiv -R.$$

Application of Green's theorem given by Equation (12.7) in Equation (12.6) yields

$$\int\int_S \frac{\partial U}{\partial t} dS + \oint_{\partial S} (Fdy - Gdx) = \oint_{\partial S} (Qdy - Rdx). \tag{12.8}$$

The whole domain is split into small cells and the above expression of conservation laws are satisfied at each cell to produce discrete algebraic equations for the unknowns.

12.3 Geometric Constraints of FVM

For the Greens theorem given by Equation (12.7), if we define

$$\vec{V} = A\hat{i} + B\hat{j},$$

then the right hand side of Equation (12.7) can be written as $\oint \vec{V}.\hat{n}dl$, where $\hat{n}$ is the unit normal along the edges of the finite volume cell whose differential length element is indicated by dl.

If the $\vec{V}$ vector denies a smooth flow field then Green's theorem gives

$$\oint \hat{n}dl = 0 \tag{12.9}$$

because the left hand side of Equation (12.7) is identically zero for the constant vector $\vec{V}$.

Equation (12.9) therefore implies that each of the small cells must be closed. This is a basic requirement for the construction of finite volume grids and is not a particularly difficult one to satisfy. Secondly, one require that the total area spanned by all the elementary cells equals the area of the flow domain, i.e.

$$S = \sum_i (dx.dy)_i.$$

12.4 Finite Volume Method for Three-Dimensional Flows

For three-dimensional flow problems, let us rewrite the equation written in Section 9.7 as

$$\frac{\partial Q}{\partial t} + \frac{\partial E}{\partial x} + \frac{\partial F}{\partial y} + \frac{\partial G}{\partial z} = \frac{1}{Re}\left[\frac{\partial E_v}{\partial x} + \frac{\partial F_v}{\partial y} + \frac{\partial G_v}{\partial z}\right]. \tag{12.10}$$

Once again, one can define the following convective and viscous fluxes in short-hand notation by

$$\left.\begin{aligned} \vec{V} &= E\hat{i} + F\hat{j} + G\hat{k} \\ \vec{W} &= E_v\hat{i} + F_v\hat{j} + G_v\hat{k} \end{aligned}\right\}. \tag{12.11}$$

Then Equation (12.10) can be written in the short notation by

$$\frac{\partial Q}{\partial t} + \nabla.\vec{V} = \frac{1}{Re}\nabla.\vec{W}. \tag{12.12}$$

Now integrating this equation over the (i, j, k)th control volume Ω_{ijk}, we get

$$\frac{\partial}{\partial t}\int_{\Omega_{ijk}} Q d\Omega + \int_{\Omega_{ijk}} \nabla.\vec{V} d\Omega = \frac{1}{Re}\int_{\Omega_{ijk}} \nabla.\vec{W} d\Omega \tag{12.13}$$

The explicit integration for the (i, j, k)-th cell is not directly performed in Equation (12.13). Instead Gauss's divergence theorem is used where the volume integral of fluxes is transformed to a surface integral of efflux through the control surface by

$$\int_{\Omega_{ijk}} \nabla.\vec{V} d\Omega = \oint_{\delta\Omega_{ijk}} \vec{V}.d\vec{S}, \tag{12.14}$$

where the control volume Ω_{ijk} has control surfaces of total area $\delta\Omega_{ijk}$. A similar transformation can be performed on viscous flux terms. Finally one solves the equation

$$\frac{\partial}{\partial t}\int_{\Omega_{ijk}} Q d\Omega + \int_{\delta\Omega_{ijk}} \vec{V}.d\vec{S} = \frac{1}{Re}\int_{\delta\Omega_{ijk}} \vec{W}.d\vec{S}. \tag{12.15}$$

If one defines by $\bar{Q}^n_{ijk}$ as the cell-averaged value of the conservative variable at time t^n, i.e.,

$$\bar{Q}^n_{ijk} = \frac{\int_{\Omega_{ijk}} Q(x_i, y_j, z_k, t^n) dx dy dz}{\int_{\Omega_{ijk}} dx dy dz}, \tag{12.16}$$

then this brings in the additional problem of fixing the position of variables within the (i, j, k)-th cell.

If (n_x, n_y, n_z) are the outward normal unit vector of $\delta\Omega_{ijk}$ and dS its surface element, then Equation (12.15) can be written as

$$(\bar{Q}^{n+1}_{ijk} - \bar{Q}^n_{ijk})\Omega_{ijk} + \int_{t^n}^{t^{n+1}}\int_{\delta\Omega_{ijk}} \{En_x + Fn_y + Gn_z\} dS dt$$

$$= \frac{1}{Re}\int_{t^n}^{t^{n+1}}\int_{\delta\Omega_{ijk}} \{E_v n_x + F_v n_y + G_v n_z\} dS dt. \tag{12.17}$$

In order to evaluate surface fluxes it is convenient to think of $\bar{Q}_{ijk}$ as the value of Q at some representative point within the cell. This is easily understood if we want to evaluate for two-dimensional problem the following quantity:

$$A = \int_{\delta\Omega_{ij}} \frac{\partial Q}{\partial x}(x, y) dS. \tag{12.18}$$

In the above equation, one can expand $\frac{\partial Q}{\partial x}$ about a nominal point (x_0, y_0) by a Taylor series expansion. Upon doing this, it is easy to express

$$A = \int \frac{\partial Q}{\partial x}(x_0, y_0) dS + \int (x - x_0)\frac{\partial^2 Q}{\partial x^2}(x_0, y_0) dS + \int (y - y_0)\frac{\partial^2 Q}{\partial x \partial y}(x_0, y_0) dS$$

$$+ \int \frac{\partial^3 Q}{\partial x^2 \partial y}(x - x_0)(y - y_0) dS + \frac{1}{2}\int (x - x_0)^2 \frac{\partial^3 Q}{\partial x^3}(x_0, y_0) dS$$

$$+ \frac{1}{2}\int \frac{\partial^3 Q}{\partial x \partial^2 y}(y - y_0)^2 dS + \frac{1}{6}\int \frac{\partial^4 Q}{\partial x \partial^3 y}(y - y_0)^3 dS + \ldots \tag{12.19}$$

If the nominal point (x_0, y_0) is chosen as the centroid of the cell, then by definition of the centroid

$$\int\int (x - x_0)^m (y - y_0)^n dS \equiv 0$$

for all positive integral values of m and n. Thus, from Equation (12.19),

$$A = \frac{\partial Q}{\partial x}(x_0, y_0) \int dS.$$

Therefore,

$$\frac{\partial Q}{\partial x}(x_0, y_0) = \frac{1}{\delta\Omega_{ij}} \int_{\delta\Omega_{ij}} \frac{\partial Q}{\partial x} dS, \tag{12.20}$$

where $\int dS = \delta\Omega_{ij}$.

For simple geometrical shapes like parallelogram, square, rectangle, etc, the geometric center and centroids coincide. Vinokur (1989) suggested taking the nominal point (x_0, y_0) as the cell centre and using Equation (12.20), where the cell center is defined as the vectorial average of the cell vertices. Kallinderis & Baron (1990) states that "the error that results when it is assumed the two coincide, is of higher order in most cases".

12.5 Evaluating Viscous Terms

For viscous terms second derivatives are involved, and these cannot be evaluated the way the convective flux terms were handled in two- and three-dimensions by using Green's and Gauss divergence theorems respectively. These terms are instead, handled in two steps. Firstly the viscous stresses are computed. Using these stresses, in the second step the viscous terms are developed.

Thus the discretization of viscous terms would require taking auxiliary cells as compared to evaluating the convective flux terms on primary cells. These auxiliary cells are the staggered version of the primary cells themselves. Thus, the auxiliary cells help keeping the stencils small and compact to minimize boundary closure problems. Also, such compact computational schemes retain higher accuracy. This is illustrated below with the help of the formulation for a two-dimensional problem.

Consider the primary cell ABCD surrounding the point P, used for evaluating the inviscid fluxes through the cell faces in Fig. 12.1. If we identify this cell as the (ij)-th cell with the point P as its centroid, then to evaluate second derivatives at P, we follow the procedures outlined below.

Referring the Fig. 12.1, let us evaluate u_{xx} at P. Let S_{ij} be the area of the cell ABCD.

Any quantity at the centroid is considered as a cell-averaged quantity defined in Equation (12.16). Hence,

$$(u_{xx})_P = \frac{1}{S_{ij}} \int\int u_{xx} dx dy. \tag{12.21}$$

Applications of Green's theorem to the right hand side yields

$$(u_{xx})_P = \frac{1}{S_{ij}} \oint_{\text{ABCD}} u_x dy. \tag{12.22}$$

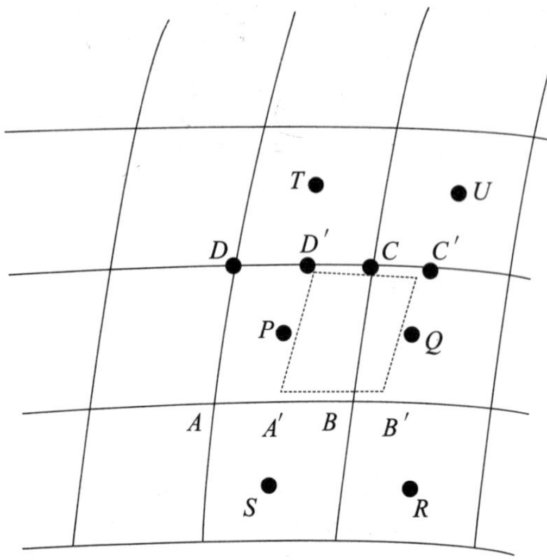

Fig. 12.1 *Various cells used in finite volume formulations. ABCD is the primary cell used to calculate inviscid fluxes. Viscous flux evaluation requires other points and a cell in the figure.*

The line-integral on the right hand side can be evaluated by considering mean values of u_x along the edges of the cell to give

$$(u_{xx})_P = \frac{1}{S_{ij}}\{(u_x)_{\text{BA}}(y_B - y_A) + (u_x)_{\text{BC}}(y_C - y_B) + (u_x)_{\text{CD}}(y_D - y_C) + (u_x)_{\text{AD}}(y_A - y_D)\}. \tag{12.23}$$

Starting from point A, the integral has been performed in anti-clockwise direction to perform a consistent line integral. $(u_x)_{\text{AB}}, (u_x)_{\text{BC}} \dots$ are interpreted as the value of the quantity evaluated using mean value theorem or simply the value at some representative value along that edge. Once again, treating this quantity as an area-average, one can construct another auxiliary cell around the chosen representative point. For example, $(u_x)_{\text{CB}}$ can be obtained from

$$(u_x)_{\text{CB}} = \frac{1}{S_{i+1/2j}}\int\int_{S_{i+1/2j}} u_x dxdy, \tag{12.24}$$

where $S_{i+1/2j}$ is the secondary cell A′ B′ C′ D′ shown in Fig. 12.1. Thus,

$$(u_x)_{\text{CB}} = \frac{1}{S_{i+1/2j}}\oint_{\text{A'B'C'D'}} udy = \frac{1}{S_{i+1/2j}}[u_B(y_{B'} - y_{A'}) + u_Q(y_{C'} - y_{B'}) + u_C(y_{D'} - y_{C'}) + u_P(y_{A'} - y_{D'})]. \tag{12.25}$$

As the variables are stored at the cell-centers, the value of u at B and C can be evaluated in terms of nearest cell-center values. For example

$$u_B = \frac{1}{4}[u_P + u_Q + u_R + u_S],$$

$$u_C = \frac{1}{4}[u_P + u_Q + u_T + u_U].$$

Of course, these are not unique and there are many other alternative ways of evaluating the viscous terms. These methods seem all right while working in Cartesian meshes. Non-Cartesian and non-uniform meshes will bring about significant non-trival problems that are not discussed here. There is another way that involves mapping the physical plane to a transformed plane and then applying the FV discretization in the transformed plane. However, as discussed in Chapter 10, this can bring in significant aliasing error. If this problem is circumvented, then one can develop higher order accurate FV methods on uniform meshes. This is discussed next.

12.6 High Resolution FV Upwind Schemes

In this section, the use of higher order projection (in space) for FV methods are discussed for incompressible flows. No efforts have been made to discuss FV methods that are currently used for compressible flows, where the projection stage is considered as the first stage of computations. The projection stage is usually followed by introducing time discretization schemes that help in preventing unwanted oscillations while satisfying positive change of entropy condition. This is referred to as satisfying the Total Variation Diminishing (TVD) or Local Extremum Diminishing (LED) condition. It is also often stated (see Hirsch (1990) for a detailed exposition) that satisfaction of the physical requirements of entropy increases in a flow process and TVD conditions cannot be satisfied when higher order projections are used. Following the work of Van Leer (1973) and Boris & Book (1976) it is claimed that one needs to resort to nonlinear discretization as the third step of the solution process. This nonlinear discretization appears in the form of *limiters* that in essence prevent Gibbs' phenomenon associated with discontinuous solution. These *limiters* control the gradient of the solution to prevent under- and over-shoots associated with Gibbs' phenomenon. A high resolution slope limiter scheme that works directly on limiting the slope of computed solution is proposed in Causon et al. (1998). This scheme is claimed to perform all the three steps simultaneously. This is an interesting observation because the results of linear convection equations shown in Chapter 11 also demonstrated that the compact schemes are capable of capturing discontinuity with first order time discretization without requiring any TVD time discretization or nonlinear space discretization. In a sense, the use of higher order numerical dissipation to control Gibbs' phenomenon is similar to the selective addition of different order dissipation in Jameson et al. (1981). However, the presence of nonlinear terms can bring in additional sources of dispersion which can create spurious waves traveling in all directions from the point of discontinuity. This cannot be controlled easily by higher order dissipation alone, and will not be discussed here any further.

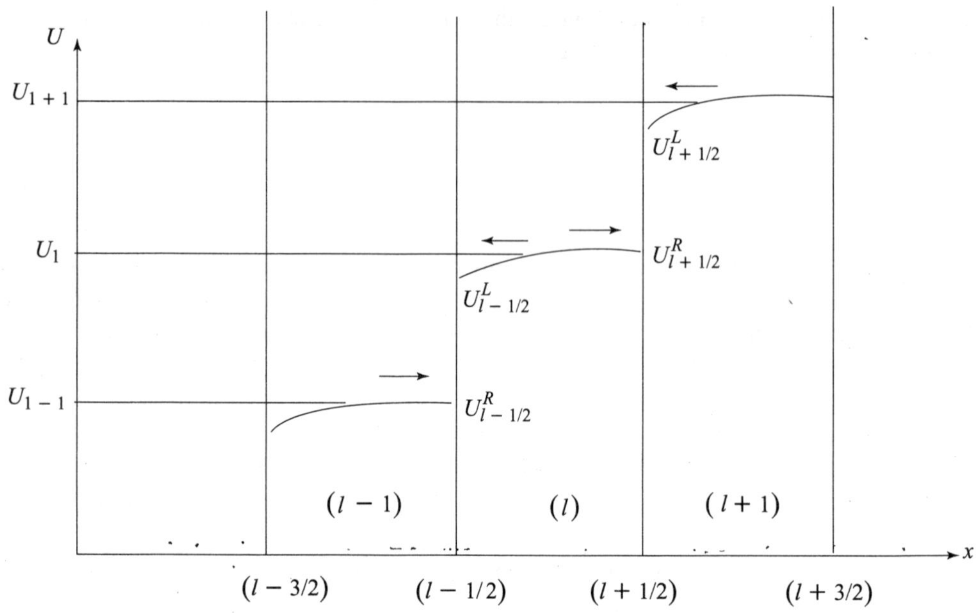

Fig. 12.2 *Piecewise continuous representation of solution used for higher order reconstruction scheme.*

With these introductory remarks, we discuss the higher order projections used in FV methods for incompressible flows alone, which uses flux vector splitting as the basic building block to represent the physical processes. We follow the work of Van Leer (1979) introduced in developing what is known as the MUSCL (Monotone Upstream-Centered Schemes of Conservations Laws) scheme. In this approach the solution is considered piecewise continuous, as shown in Fig. 12.2.

Consider the following interpolation valid for $x_{l-1/2} \leq x \leq x_{l+1/2}$:

$$U(x) = U_l + \frac{\Delta_l U}{h}(x - x_l) + \frac{3\Delta_l^2 U}{2h^2}\kappa\left[(x - x_l)^2 - \frac{h^2}{12})\right], \tag{12.28}$$

with $h = x_{l+l/2} - x_{l-l/2}$

where U_l is the cell-averaged value defined by

$$U_l = \frac{1}{h}\int_{x_{l-1/2}}^{x_{l+1/2}} U(x)dx. \tag{12.29}$$

Let us first show the equivalence of equations (12.28) and (12.29). Note that the interpolation equation (12.28) is not the Taylor series representation that is usually applied even if we choose $\kappa = 1/3$, and when $\frac{\Delta_l U}{h}$ and $\frac{\Delta_l^2 U}{h^2}$ represent the first and second derivatives, respectively.

Plugging Equation (12.28) into Equation (12.29), one gets

$$hU_l = \int_{x_{l-1/2}}^{x_{l+1/2}} \left\{ U_l + \frac{\Delta_l U}{h}(x - x_l) + \frac{3\kappa \Delta_l^2 U}{2h^2}\left[(x - x_l)^2 - \frac{h^2}{12}\right]\right\} dx. \tag{12.30}$$

Since $\int_{x_{l-1/2}}^{x_{l+1/2}} (x - x_l)dx \equiv 0$
and

$$\int_{x_{l-1/2}}^{x_{l+1/2}} (x - x_l)^2 dx = \frac{h^3}{3} + h(x_{l+1/2} - x_l)(x_{l-1/2} - x_l),$$

the right hand side of Equation (12.30) simplifies to

$$\begin{aligned} &= hU_l + \frac{\kappa \Delta_l^2 U}{2}h + \frac{3\kappa \Delta_l^2 U}{2h}(x_{l+1/2} - x_l)(x_{l-1/2} - x_l) - \frac{\kappa \Delta_l^2 U}{8}h \\ &= hU_l + \frac{3\kappa \Delta_l^2 U}{8}h + \frac{3\kappa \Delta_l^2 U}{2h}(x_{l+1/2} - x_l)(x_{l-1/2} - x_l) \end{aligned}$$

as $(x_{l+1/2} - x_l) = (x_l - x_{l-1/2}) = \frac{h}{2}$.

So the right hand side of Equation (12.30) simplifies to hU_l. Thus the interpolation formula given by Equation (12.28) is a consequence of the definition given by Equation (12.29). Also in proposing Equation (12.28) no specific expression for $\Delta_l U$ and $\Delta_l^2 U$ have been used.

Since $x_l = \frac{x_{l+1/2} + x_{l-1/2}}{2}$; using this in Equation (12.28) one gets

$$U(x_l) = U_l - \frac{\kappa}{8}\Delta_l^2 U. \tag{12.31}$$

Also from Equation (12.28):

$$U_{l+1/2}^R = U(x_{l+1/2}) = U_l + \frac{1}{2}\Delta_l U + \frac{\kappa}{4}\Delta_l^2 U \tag{12.32}$$

and

$$U_{l-1/2}^L = U(x_{l-1/2}) = U_l - \frac{1}{2}\Delta_l U + \frac{\kappa}{4}\Delta_l^2 U. \tag{12.33}$$

In proposing a class of reconstruction formulae for the state vectors at cell-boundaries—as depicted in Fig. 12.2, Hirsch (1990) suggests the following centered differences in Equations (12.32) and (12.33).

$$\Delta_l U = \frac{1}{2}(U_{l+1} - U_{l-1}). \tag{12.34}$$

$$\Delta_l^2 U = (U_{l+1} - 2U_l + U_{l-1}). \tag{12.35}$$

These are nothing but the central differences calculated based on the cell averaged values of state vector. When these are used in Equations (12.32) and (12.33) one gets,

$$U^R_{l+1/2} = U_l + \frac{1-\kappa}{4}(U_l - U_{l-1}) + \frac{1+\kappa}{4}(U_{l+1} - U_l), \tag{12.36}$$

and

$$U^L_{l-1/2} = U_l - \frac{1+\kappa}{4}(U_l - U_{l-1}) - \frac{1-\kappa}{4}(U_{l+1} - U_l). \tag{12.37}$$

In Equations (12.36) and (12.37), the first terms on the right hand side are nothing but the values obtained from first order upwinding scheme. A first order representation like this is akin to solving a Riemann problem at each cell interface, a procedure often used in solving high Mach number flow equations. The second and third set of terms in the equations above, impart higher order accuracy to the reconstruction scheme. Thus, Hirsch (1990) recommends rewriting the above equations as

$$U^R_{l+1/2} = U_l + \frac{\epsilon}{4}[(1-\kappa)(U_l - U_{l-1}) + (1+\kappa)(U_{l+1} - U_l)], \tag{12.38}$$

$$U^L_{l-1/2} = U_l - \frac{\epsilon}{4}[(1+\kappa)(U_l - U_{l-1}) + (1-\kappa)(U_{l+1} - U_l)]. \tag{12.39}$$

The reconstruction schemes given by Equations (12.38) and (12.39) will be referred to as $(\epsilon - \kappa)$-scheme in future usage. For first order upwinding one simply sets $\epsilon \equiv 0$. For other schemes, one sets $\epsilon = 1$.

12.7 Properties of the Reconstruction Scheme

From Fig. 12.2, one can also estimate $U^L_{l+1/2}$ and $U^R_{l-1/2}$ by obtaining equivalent interpolation formulae (as given by Equation (12.28) for the lth cell) for $(l-1)$th and $(l+1)$th cells. These can be written down by inspection of Equations (12.38) and (12.39) as,

$$U^R_{l-1/2} = U_{l-1} + \frac{\epsilon}{4}[(1-\kappa)(U_{l-1} - U_{l-2}) + (1+\kappa)(U_l - U_{l-1})] \tag{12.40}$$

and

$$U^L_{l+1/2} = U_{l+1} - \frac{\epsilon}{4}[(1-\kappa)(U_{l+1} - U_l) + (1+\kappa)(U_{l+2} - U_{l+1})]. \tag{12.41}$$

This reconstruction scheme has the following properties for $\epsilon = 1$:

(a) If $\kappa = -1$, then we have a one-sided linear (OSL) reconstruction scheme.

(b) If $\kappa = 0$, then we get a linear interpolation (LI) scheme.

(c) If we assign $\kappa = 1$, then the reconstruction scheme corresponds to second order central difference scheme (CD_2).

(d) If we choose $\kappa = 1/2$ then we recover the QUICK scheme of Leonard (1979).

(e) If we take $\kappa = 1/3$, then we have the monotone upstream-centered scheme for conservation laws (MUSCL) given by Van Leer (1979).

(f) Equation (12.38) and (12.41) represents the same function at $(l + 1/2)$th node admitting a discontinuity at the cell-face. The discontinuity value is given by

$$U^R_{l+1/2} - U^L_{l+1/2} = -\frac{1-\kappa}{4}(U_{l+2} - 3U_{l+1} + 3U_l - U_{l-1}). \tag{12.42}$$

The quantity in the parenthesis on the right hand side of Equation (12.42) is a difference expression for $\left(\frac{\partial^3 U}{\partial x^3}\right)_{l+1/2}(\Delta x)^3$. Thus, these schemes are called third order schemes except for $\kappa = 1$, where the leading term is a fourth derivative term. Thus, except for $\kappa = 1$, the reconstruction scheme gives rise to a source of dispersion at each cell face. This spurious dispersion is not seen in the smooth part of the flow. However, at discontinuities like a shock or contact discontinuity, this term is the source of Gibbs' phenomenon. One must also note that the actual discretization error for a differential equation will be different and the above observation has to be related to that particular equation. This will be attempted later. These reconstruction schemes, discussed above, can be directly used for incompressible flows. However, before using these schemes one would like to analyse such schemes for accuracy, stability and DRP property.

12.8 Spectral Resolution of Flux Vector Splitting (FVS) Scheme

If we define the cell-averaged state vectors in terms of Fourier spectral representation given by

$$U_l = \int \hat{U}(k) e^{ikx_l} dk, \tag{12.43}$$

then it is possible to spectrally analyze the FV schemes for the generic PDE given by

$$\frac{\partial U}{\partial t} + \frac{\partial f}{\partial x} = 0 \tag{12.44}$$

where it is customary to relate the flux term in terms of Jacobian (A) by

$$\frac{\partial f}{\partial x} = A\frac{\partial U}{\partial x}. \tag{12.45}$$

To solve the 1D conservation law given by Equation (12.44), one evaluates the flux term AU at the cell faces in terms of the information propagation in either direction dictated by the eigenvalue and eigenvectors of A matrix. This procedure is known as flux vector splitting and this is performed by

$$AU = A^+U^+ + A^-U^-, \tag{12.46}$$

where U^+ and U^- are the components of the state vector propagating in downstream and upstream direction from any cell interface. In the same manner the flux Jacobian matrix A is split into $A^{\pm}$ depending on the direction of information propagation at the cell boundaries. $U^{\pm}$ are obtained from Equations (12.38)-(12.41). For example, for all the cell interfaces located at $(l - 1/2)$, the split state vectors are given by

$$U^+_{l-1/2} = U_{l-1} + \frac{\varepsilon}{4}[(1-\kappa)(U_{l-1} - U_{l-2}) + (1+\kappa)(U_l - U_{l-1})], \tag{12.47}$$

$$U^-_{l-1/2} = U_l - \frac{\varepsilon}{4}[(1+\kappa)(U_{l+1} - U_l) + (1-\kappa)(U_{l+2} - U_{l+1})]. \tag{12.48}$$

Thus Equation (12.47) represents $U^g_{l-1/2}$ and Equation (12.48) represents $U^l_{l-1/2}$. The split Jacobian matrices $A^\pm$, can be defined in terms of left and right eigenvectors (R) and eigenvalues $(\Lambda^\pm_A)$ from

$$A^\pm = R\Lambda^\pm_A R^{-1}. \tag{12.49}$$

Thus, the performance of FVS scheme can be represented by the following ratio which we term as the effectiveness ratio. It defines the actual effectiveness of the numerical flux calculated by a particular choice of the reconstruction scheme with respect to the actual flux calculated by a global spectral representation.

$$E_f = \frac{\text{Equivalent FV flux}}{\text{Actual flux}} = \frac{(A^+U^+ + A^-U^-)_i}{(AU)_i}. \tag{12.50}$$

Using Equation (12.49) in Equation (12.50) and noting that $A = R(\Lambda^+_A + \Lambda^-_A)R^{-1}$,

$$E_f = \frac{\Lambda^+_A}{\Lambda^+_A + \Lambda^-_A}\left(\frac{U^+}{U}\right) + \frac{\Lambda^-_A}{\Lambda^+_A + \Lambda^-_A}\left(\frac{U^-}{U}\right).$$

Now if we evaluate this quantity at $(l - 1/2)$th node and use the expressions for U^+ and U^-, then

$$E_f = \frac{\Lambda^+_A}{\Lambda^+_A + \Lambda^-_A}\left\{\frac{U^+_{l-1/2}}{U_{l-1/2}}\right\} + \frac{\Lambda^-_A}{\Lambda^+_A + \Lambda^-_A}\left\{\frac{U^-_{l-1/2}}{U_{l-1/2}}\right\}. \tag{12.51}$$

Using expressions for $U^+_{l-1/2}$ and $U^-_{l-1/2}$ from Equations (12.47) and (12.48) and the Fourier-Laplace representation

$$U_{l-1/2} = \int \hat{U}(k)e^{ik\Delta x(l-1/2)}dk$$

obtained from Equation (12.43), and substituting it in Equation (12.51), one gets,

$$E_f = \frac{\Lambda^+_A}{\Lambda^+_A + \Lambda^-_A}\left[e^{\frac{-ik\Delta x}{2}} + \frac{\varepsilon(1-\kappa)}{4}\left\{e^{\frac{-ik\Delta x}{2}} - e^{\frac{-3ik\Delta x}{2}}\right\}\right.$$
$$\left.+\frac{\varepsilon(1+\kappa)}{4}\left\{e^{\frac{ik\Delta x}{2}} - e^{\frac{-ik\Delta x}{2}}\right\}\right] + \frac{\Lambda^-_A}{\Lambda^+_A + \Lambda^-_A}$$
$$\left[e^{\frac{ik\Delta x}{2}} - \frac{\varepsilon(1+\kappa)}{4}\left\{e^{\frac{ik\Delta x}{2}} - e^{\frac{ik\Delta x}{2}}\right\} - \frac{\varepsilon(1-\kappa)}{4}\left\{e^{\frac{3ik\Delta x}{2}} - e^{\frac{-ik\Delta x}{2}}\right\}\right]. \tag{12.52}$$

Ideally, the left hand side should be equal to one. Used reconstruction scheme shows that the right hand side is a combination of a real and an imaginary quantity. The departure of the real part from unity represents the phase error. The imaginary part contributes to the added dissipation necessary for numerical stability.

In Equation (12.52) the factor $\frac{\Lambda_A^+}{\Lambda_A^+ + \Lambda_A^-}$ is always positive if Λ is positive and greater than one. If this factor is denoted by α, then,

$$\frac{\Lambda_A^-}{\Lambda_A^+ + \Lambda_A^-} = (1 - \alpha).$$

Irrespective of the value of α, the real and imaginary part of $E_j(= k_{\text{eq}}/k)$ will always be same. In plotting the Fig. 12.3, this observation is made use of. In these figures, these values are shown for all possible resolved length scales defined up to Nyquist limit.

From the plot of the real part shown in Fig. 12.3(a), it is evident that the one-sided linear (OSL) reconstruction scheme does not match with the exact representation for any length scale and hence should not be used at all. Interestingly, the MUSCL scheme also displays a small overshoot over one at intermediate wave numbers. However, as this occurs at a relatively small scale and such distortions are small, it will not cause major phase error problem. The QUICK scheme does not have this attribute. Also the QUICK scheme adds lesser amount of numerical dissipation as compared to MUSCL scheme, although both the schemes are third order accurate, as noted from Equation (12.42). The observation that these schemes add dispersion to fluxes is correct. However, when we view it from the governing differential equation, we actually add a fourth derivative term that provides equivalent negative feedback in providing numerical stability. These aspects can only be investigated for an equation that has to be solved. However, we can observe the property of the schemes by studying a particular equation to calibarate the methods. This is attempted next with respect to the 1D wave equation, given by

$$\frac{\partial u}{\partial t} + c\frac{\partial u}{\partial x} = 0 \text{ for } c > 0. \tag{12.53}$$

In particular, the numerical stability, DRP property will be studied for the spatial discretization using the flux vector splitting approach for the above equation. The information can be compared with the similar results for the explicit finite difference and compact difference schemes obtained in the previous chapters. These results were reported in Sengupta et al. (2001) for finite volume methods.

Since disturbance only propagates in the positive x-direction, the discrete finite volume versions of Equations (12.53) is given for the lth cell by

$$\frac{U_l^{n+1} - U_l^n}{\Delta t}\Delta x + c(U_{l+1/2}^+ - U_{l-1/2}^+) = 0. \tag{12.54}$$

The choice of Euler explicit time splitting is deliberate, although it may appear arbitrary at this point. We will also investigate other time integration strategies later on and show the effectiveness

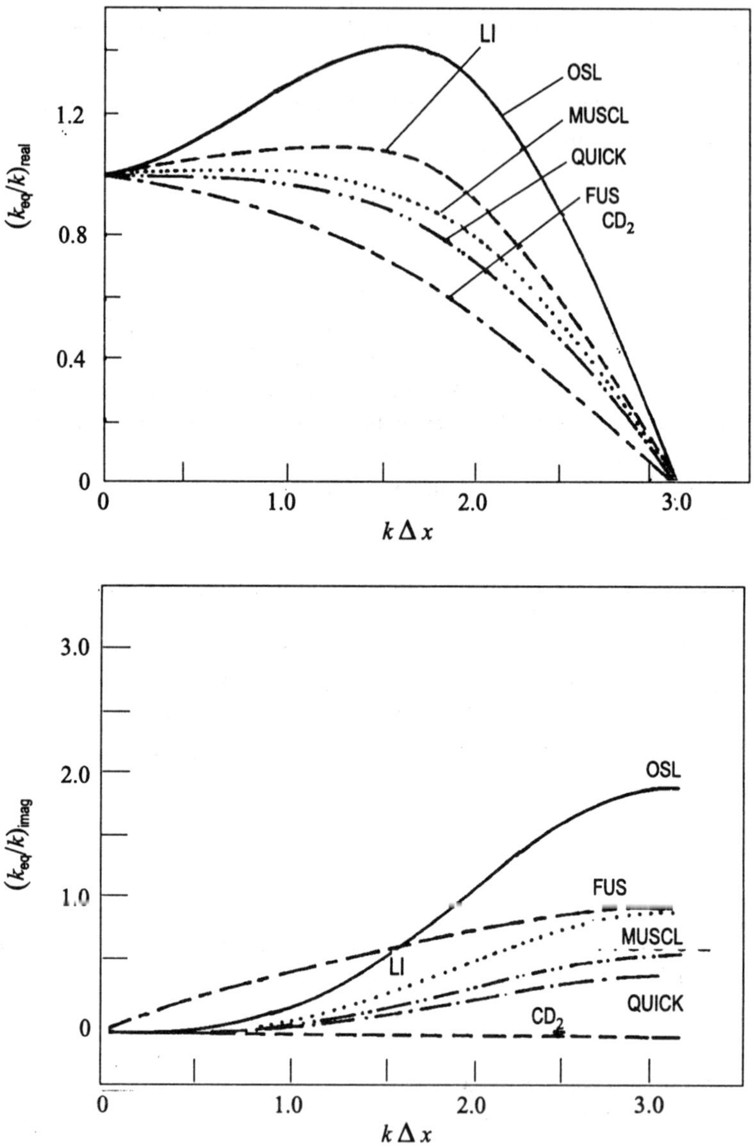

Fig. 12.3 *Spectral resolution of higher order reconstruction schemes. (a) Spectral resolution, (b) added numerical dissipation.*

of Euler time integration strategy over higher order methods in terms of storage requirments and computational speed.

The second set of terms in Equation (12.54) represents the balance of outgoing and incoming fluxes through cell interfaces. Once again, the vectors with superscript ($^+$) in Equations

(12.54) relates to values obtained from Equations (12.41) and (12.37) respectively, in terms of the cell-averaged values. The first set of terms involve the cell-averaged values themselves. Thus, substituting these expressions in Equations (12.54) and using the Fourier spectral representation of the function

$$U(x,t) = \int\int \hat{U}(k,\omega)e^{i(kx-\omega t)}dkd\omega \tag{12.55}$$

yields the following relation,

$$\frac{e^{-i\omega\Delta t}-1}{\Delta t}\Delta x + c\left\{(e^{-ik\Delta x}-1) + \frac{\varepsilon(1-\kappa)}{4}[2e^{-ik\Delta x} - 1 - e^{-2ik\Delta x}]\right.$$
$$\left. + \frac{\varepsilon(1+\kappa)}{4}[2 - e^{-ik\Delta x} - e^{ik\Delta x}]\right\} = 0. \tag{12.56}$$

This equation represents the numerical dispersion relation for the flux vector splitting finite volume scheme. Once again, if one defines an amplification factor by

$$\left.\begin{aligned} G(k) &= \tfrac{\hat{U}(k,t+\Delta t)}{\hat{U}(k,t)} \\ &= G_{\text{real}} + iG_{\text{imag}} \end{aligned}\right\},$$

then

$$G_{\text{real}} = 1 - 2N_c\sin^2\left(\frac{k\Delta x}{2}\right)\left[1 - \varepsilon\left(\cos^2\frac{k\Delta x}{2} + \kappa\sin^2\frac{k\Delta x}{2}\right)\right] \tag{12.57}$$

and

$$G_{\text{imag}} = -N_c\sin(k\Delta x)\left[1 + \varepsilon(1-\kappa)\sin^2\frac{k\Delta x}{2}\right], \tag{12.58}$$

where $N_c = \frac{c\Delta t}{\Delta x}$ is the CFL number.

Note that in defining the amplification factor we have not used the Fourier representation for time; the definition used is going to inform us about the amplification factors for each and every wave number seperately. The amplification factor, $G(k)$, is shown in Fig. 12.4 for $N_c = 0.1$.

Excepting the second order central difference scheme, all the other schemes show numerical stability. As any physical system demands neutral stability from numerical methods, the adopted numerical scheme must have the same attribute. As the CFL number is reduced, all the schemes show a move towards the right direction and it is only the central difference scheme that approaches neutral stability from the unstable side and hence cannot be used for even solving the one-dimensional wave equation. Among the other schemes, OSL and first-order upwind schemes are the most dissipative at all wave numbers. Next in the order of added dissipation is the linear interpolaration (LI) scheme. Between the two higher order upwind schemes, MUSCL adds more dissipation when compared to QUICK.

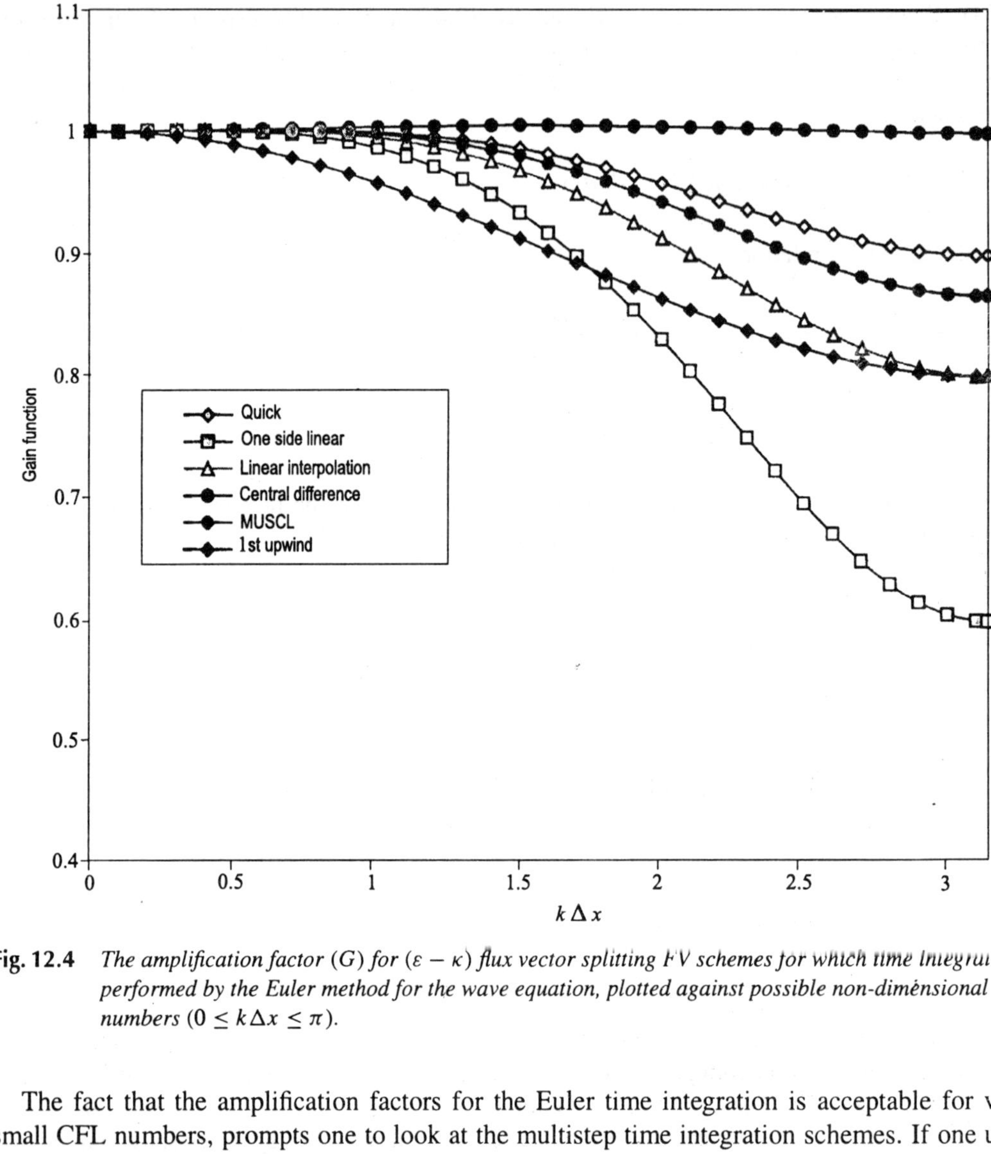

Fig. 12.4 *The amplification factor (G) for (ε − κ) flux vector splitting FV schemes for which time integration is performed by the Euler method for the wave equation, plotted against possible non-dimensional wave numbers* $(0 \le k\Delta x \le \pi)$.

The fact that the amplification factors for the Euler time integration is acceptable for very small CFL numbers, prompts one to look at the multistep time integration schemes. If one uses second-order Adams-Bashforth time integration scheme given by

$$U_l^{n+1} = U_l^n + \frac{\Delta t}{2}\left[3\left.\frac{\partial u}{\partial t}\right|_l^n - \left.\frac{\partial u}{\partial t}\right|_l^{n-1}\right], \tag{12.59}$$

then the equivalent of Equation (12.54) is obtained as

$$U_l^{n+1} = U_l^n - \frac{N_c}{2}\left[3U_{l+1/2}^+\big|^n - 3U_{l-1/2}^+\big|^n - U_{l+1/2}^+\big|^{n-1} + U_{l-1/2}^+\big|^{n-1}\right]. \tag{12.60}$$

Once again using the cell-face quantities given by Equations (12.41) and (12.37) and using the Fourier representation (Equation (12.55)), one gets the numerical dispersion relation

$$\sin(\omega\Delta t) = \frac{N_c}{2}\sin(k\Delta x)\left\{\frac{3\varepsilon}{2} + 3 - \frac{3\varepsilon}{4}[(1-\kappa)(2\cos k\Delta x - 1) + 1 + \kappa]\right\}$$

$$+\frac{N_c}{2}\cos(\omega\Delta t)\sin(k\Delta x)\left\{\frac{\varepsilon}{2} + 1 - \frac{\varepsilon}{4}[(1-\kappa)(2\cos k\Delta x - 1) + 1 + \kappa]\right\}$$

$$-\sin(\omega\Delta t)\left\{\left(1 - \frac{\varepsilon}{2}\kappa\right)(1 - \cos(k\Delta x))\right.$$

$$\left.-\frac{\varepsilon}{4}[1 + \kappa - 2\kappa\cos(k\Delta x) - (1-\kappa)\cos(2k\Delta x)]\right\}. \tag{12.61}$$

And the amplification factors are obtained as the roots of the quadratic equations

$$G^2 + aG - b = 0, \tag{12.62}$$

where,

$$b = N_c\left\{\sin^2\frac{k\Delta x}{2}\left[1 - \varepsilon\kappa\sin^2\frac{k\Delta x}{2} - \cos^2\frac{k\Delta x}{2}\right]\right.$$

$$\left.+\frac{i}{2}\sin(k\Delta x)\left[1 + \varepsilon(1-\kappa)\sin^2\frac{k\Delta x}{2}\right]\right\}$$

and $a = -1 + 3b$.

In Fig. 12.5, G_1 and G_2, the two roots of the above quadratic are shown for $N_c = 0.1$ for the FVS scheme.

G_1 is the severely attenuated factor while G_2 resembles the amplification factors of the Euler time marching scheme shown in Fig. 12.4 except for the second order central differencing scheme. While for the Euler time marching scheme, a second order central difference scheme was found to be unconditionally unstable, here it is conditionally stable, i.e. it is stable for the low values of $k\Delta x$. Once again, for G_2, the OSL, first-order upwind scheme and LI schemes are excessively dissipative. Also for G_2, QUICK is less dissipative than MUSCL. However, the amplification factor (G_2) is not the reason for which this time integration scheme should not be used. The main reason why this scheme is not recommended for time accurate simulation is the presence of the other amplification factor G_1, which is overtly dissipative. The same feature was noted for finite difference scheme, as well for Adams-Bashforth time integration scheme. For the finite difference scheme, G_2 was identified with the *physical mode*, while G_1 was identified as the *computational mode*. Here one notices also an interesting feature of CD_2 scheme with respect to the physical mode (G_2).

For the second order central scheme, the physical mode displays near-neutral behaviour at the largest and smallest length scales. For finite volume computations, Jameson et al. (1981) have used this spatial discretization scheme, but with explicit addition of second and fourth order dissipation terms based on pressure sensitive switches in the code. However, Jameson et al.'s method is mostly used for steady state calculations.

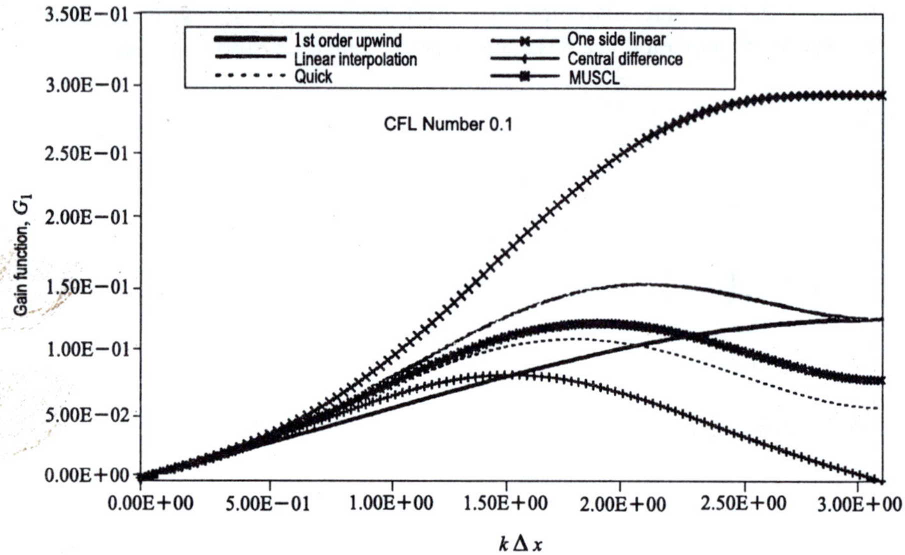

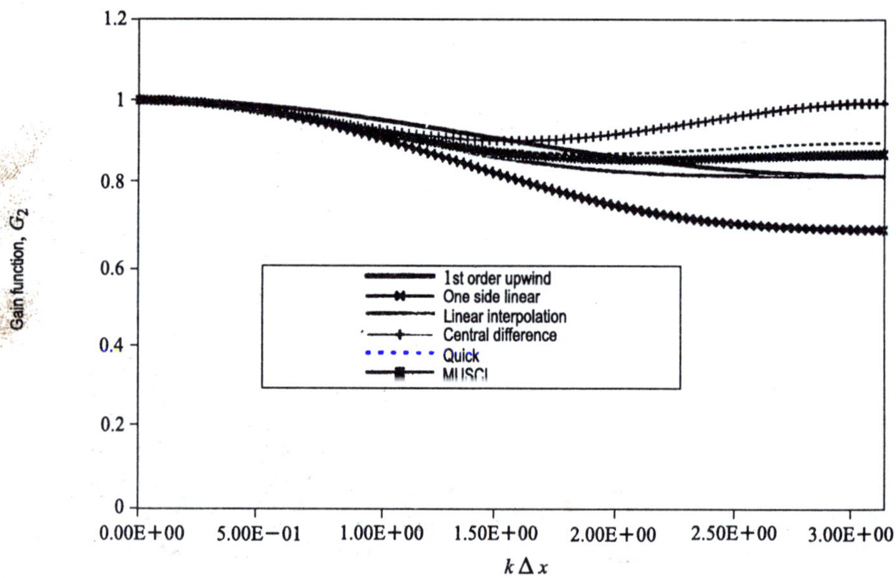

Fig. 12.5 *The amplification factor (G) for $(\varepsilon - \kappa)$ FVS-FV schemes for which time integration is performed by an Adams-Bashforth scheme plotted against non-dimensional wave number. Shown on top is the severely attenuated root; the neutral root is shown below.*

Next, some of the multi-time-step integration schemes designed to posses total variation diminishing (TVD) property are discussed. Some such Runge-Kutta schemes from Gottlieb & Shu (1996) are given below.

The two-stage Runge-Kutta time integration scheme of this class for the equation

$$\frac{\partial u}{\partial t} = L(u) \tag{12.63}$$

are given by following two stages:

$$U^{(1)} = U^n + \Delta t L(U^n)$$

and

$$U^{n+1} = \frac{U^n + U^{(1)}}{2} + \frac{\Delta t}{2} L(U^{(1)}).$$

This scheme replaces the time integration scheme in Equation (12.54), and the resultant amplification factor is obtained from

$$G_{\mathrm{RK}_2} = 1 - N_c g + \frac{N_c^2}{2} g^2, \tag{12.64}$$

where $g = a + ib = 2\sin^2\frac{k\Delta x}{2} + \frac{\varepsilon(1-\kappa)}{2}(\cos k\Delta x - 1)\cos k\Delta x - \varepsilon(1-\kappa)\sin^2\frac{k\Delta x}{2} + i\left[1 + \frac{\varepsilon(1-\kappa)}{2}(1 - \cos k\Delta x)\right]\sin k\Delta x$.

In Fig. 12.6 the amplification factors for the six schemes given by the $(\varepsilon - \kappa)$ FVS scheme are plotted for three CFL numbers (= 0.01, 0.2 and 0.5).

For the RK_2 time intregation scheme, the OSL interpolation is again excessively dissipative, and when the CFL number is increased from 0.2 to 0.5, the phase error is so large that for a physical system for which the wave propogates downstream, the numerical scheme would predict a disturbance field whose wave numbers components given by $k\Delta x \geq \pi/2$ would travel upstream.

The other interpolation schemes are such that the intermediate wave numbers are the most damped ones, and for first-order upwinding it is the lowest and the highest wave numbers that are least damped; in that sense it is an interesting scheme. Between the two higher-order schemes, the QUICK is the least dissipative.

Similarly, for Equation (12.62) the three-stage TVD Runge-Kutta scheme is given by

$$\left.\begin{aligned} U^{(1)} &= U^n + \Delta t L(U^{(n)}) \\ U^{(2)} &= \tfrac{3}{4}U^{(n)} + \tfrac{U^{(1)}}{4} + \tfrac{\Delta t}{4}L(U^{(1)}) \\ U^{(n+1)} &= \tfrac{U^{(n)}}{3} + \tfrac{2U^{(2)}}{3} + \tfrac{2\Delta t}{3}L(U^{(2)}) \end{aligned}\right\}. \tag{12.65}$$

For this time integration scheme, the amplification factor is given by

$$G_{\mathrm{RK}_3} = 1 - N_c g + \frac{N_c^2 g^2}{2} - \frac{N_c^3 g^3}{6}. \tag{12.66}$$

In Fig. 12.7 the amplification factors for all the six spatial discretization schemes are shown for the three indicated CFL numbers for the RK_3 time integration scheme. The features of this time integration scheme is similar to two-stage Runge-Kutta scheme.

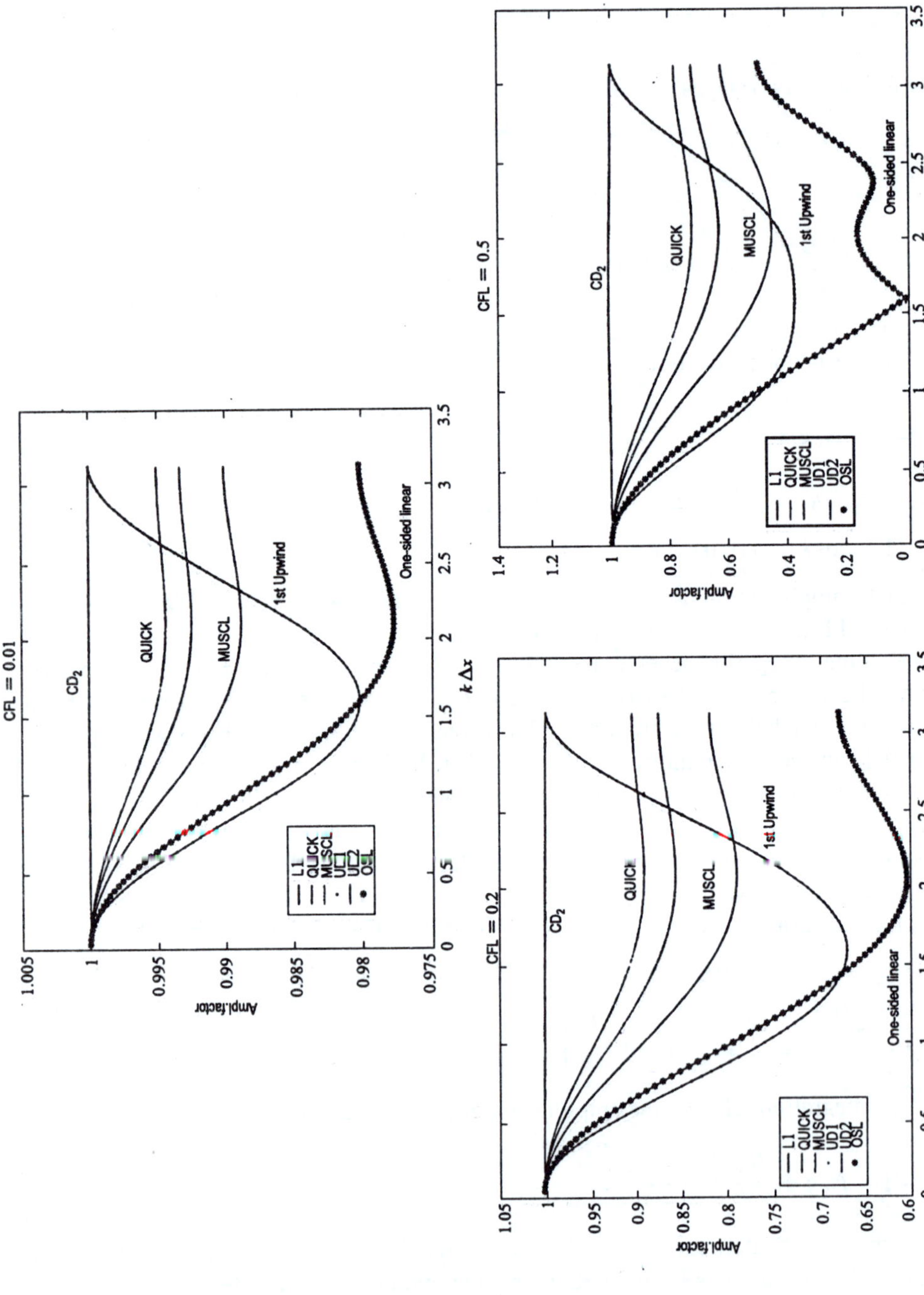

Fig. 12.6 *The amplification factor shown for (ε−κ) FVS-FV schemes for which time integration is performed by two stage Runge-Kutta scheme plotted against non-dimensional wave number for three different CFL numbers.*

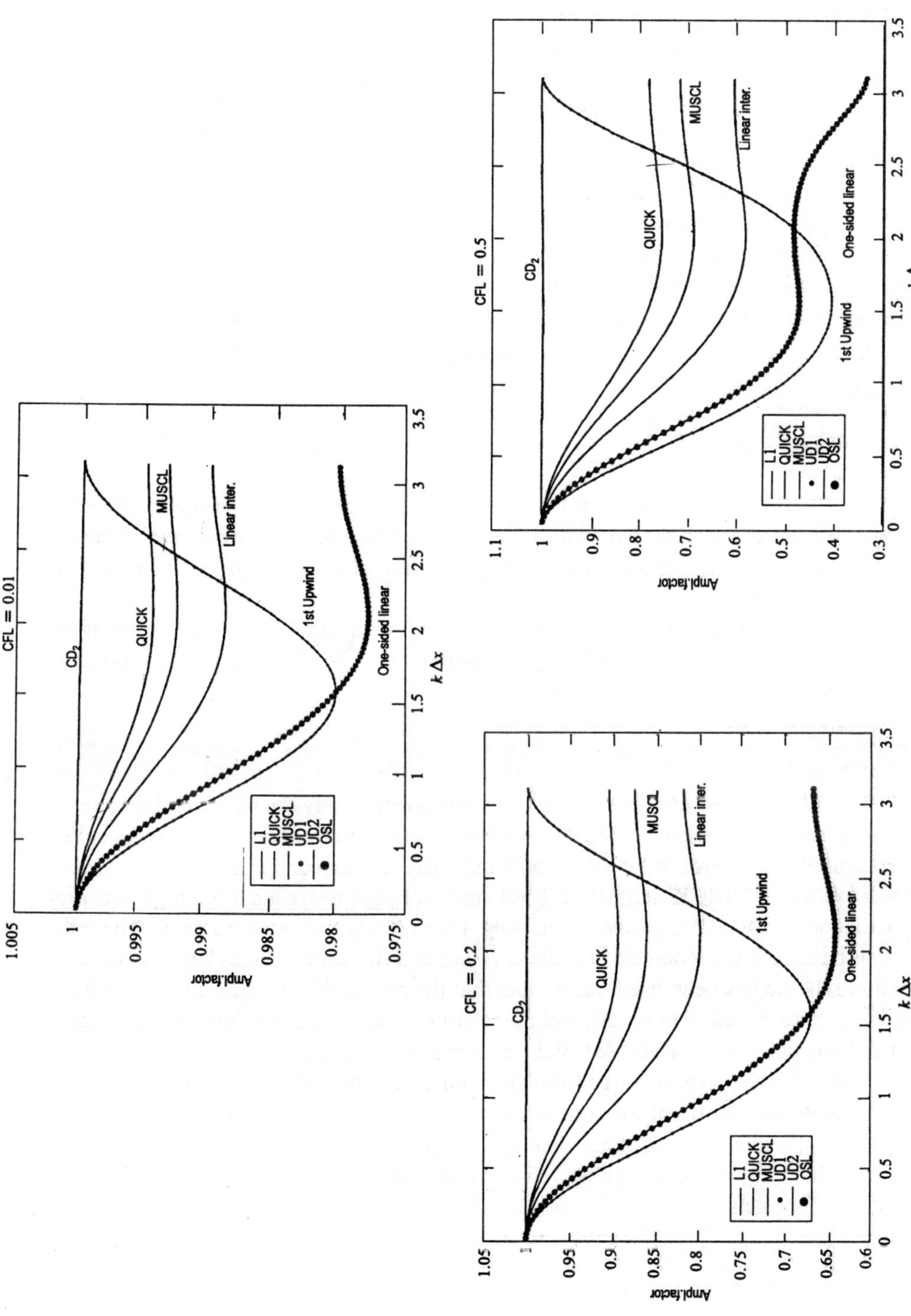

Fig. 12.7 *The amplification factor (G) for ($\varepsilon - \kappa$) FVS-FV methods for which time integration is performed by three stage Runge-Kutta scheme (Eqn. (12.65)) plotted against non-dimensional wave number for three different CFL numbers.*

12.9 Dispersion Relation Preservation Property

This property is specific to the particular differential equation that has to be solved. Once again, the efficiency and other properties of various numerical schemes will be calibrated with respect to the model one-dimensional wave equation given by Equation (12.53). As this equation represents a non-dispersive physical phenomenon, the dispersive property of the used numerical scheme will be quantified by comparing the exact group velocity (in this case equal to the phase speed c) with the equivalent group velocity of the space-time discretization of the numerical scheme.

For the (ε–κ) schemes given by Equations (12.38)–(12.41) for spatial discretization using Euler time integration for one dimensional wave equation, the numerical group velocity is obtained from the numerical dispersion relation in Equation (12.56). This can be simplified as,

$$V_{gn} = \frac{d\omega}{dk} = \frac{c}{\cos\omega\Delta t}\left[\left\{1+\varepsilon(1-\kappa)\sin^2\frac{k\Delta x}{2}\right\}\cos k\Delta x + \frac{\varepsilon(1-\kappa)}{2}\sin^2\frac{k\Delta x}{2}\right] \tag{12.67}$$

For the present purpose of comparision between physical and numerical group velocities, a $\pm 5\%$ tolerance is consider as acceptable. In the ($k\Delta x$–$\omega\Delta t$) plane, the points where the numerical group velocity is within the above acceptable limits are marked by a cross in Fig. 12.8 for four spatial reconstruction schemes.

For the CD_2 scheme, one notices in the above figure a region diagonally opposite to the origin where numerical group velocity is close to the actual group velocity. For this scheme, the expression for numerical group velocity can be simplified to

$$V_{gn} = c\frac{\cos k\Delta x}{\cos\omega\Delta t}.$$

Thus, although locally the phase error is maximum at high wave numbers, for these combinations of wave numbers and circular frequencies, the dispersion error is minimum. The patch of region close to the origin should be of interest to us where both the phase and dispersion error are minimum. From this point of view, QUICK and MUSCL schemes are seen to be ideal for high accuracy FVS-FV calculation, with MUSCL scheme showing a larger range of wave numbers where the DRP property is better. All the schemes show the same circular frequency ranges because the same time integration scheme has been used. Also, note that the results of the semi-discrete analysis (when time is not discretized) can be directly noted from these figures by looking at the limit $\omega\Delta t \rightarrow 0$; from this point of view also MUSCL scheme shows better property.

One can investigate the same property for higher order time discretization schemes. For the second order accurate Adams-Bashforth time integration scheme applied to the wave equation in conjuction with (ε–κ) spatial discretization scheme, the dispersion relation was derived in Equation (12.60). One can obtain the numerical group velocity from this equation by differentiating it, and upon simplification one gets

$$V_{gn} = \frac{i\frac{db}{dk} - \frac{da}{dk}\sin(\omega\Delta t)}{\Delta t[2\cos 2\omega\Delta t + a\cos\omega\Delta t]}, \tag{12.68}$$

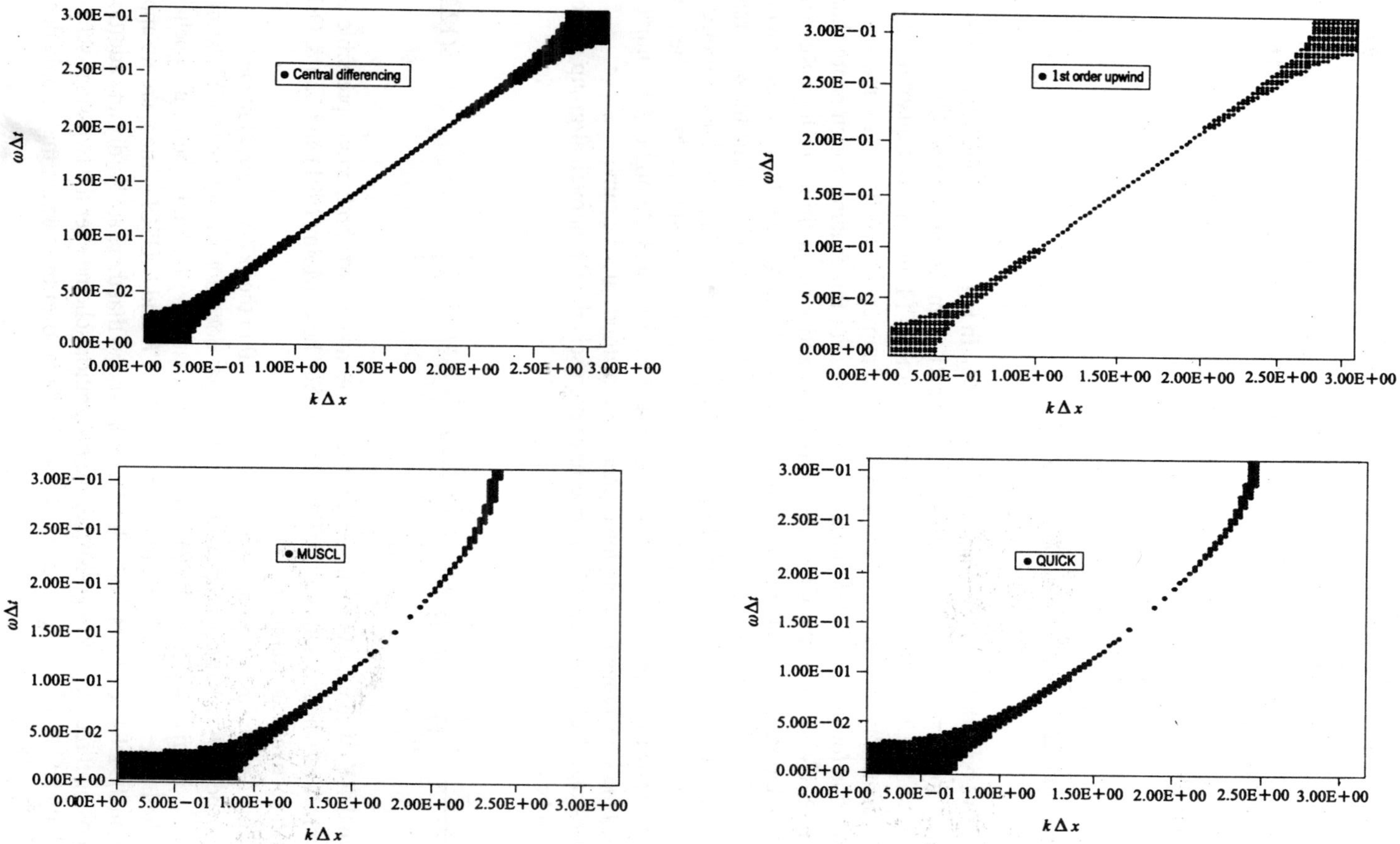

Fig. 12.8 *DRP region in $(k\Delta x - \omega\Delta t)$ plane for $(\varepsilon - \kappa)$ scheme for (a) CD_2; (b) 1st order upwinding; (c) MUSCL; and (d) QUICK interpolation schemes.*

where a and b have been defined in Equation (12.62). However, this method is already seen to be unsuitable for providing near-neutral numerical stability, and hence its DRP property is not investigated. Also, in analyzing compact schemes, we have noted the inferiority of Adams-Bashforth time integration scheme for the DRP region in $(\kappa\Delta x-\omega\Delta t)$ plane, as compared to other time integration schemes.

For the two-stage Runge-Kutta (RK_2) time integration scheme, one can similarly calculate the numerical group velocity as

$$V_{gn} = \frac{c}{\Delta x \cos \omega\Delta t}\left[\frac{db}{dk} - N_c\left(b\frac{da}{dk} + a\frac{db}{dk}\right)\right], \tag{12.69}$$

where a and b have been defined following Equation (12.64). In Fig .12.9 the DRP region of some of these schemes are displayed in $(k\Delta x-\omega\Delta t)$ plane for RK_2 time integration scheme.

For the CD_2 scheme, the DRP region is independent of the CFL number in Equation (12.69), since $\frac{d}{dk}(ab)$ is identically zero for $\varepsilon = 1$ and $\kappa = 1$. Thus if numerical stability is ensured, one can use the CD_2 scheme for as high a CFL number as possible. For this feature, which is not shared by any other FVS-FV scheme, the CD_2 scheme is an attractive proposition for high Reynolds number flow simulation. Comparatively the first-order upwind scheme results show that the $k\Delta x$ range over which DRP property is valid, keeps reducing with CFL number, while the $\omega\Delta t$ range does not change. Similarly for QUICK and MUSCL schemes, the $\omega\Delta t$ range for DRP region does not change with CFL number. However, the $k\Delta x$ range is a strong function of CFL number, with MUSCL scheme showing superior $k\Delta x$ range for lower CFL numbers. At higher CFL numbers (larger than 0.2) the $k\Delta x$ range for both the higher order methods are the same.

For the three-stage Runge-Kutta (RK_3) time integation scheme the numerical group velocity can be shown to be as

$$V_{gn} = \frac{c}{\cos \omega\Delta t}\left[\frac{db}{dk} - \frac{N_c}{2}\frac{d}{dk}(ab) + \frac{N_c^2}{6}\left(4ab\frac{da}{dk} - 3b^2\frac{db}{dk}\right)\right]. \tag{12.70}$$

The RK_2 and RK_3 time integration schemes have been used for both incompressible and compressible flows (as shock-capturing TVD schemes). See Le & Moin (1991) and Shu & Osher (1988, 1989) respectively as examples.

The DRP plots for the RK_3 scheme is shown in Fig. 12.10 in the $(k\Delta x - \omega\Delta t)$ plane.

Once again, the CD_2 scheme shows a DRP region that does not change with CFL number and which is indistinguishable from the RK_2 results. The first order upwind scheme also behaves similarly. For the higher-order interpolation schemes (QUICK and MUSCL), there are no appreciable changes in the DRP region as well. This therefore shows that there are hardly any additional benefits in choosing three-stage Runge-Kutta time integration scheme over the two-stage scheme. It is the simplicity and lesser storage requirement of the latter over the former, that should decide in favor of the RK_2 scheme.

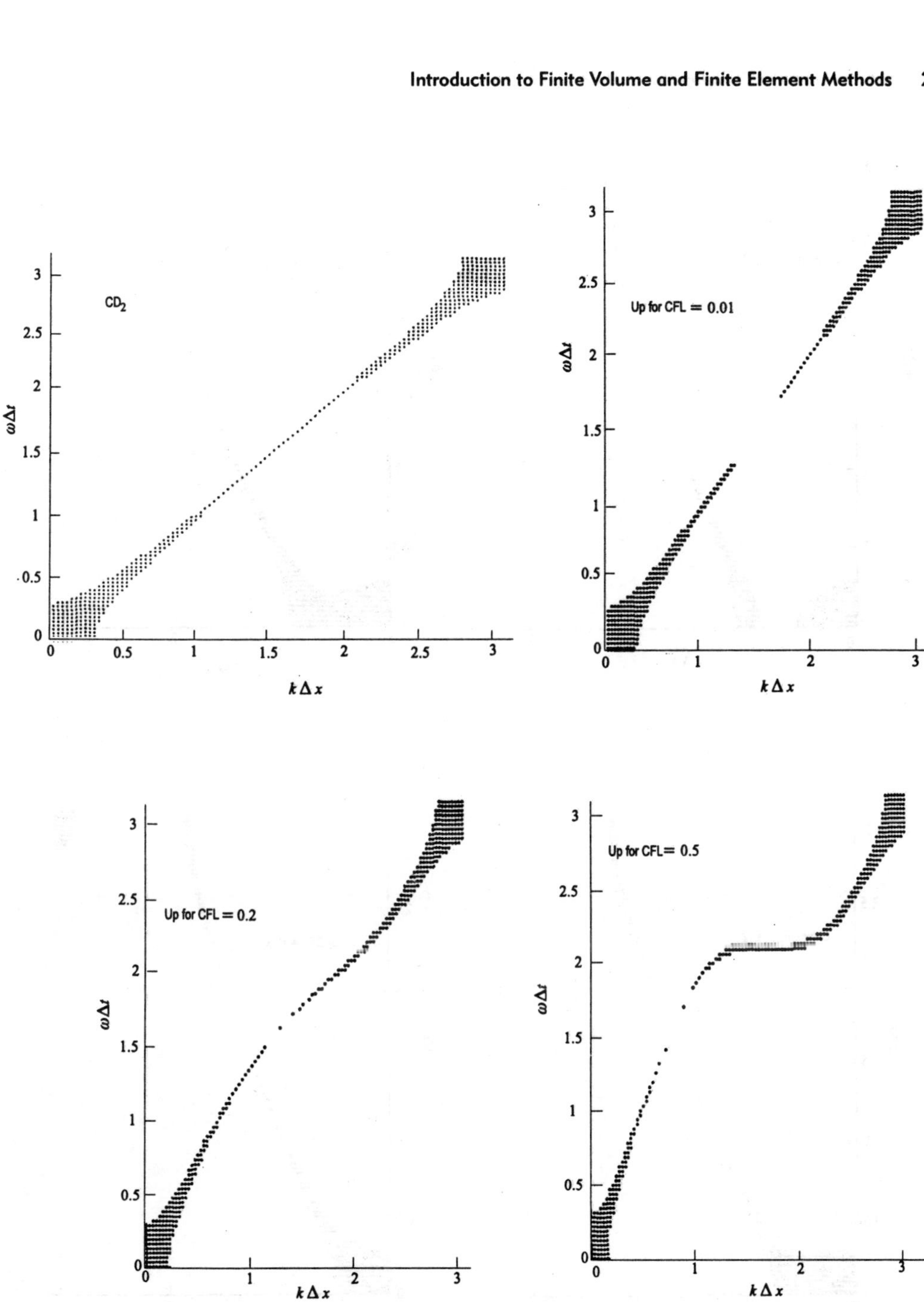

Fig. 12.9 *(a) The DRP region of RK_2 time integration scheme for CD_2 and 1st order upwind spatial discretization scheme for the indicated CFL numbers. Note that CD_2 results are independent of CFL number.*

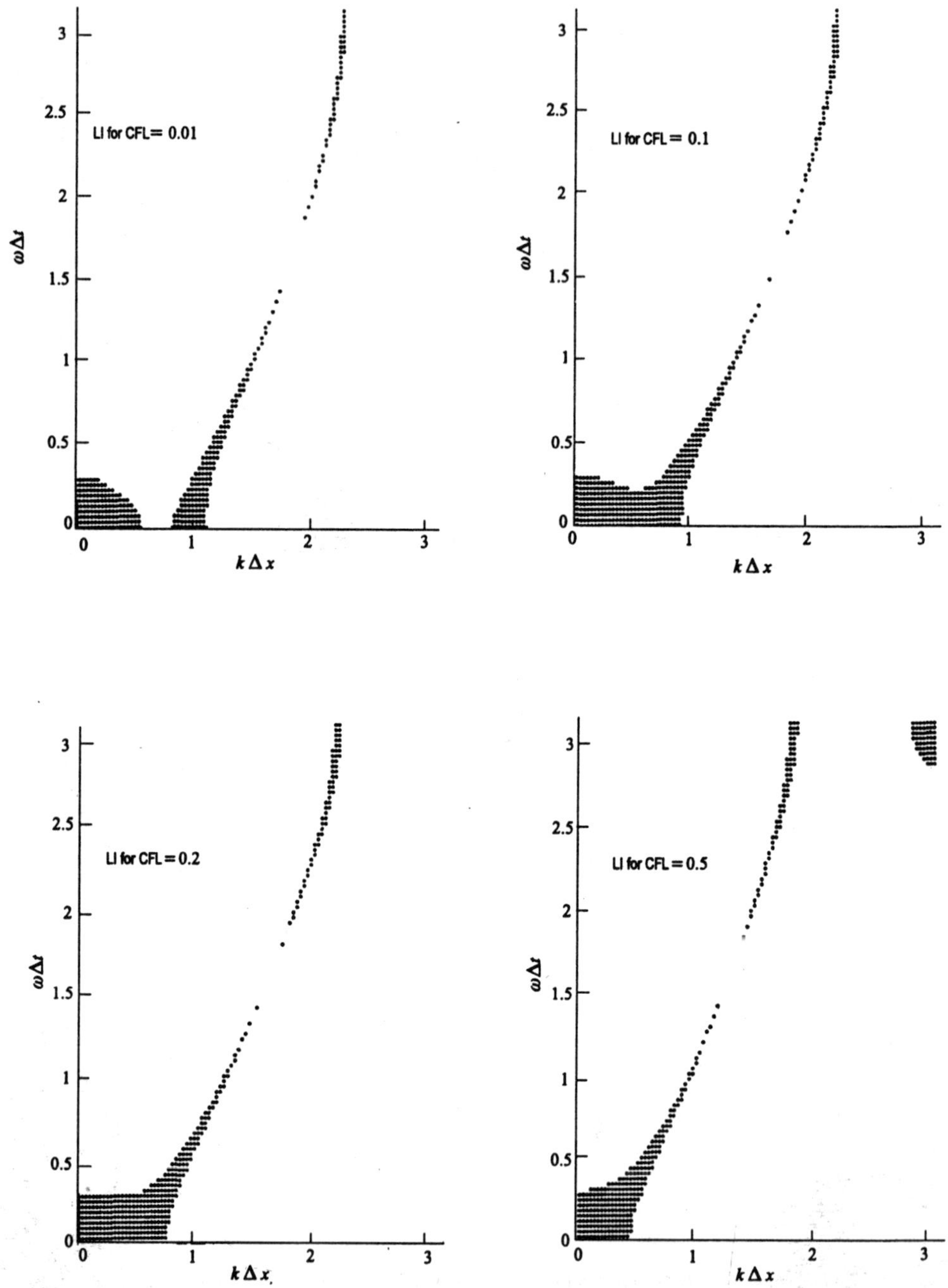

Fig. 12.9 *(b) The DRP region of RK_2 time integration scheme ($\varepsilon = 1$ and $\kappa = 0$) for linear interpolation spatial discretization scheme for the indicated CFL numbers.*

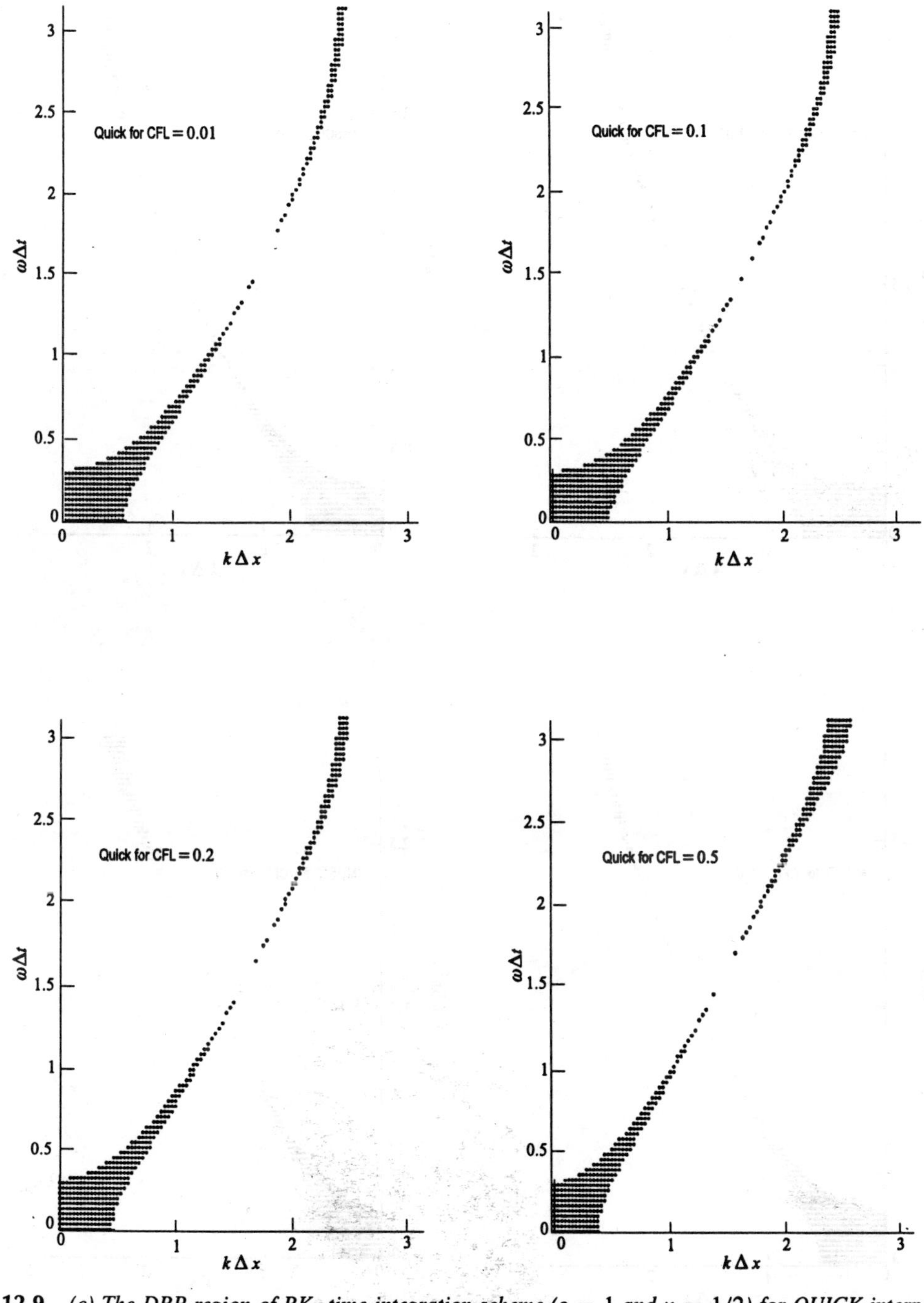

Fig. 12.9 *(c) The DRP region of RK_2 time integration scheme ($\varepsilon = 1$ and $\kappa = 1/2$) for QUICK interpolation spatial discretization scheme for the indicated CFL numbers.*

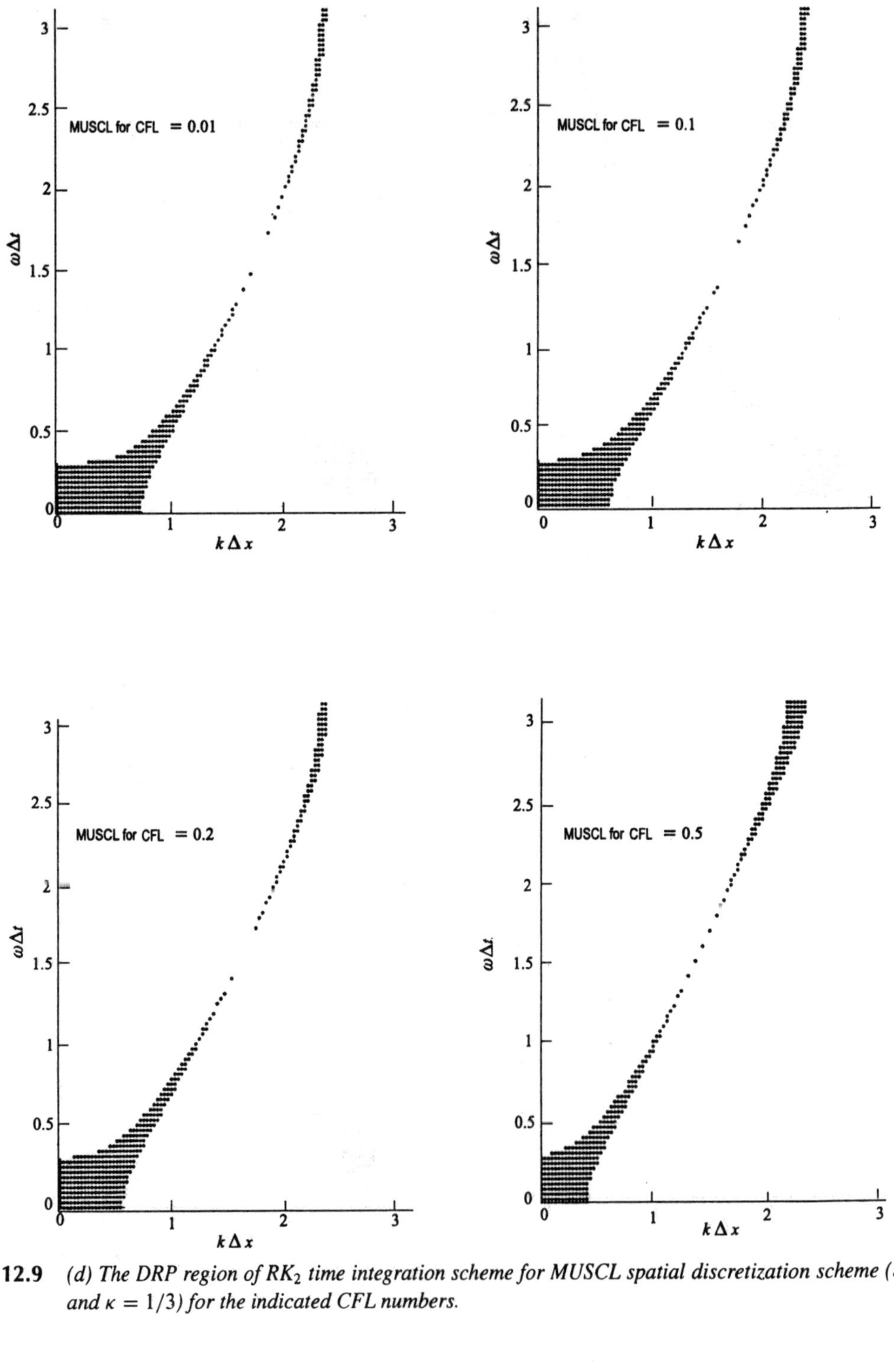

Fig. 12.9 *(d) The DRP region of RK_2 time integration scheme for MUSCL spatial discretization scheme ($\varepsilon = 1$ and $\kappa = 1/3$) for the indicated CFL numbers.*

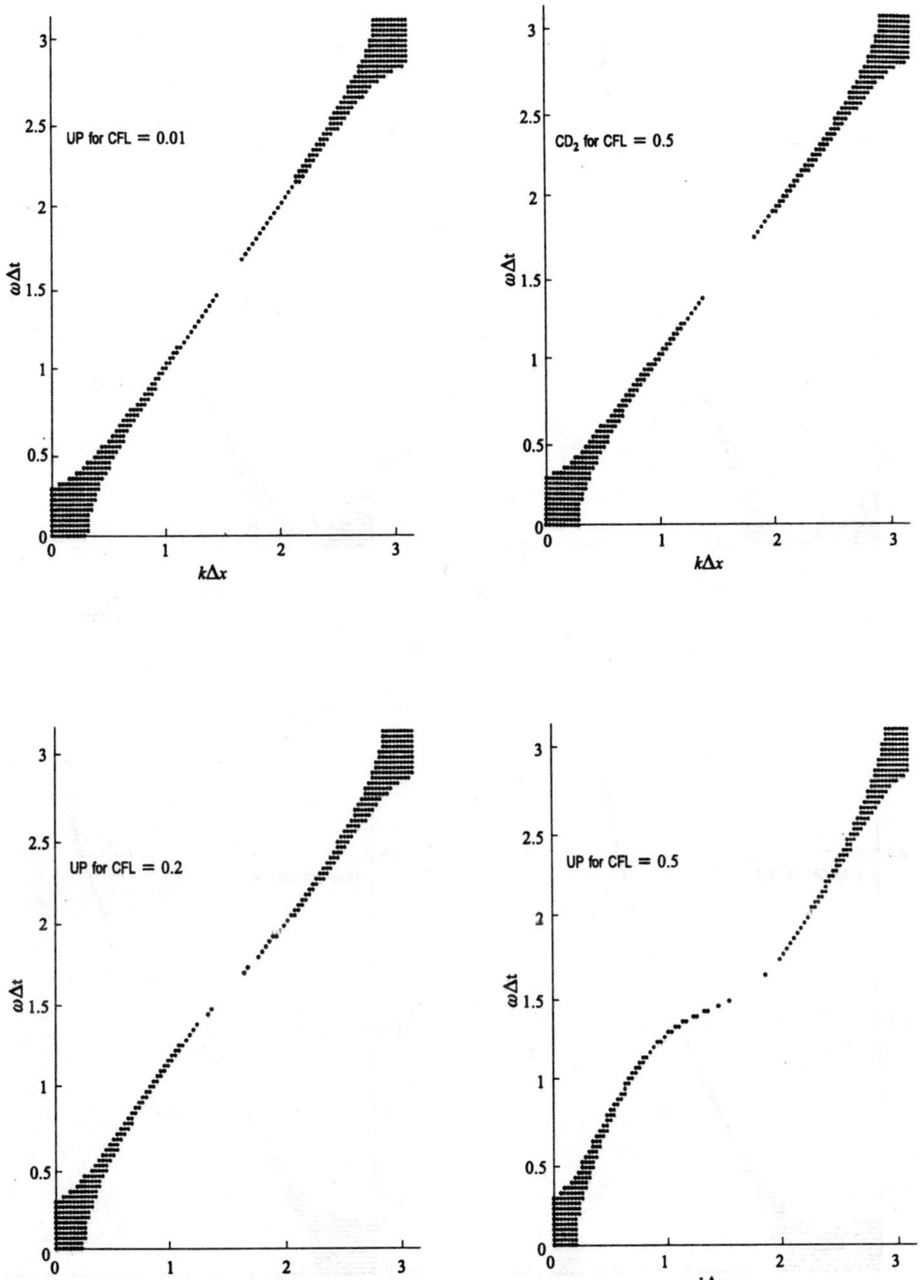

Fig. 12.10 *(a) The DRP region of RK_3 time integration scheme for $CD_2(\varepsilon = 1$ and $\kappa = 1)$ and 1st order upwind spatial discretization scheme ($\varepsilon = 0$) for the incidicated CFL numbers. Note that the CD_2 results do not depend on CFL number.*

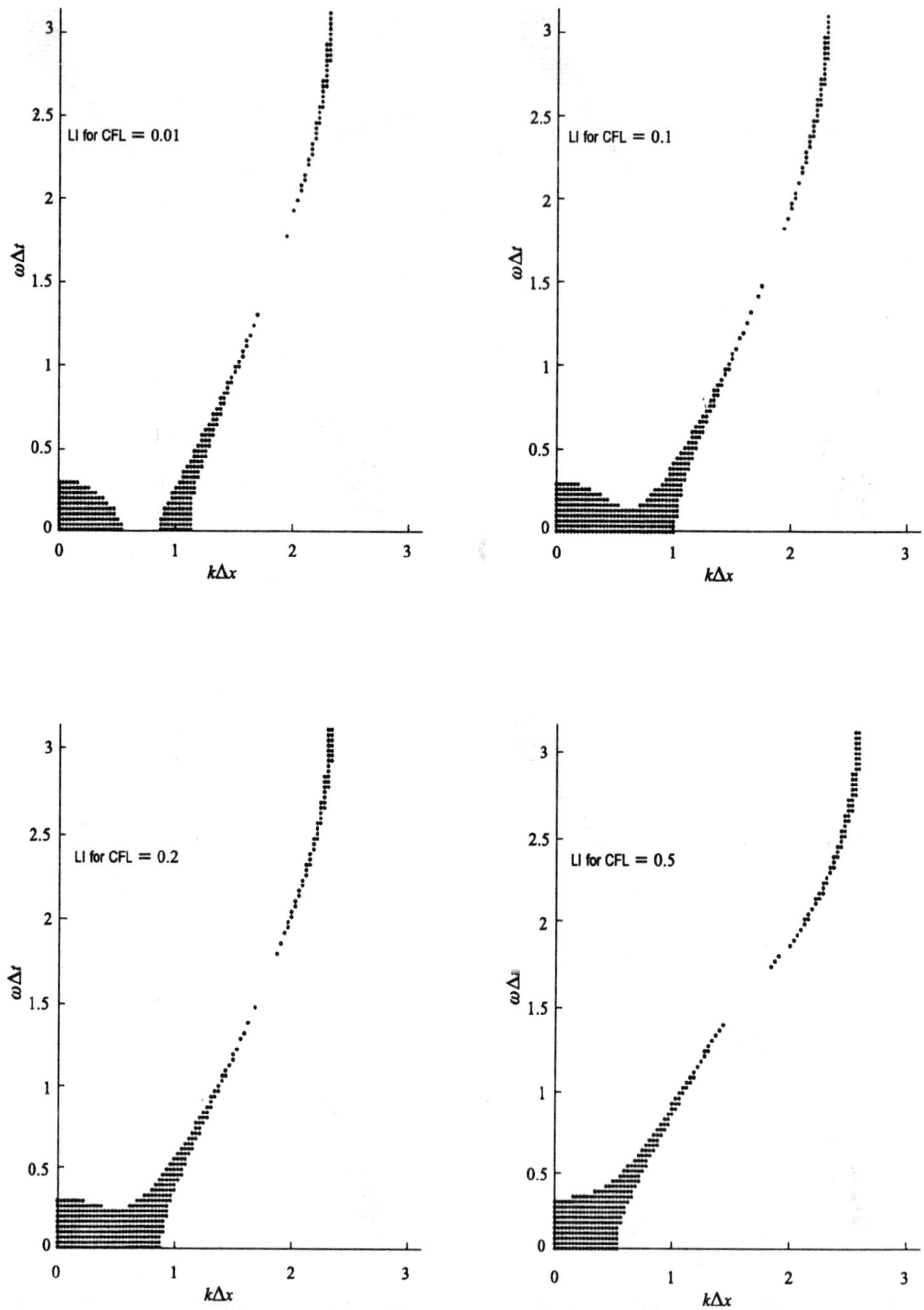

Fig. 12.10 *(b) The DRP region of RK_3 time integration scheme for linear interpolation spatial discretization scheme ($\varepsilon = 1$ and $\kappa = 0$) for the incidicated CFL numbers.*

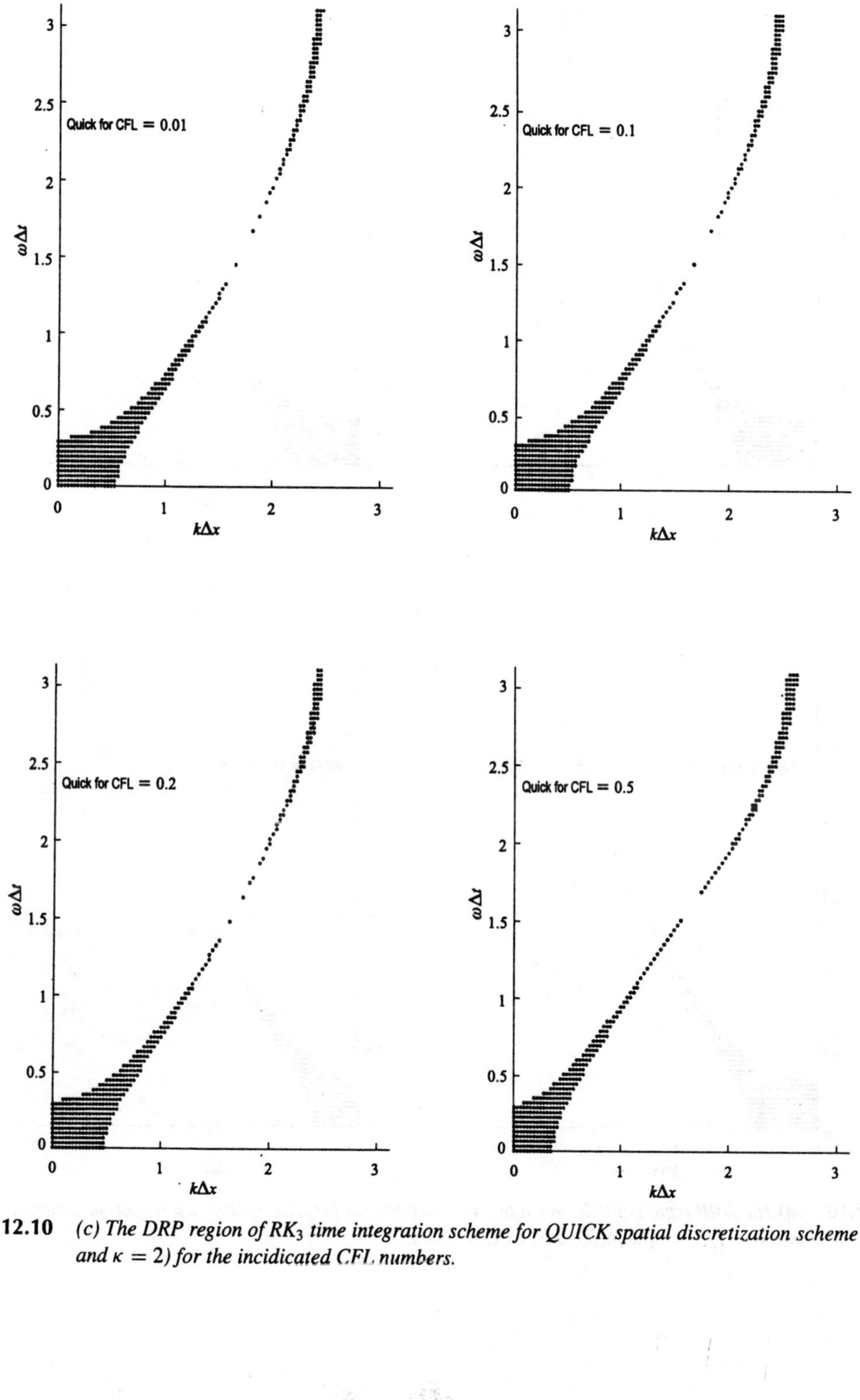

Fig. 12.10 *(c) The DRP region of RK_3 time integration scheme for QUICK spatial discretization scheme ($\varepsilon = 1$ and $\kappa = 2$) for the incidicated CFL numbers.*

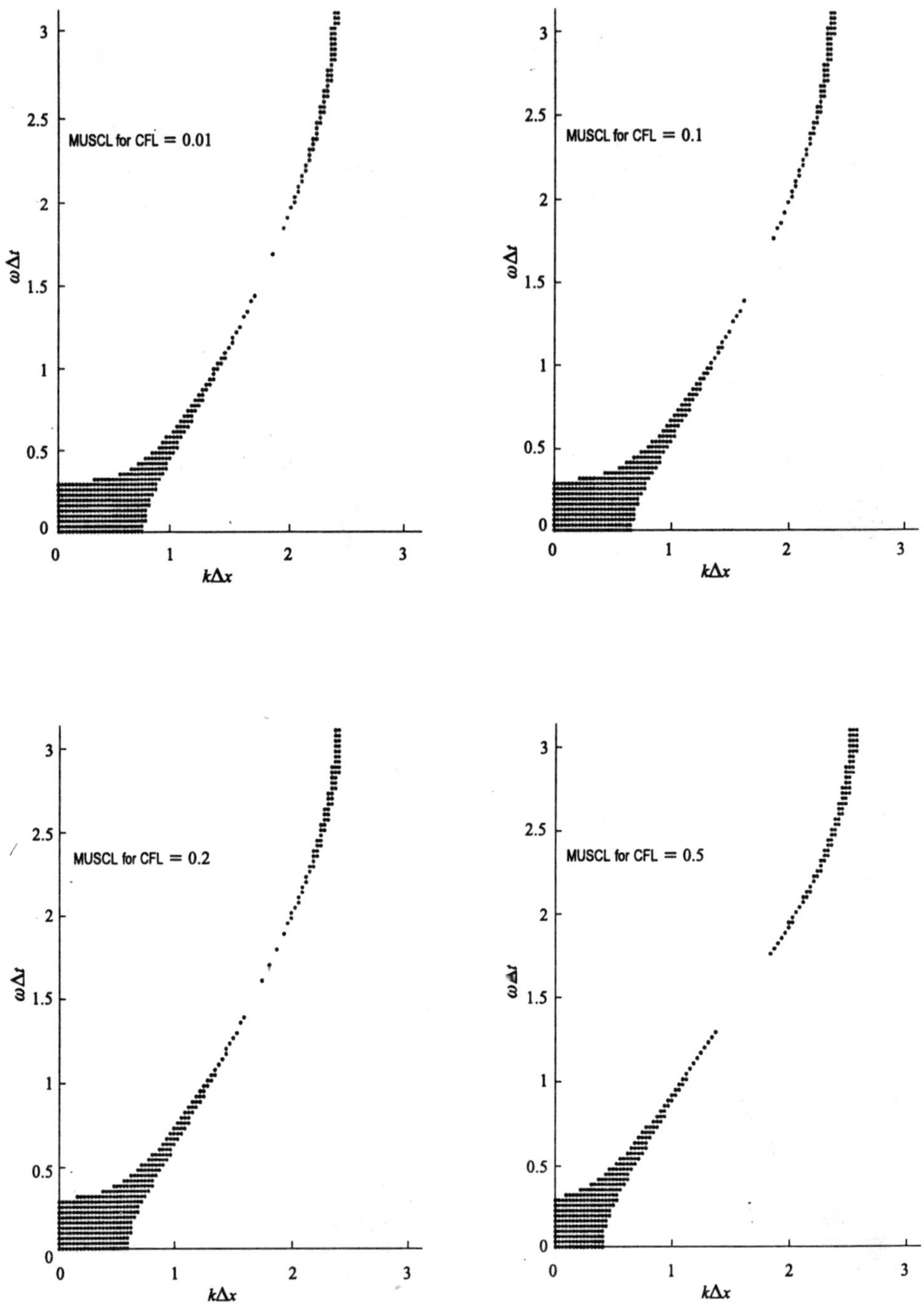

Fig. 12.10 *(d) The DRP region of RK_3 time integration scheme for MUSCL spatial discretization scheme ($\varepsilon = 1$ and $\kappa = 1/3$) for the incidicated CFL numbers.*

12.10 Solution of 1D Wave Equation

The ability of FVS-FV schemes as a tool for high accuracy computing can be tested by solving 1D wave equation with initial condition that requires large spectral resolution and ability to capture discontinuities. For this purpose, Equation (12.53) would be solved using Euler time integration and $(\varepsilon - \kappa)$ schemes for spatial reconstruction for the FVS representation given in Equation (12.54). In particular, the linear interpolation $(\varepsilon = 1, \kappa = 0)$, QUICK $(\varepsilon = 1, \kappa = 1/2)$ and MUSCL schemes $(\varepsilon = 1, \kappa = 1/3)$ are evaluated. The time accurate solutions are presented in Fig. 12.11 to 12.13. The initial condition is shown in the top panel of these figures and one notices that it has discontinuities at $x = 0$ and 1, a region over which Equation (12.53) is solved. Such a case with a wave packet as initial condition was also solved with the higher order finite and compact difference schemes in Chapter 11.

For LI case, solution is displayed in Figure 12.11. This scheme is practically non-dissipative wherever the solution is continuous. But its excessive dissipation is evident at the discontinuities and for the same reason the Gibbs phenomenon is not evident. Also, another notable feature of this scheme is that the dispersion of the numerical solution with respect to the exact solution is clearly visible for $t \geq 10$. Hence such a scheme is not recommended for long-time high-accuracy solutions. The dispersion is caused by poor spectral resolution and small DRP region for this scheme. In contrast, QUICK and MUSCL schemes have high spectral resolution and low numerical dissipation. This is evident from numerical solution shown in Figs. 12.12 and 12.13. Both these schemes have poor ability to capture discontinuity. Also, the low dissipation of these schemes is responsible for the presence of Gibbs' phenomenon — more for QUICK than MUSCL scheme, as explained in the next section.

12.11 An Explanation for Gibbs Phenomenon in FVS-FV Schemes

Consider the solution of Equation (12.53) using the FVS-FV schemes as given by Equation (12.54). In this equation

$$U^{+}_{l+1/2} - U^{+}_{l-1/2} = U_l + \frac{1-k}{4}(U_l - U_{l-1}) + \frac{1+k}{4}(U_{l+1} - U_l) - U_{l-1} - \frac{1-k}{4}(U_l - U_{l-2}) - \frac{1+k}{4}(U_l - U_{l-1}). \quad (12.71)$$

The central idea behind FVS-FV schemes, as given here, consists of treating the state vectors in a piecewise-continuous fashion in terms of cell-averaged values. The Fourier spectral representation, as expressed by Equation (12.55), admits such discontinuities. At the same time the various terms in Equation (12.71) are considered piecewise-continuous expressible by Taylor series. Therefore,

$$(U_{l+2} - U_l) - (U_{l+1} - U_{l-1}) = 2h^2 U_l'' + h^3 U_l''' + \frac{2h^4}{3} U_l^{iv} + \frac{h^5}{4} U_l^{v}$$

and $$(U_{l+2} - 3U_{l+1} + 3U_l - U_{l-1}) = h^3 U_l''' + \frac{h^4}{2} U_l^{iv} + \frac{h^5}{4} U_l^{v}.$$

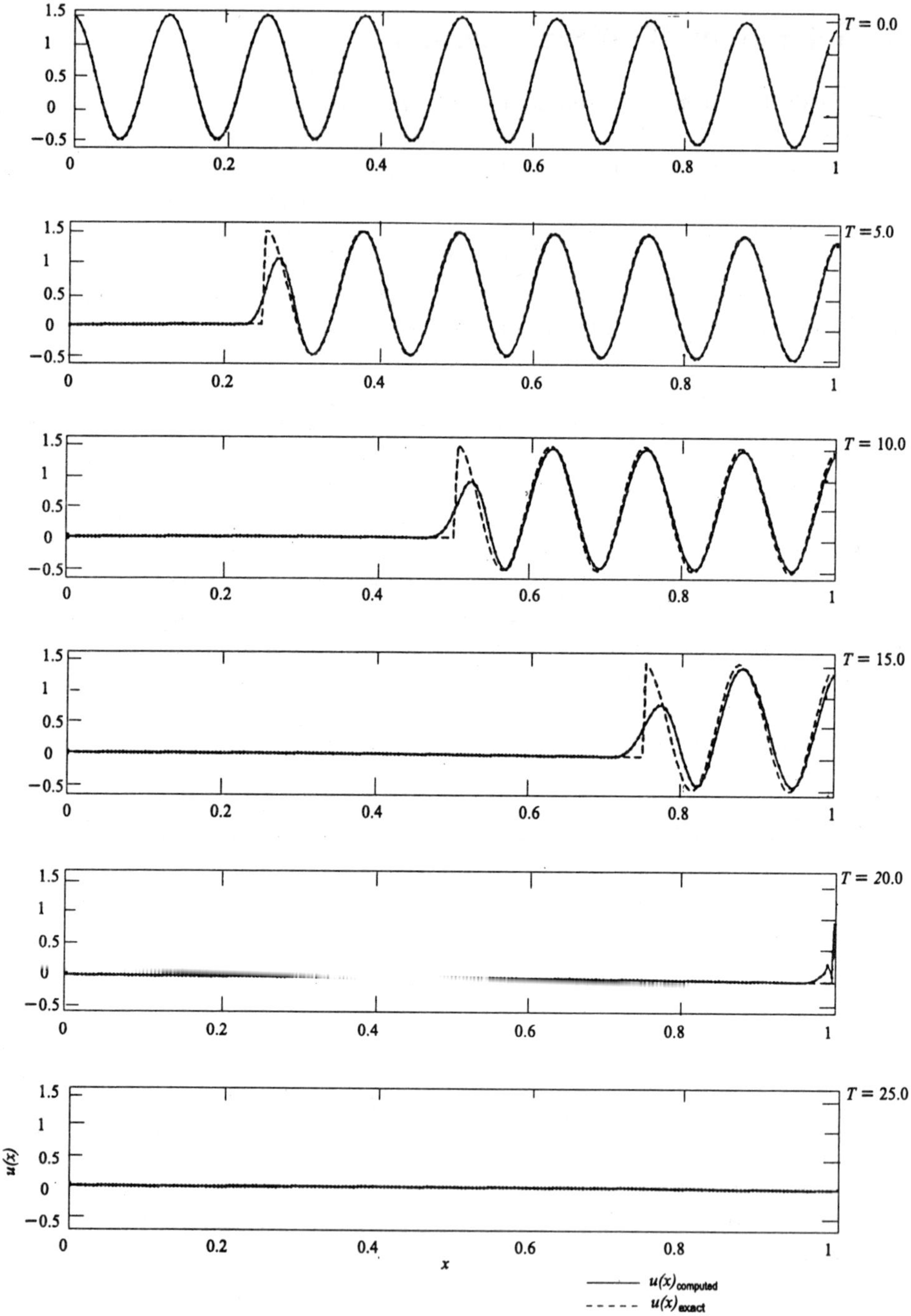

Fig. 12.11 *Solution of Equation (12.53) obtained numerically by LI spatial reconstruction scheme and Euler time integration scheme and compared with the analytic solution at the indicated non-dimensional times. Here, c = 0.05, the number of points used are 201 and time step for integration is 0.0001.*

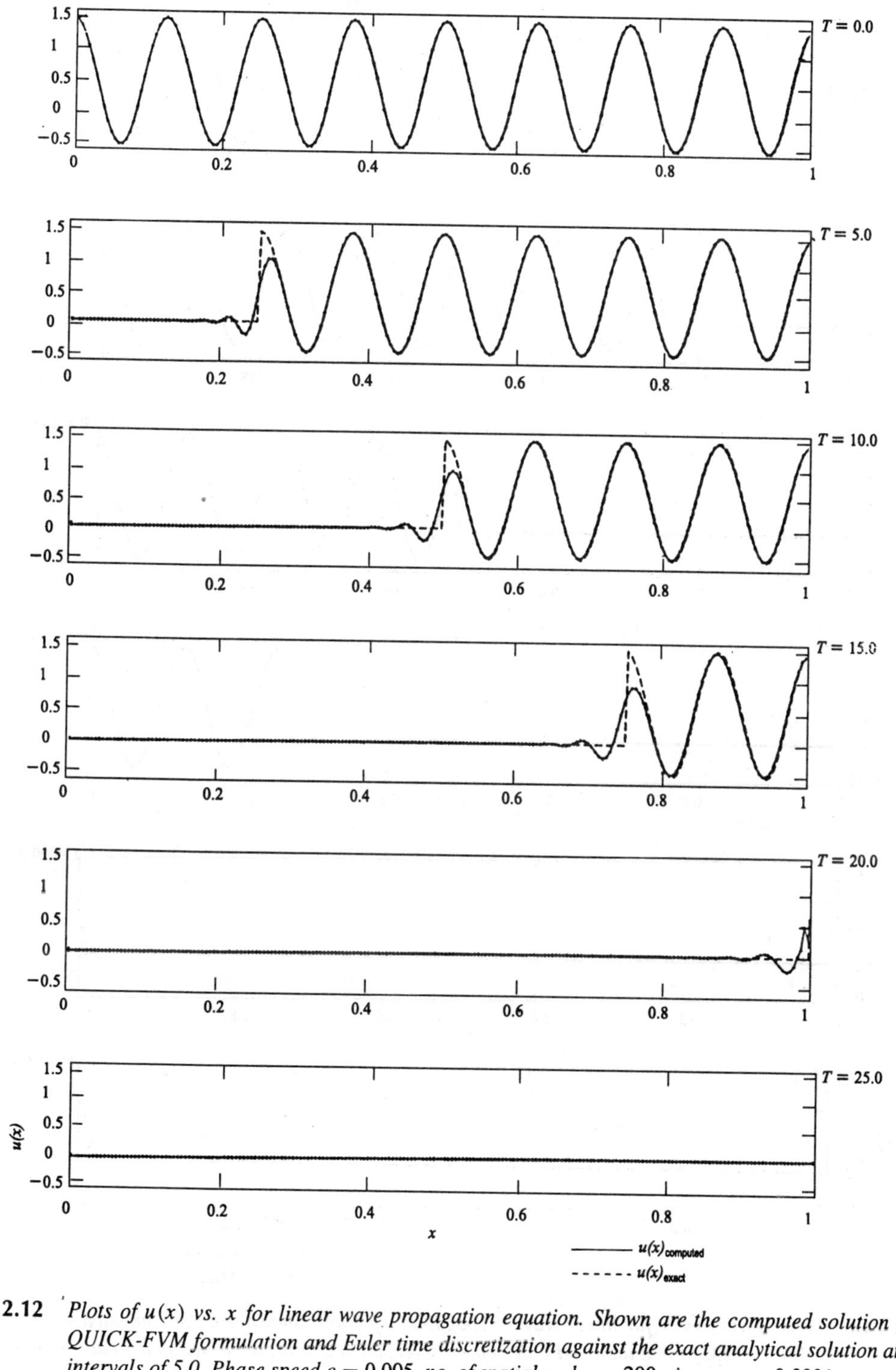

Fig. 12.12 *Plots of $u(x)$ vs. x for linear wave propagation equation. Shown are the computed solution using QUICK-FVM formulation and Euler time discretization against the exact analytical solution at time intervals of 5.0. Phase speed $c = 0.005$, no. of spatial nodes = 200, time step = 0.0001.*

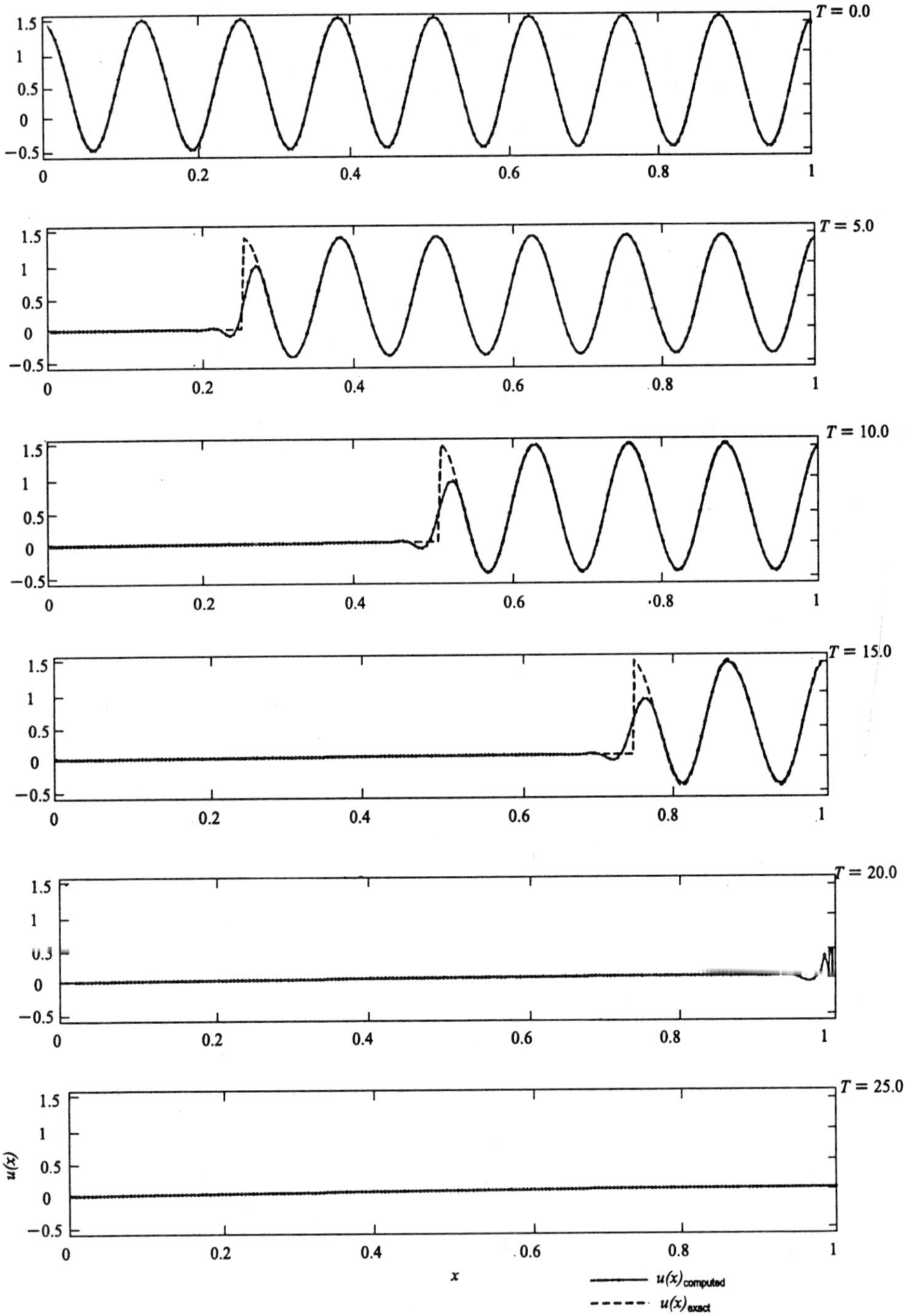

Fig. 12.13 *Plots of $u(x)$ vs. x for linear wave propagation equation. Shown are the computed solution using MUSCL-FVM formulation and Euler time discretization against the exact analytical solution at time intervals of 5.0. Phase speed $c = 0.005$, no. of spatial nodes $= 200$, time step $= 0.0001$.*

Thus from Equation (12.71)

$$U^{+}_{l+1/2} - U^{+}_{l-1/2} = hU_l' + \frac{h^2}{2}U_l''\left(\frac{k-1}{2}\right)$$

$$+\frac{h^3}{6}\left(\frac{3k-1}{2}\right)U_l''' + \frac{h^4}{24}(2-4k)U_l^{iv}$$

Thus the solution of Equation (12.53) by the algorithm of Equation (12.54) is the same as the that would be obtained by solving the following equation computationally.

$$\frac{\partial u}{\partial t} + c\frac{\partial u}{\partial x} = -c\frac{h^2}{12}(1-3\kappa)\frac{\partial^3 u}{\partial x^3}$$

$$-\frac{ch^3}{8}(\kappa-1)\frac{\partial^4 u}{\partial x^4} - \frac{ch^4}{240}(15\kappa-13)\frac{\partial^5 u}{\partial x^5}. \tag{12.72}$$

A close inspection of Equation (12.72) will help us in understanding the *computational physics* revealed by the numerical solution of 1D wave equation in Figs. (12.11) to (12.13). The equation that is actually solved by the algorithm of Equation (12.54) reveals that in essence a host of terms are added. The leading odd derivative terms are a source of spurious *dispersive waves*, while the dissipation terms tend to attenuate not only the numerical instabilities, but also the spurious waves created by Gibbs' phenomenon. In the regular part of the flow these spurious terms are negligibly small and their effects are not seen. However, at the discontinuities these error terms are not negligible and their effects are clearly visible as Gibbs' phenomenon. Unfortunately, the truncation error terms at the discontinuity should not be viewed by the leading truncation error terms. This is explained below.

For the three schemes of Figs. (12.11) to (12.13) the leading truncation error terms are given by:

Linear interpolaration scheme: $\dfrac{ch^2}{12}\dfrac{\partial^3 u}{\partial x^3}$

QUICK scheme: $-\dfrac{ch^2}{24}\dfrac{\partial^3 u}{\partial x^3}$

MUSCL scheme: $+\dfrac{ch^3}{12}\dfrac{\partial^4 u}{\partial x^4}$

If these terms only contribute to Gibbs' phenomenon, then for the linear interpolaration scheme we have a superimposed *right-running wave* that will cause a phase mismatch to the right of discontinuities, while no *left-running waves* will be visible, as seen in the figure. Also, among these three schemes, LI scheme adds the maxmimum amount of dissipation. This is also evident from the massive smearing of the discontinuity. The situation revealed for the QUICK and MUSCL scheme is slightly more complicated. Note that the choice of κ for the MUSCL removes third derivative term altogether. Also, for the QUICK scheme the third derivative represents a right-running wave. However, for both the schemes one notices instead a left-running wave. It is related to the property of Gibbs' phenomenon itself. At the discontinuity it is seen that the oscillations increase with grid refinement. It is related to the fact that at the discontinuity the higher order derivatives become more and more important and they cannot be ignored on the basis of the order of truncation error terms. Notice that for both these schemes spurious dispersion associated with fifth and higher derivative terms are of opposite signs to that of the third derivative term. They indicate left-running waves due to the fifth derivative terms, as seen in Figures 12.12 and 12.13. The maximum contribution to the odd derivatives come from the second set of terms in Equation (12.71) and this is clearly a left-running wave. Note that the MUSCL scheme displays lesser dispersion due to the leading role played by the fourth derivative term.

12.12 Introduction to Finite Element Method

Here the essential of Galerkin finite element method is discussed and compared with other discrete methods. The dependent variables in this method are represented as a sum of functions with prescribed spatial variation. The associated coefficients with these space dependent functions are considered as a function of time. Substitution of this in the governing space-time dependent PDE converts it to a time dependent ordinary differential equation. The resulting ODE is usually solved by standard stable algorithms. The same principle can be used for steady problems, where instead of solving an ODE, one needs to solve a set of algebraic equations.

As an illustration, consider the following unsteady problem:

$$\frac{\partial u}{\partial t} + L(u) = 0 \quad \text{for} \quad a \leq x < b, \tag{12.73}$$

Now one approximates $u(x,t)$ by a finite series

$$u(x,t) = \sum_{j=1}^{N} u_j(t)\phi_j(x). \tag{12.74}$$

The functions $\phi_j(x)$ are called the basis functions and in finite element methods these are the Chapeau functions with a compact support as shown in Figure 12.14.

The jth basis function shown above extends from $(j-1)\Delta x$ to $(j+1)\Delta x$ with a maximum value of one at $x = j\Delta x$. The basis function decreases linearly to zero from this maximum in either direction till the end of the support, and is zero everywhere else.

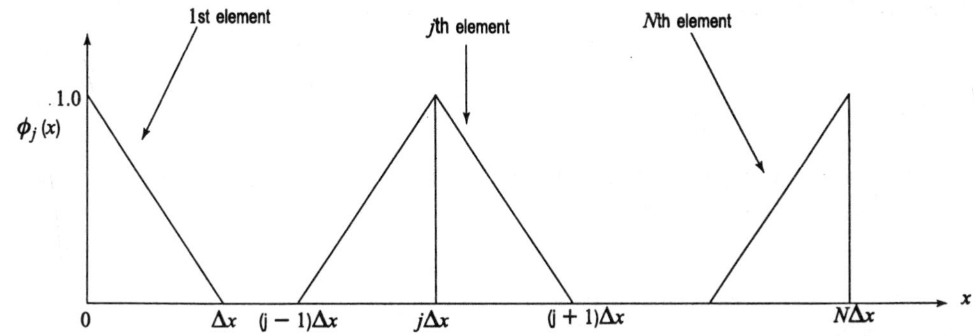

Fig. 12.14 *The commonly used Chapeau functions used for Galerkin finite element method. Note the special one-sided functions near the domain ends.*

Thus,

$$\phi_j(x) = \begin{cases} 0 \quad \text{for} \quad x > (j+1)\Delta x \quad \text{or} \quad x < (j-1)\Delta x \\ [x-(j-1)\Delta x]/\Delta x \quad \text{for} \quad (j-1)\Delta x \le x \le j\Delta x \\ [(j+1)\Delta x - x]/\Delta x \quad \text{for} \quad j\Delta x \le x \le (j+1)\Delta x \end{cases} \tag{12.75}$$

The Galerkin method involves substituting Equation (12.74) into Equation (12.73) and then multiplying the resulting equation by $\phi_i(x)$ and integrating over the whole domain as follows:

$$\sum_{j=1}^{N} \frac{du_j}{dt} \int_a^b \phi_i \phi_j dx + \int_a^b \phi_i L\left(\sum_{j=1}^{N} u_j \phi_j\right) dx = 0 \quad \text{for} \quad i = 1, 2, \ldots N. \tag{12.76}$$

Now,

$$\sum_{j=1}^{N} \frac{du_j}{dt} \int_a^b \phi_i \phi_j dx = \sum_{j=1}^{N} \frac{du_j}{dt} \int_{(i-1)\Delta x}^{(i+1)\Delta x} \phi_i \phi_j dx.$$

The integral on the right hand side will have non-zero contribution only when $j = (i-1), i$ and $(i+1)$. To facilitate evaluation of the integral, let us introduce, $\xi = x - i\Delta x$ in the above to get

$$\sum_{j=1}^{N} \frac{du_j}{dt} \int_a^b \phi_i \phi_j dx = -\frac{du_{i-1}}{dt} \int_{-\Delta x}^{0} \frac{\xi(\xi + \Delta x)}{\Delta x^2} d\xi + 2\frac{du_i}{dt} \int_{-\Delta x}^{0} \frac{(\xi + \Delta x)^2}{\Delta x^2} d\xi$$
$$+ \frac{du_{i+1}}{dt} \int_0^{\Delta x} \frac{\xi(\Delta x - \xi)}{\Delta x^2} d\xi. \tag{12.77}$$

The integrals when performed and simplified gives

$$\sum_{j=1}^{N} \frac{du_j}{dt} \int_a^b \phi_i \phi_j dx = \left[\frac{1}{6}\frac{du_{i-1}}{dt} + \frac{4}{6}\frac{du_i}{dt} + \frac{1}{6}\frac{du_{i+1}}{dt} \right] h. \tag{12.78}$$

Similarly the spatial derivative term can also be simplified, once the form of the $L(u)$ is known exactly. A typical example follows.

Example

Find out the finite element representation when $L(u) = \dfrac{\partial^2 u}{\partial x^2}$. Here,

$$\int_a^b \phi_i L(\sum u_j \phi_j) dx = \sum_{j=1}^{N} u_j \int_a^b \phi_i \frac{d^2 \phi_j}{dx^2} dx.$$

For the piece-wise linear basis function given in Equation (12.75), the above expression cannot be directly evaluated. However, this can be avoided by integrating the right hand side by parts.

$$\sum_{j=1}^{N} u_j \int_a^b \phi_i \frac{d^2 \phi_j}{dx^2} dx = \sum_{j=1}^{N} u_j \int_a^b \left[\frac{d}{dx}\left(\phi_i \frac{d\phi_j}{dx} \right) - \frac{d\phi_i}{dx}\frac{d\phi_j}{dx} \right] dx$$

The first term $\left[\phi_i \frac{d\phi_j}{dx}\right]_a^b$ vanishes as $\phi_i = 0$ at $x = a$ and b, unless $i = 1$ and /or N.
Thus ,

$$\int_a^b \phi_i \frac{\partial^2 u}{\partial x^2} dx = -\sum_{j=1}^{N} u_j \frac{d\phi_i}{dx}\frac{d\phi_j}{dx} dx. \tag{12.79}$$

From Equation (12.75) one has

$$\frac{d\phi_j}{dx} = \begin{cases} 0 \quad \text{for} \quad x > (j+1)\Delta x \quad \text{or} \quad x < (j-1)\Delta x \\ \frac{1}{\Delta x} \quad \text{for} \quad x_{j-1} < x < x_j \\ -\frac{1}{\Delta x} \quad \text{for} \quad x_j < x < x_{j+1}. \end{cases}$$

Using these expressions in Equation (12.79) and simplifying one gets the following:

$$-\sum_{j=1}^{N} u_j \int_a^b \frac{d\phi_i}{dx}\frac{d\phi_j}{dx} dx = \frac{u_{i-1}\Delta x - 2u_i \Delta x + u_{i+1}\Delta x}{\Delta x^3}. \tag{12.80}$$

This expression is valid for $i = 2$ to $(N-1)$ and is the same as one would have for finite difference formulation using second order central differencing.

12.13 Finite Element Formulation for Advection Equation

Consider the advection equation

$$\frac{\partial u}{\partial t} + c\frac{\partial u}{\partial x} = 0 \quad \text{for} \quad c > 0.$$

The time derivative discretization is already evaluated and given by Equation (12.78). The spatial derivative term is evaluated as

$$\int_a^b \phi_i c \frac{du}{dx} dx = c \sum_{j=1}^{N} u_j \int_a^b \phi_i \frac{d\phi_j}{dx} dx = \frac{c}{2\Delta x}(u_{i+1} - u_{i-1}).$$

The resulting discrete equation is then given by

$$\frac{1}{6}\left(\frac{du_{i+1}}{dt} + 4\frac{du_i}{dt} + \frac{du_{i-1}}{dt}\right) + \frac{c}{2\Delta x}(u_{i+1} - u_{i-1}) = 0. \tag{12.81}$$

In finite element formulation one does discretize various derivatives independently. However, with the help of the linear basis functions one derives a relationship among spatial and temporal derivatives as given above in Equation (12.81) for one-dimensional wave equation.

This provides a unique compact relationship among the derivatives. Equation (12.81) provides the semi-discrete *numerical dispersion relation* given by

$$\omega = \frac{3c}{\Delta x}\frac{\sin k\Delta x}{(2 + \cos k\Delta x)} \tag{12.82}$$

and the fully discretized numerical dispersion relation is given by

$$\frac{(e^{-i\omega\Delta t} - 1)}{6\Delta t}[e^{ik\Delta x} + 4 + e^{-ik\Delta x}] + \frac{c}{2\Delta x}[e^{ik\Delta x} - e^{-ik\Delta x}] = 0.$$

or

$$e^{-i\omega\Delta t} = 1 - 3iN_c\frac{\sin k\Delta x}{2 + \cos k\Delta x}. \tag{12.83}$$

Similarly, one can also obtain the amplification factor as

$$G = 1 - 3iN_c\frac{\sin k\Delta x}{2 + \cos k\Delta x}. \tag{12.84}$$

We will not use the above expressions to compare the performance of finite element discretization, as they are valid only for problems with periodic boundary condition.

From Equation (12.81), one can also obtain k_{eq} for evaluating first derivative. The time derivative is evaluated at three neighbouring nodes and this modifies the CD_2 expression for $\frac{\partial u}{\partial x}$ which is evaluated in Equation (12.81). Using the spectral representation

$$u(x,t) = \int U(k,t)e^{ikx}dk \tag{12.85}$$

in Equation (12.81), one gets

$$\int \left\{ \frac{1}{6}[e^{-ik\Delta x} + 4 + e^{ik\Delta x}]\frac{dU}{dt} + \frac{c}{2\Delta x}(e^{ik\Delta x} - e^{-ik\Delta x})U \right\} e^{ikx}dk = 0.$$

Hence one has to solve the following ordinary differential equation for each wave number.

$$\frac{dU}{dt} + \frac{6}{(4 + 2\cos k\Delta x)} c \frac{2i\sin(k\Delta x)}{2\Delta x} U = 0.$$

Therefore, $$\frac{k_{\text{eq}}}{k} = \frac{3}{(2+\cos k\Delta x)}\left(\frac{\sin k\Delta x}{k\Delta x}\right). \tag{12.86}$$

It has to be noted that the discrete equation (Equation (12.81)) cannot be used at the immediate neighbourhood of boundary points. At these points, both the time and space derivative terms needs to be modified based on the altered basis functions (see Figure 12.14).

The equivalent discrete equations for the boundary nodes can be obtained as,

$$\left.\begin{aligned} &\text{for} \quad i = 1: \frac{1}{6}\left[4\frac{du_i}{dt} + 2\frac{du_{i+1}}{dt}\right] + \frac{c}{\Delta x}(u_{i+1} - u_i) = 0 \\ &\text{for} \quad i = N: \frac{1}{6}\left[4\frac{du_i}{dt} + 2\frac{du_{i-1}}{dt}\right] + \frac{c}{\Delta x}(u_i - u_{i-1}) = 0 \end{aligned}\right\}. \tag{12.87}$$

For these points the equations corresponding to Equations (12.82) to (12.84) can be obtained.

In Fig. 12.15 the spectral resolution and numerical dissipation of the Galerkin finite element method is shown for different points with $N = 30$. All the interior points have very high spectral resolution and zero numerical dissipation. However, the boundary points show large spectral distortion at intermediate wave numbers and steep filtering at higher wave numbers.

For $j = 1$ one notices large *anti-diffusion* and for $j = 30$ there is large diffusion. For problems with Dirichlet boundary conditions, this attribute will not cause any problem. However, for high Reynolds number flow simulations, numerical dissipationless method displays wiggles. Also, numerical instability can seep through the near-boundary points. This would eventually lead to solution breakdown due to aliasing.

Due to the above mentioned problems of Galerkin finite element method, there are a few non-Galerkin methods in usage. They generally bring in numerical dissipation under different guises for numerical stabilization (and controlling aliasing unknowingly!). One of the method is the popular Petrov-Galerkin method. In developing the discrete equation for this method the test functions are not taken as the basis functions. For the one-dimensional wave equation, we show the discrete equation for the streamwise-upwind Petrov-Galerkin (SUPG) method that is given by (as in Grescho & Sani (1998)),

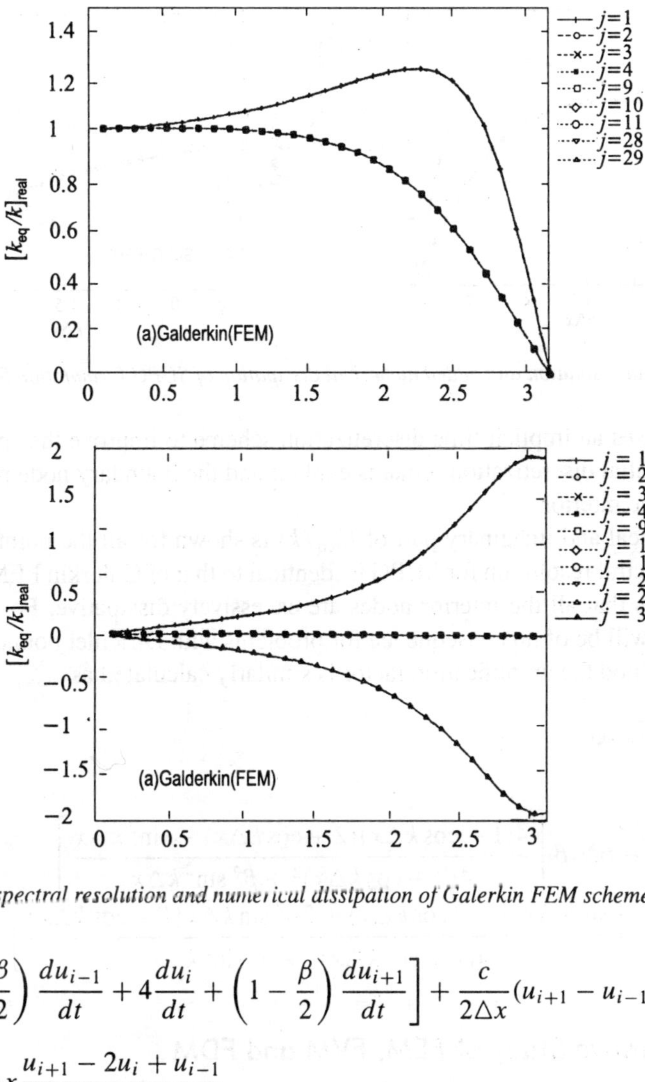

Fig. 12.15 *The spectral resolution and numerical dissipation of Galerkin FEM scheme for the whole domain.*

$$\frac{1}{6}\left[\left(1+\frac{\beta}{2}\right)\frac{du_{i-1}}{dt}+4\frac{du_i}{dt}+\left(1-\frac{\beta}{2}\right)\frac{du_{i+1}}{dt}\right]+\frac{c}{2\Delta x}(u_{i+1}-u_{i-1})$$

$$=\beta c\Delta x\frac{u_{i+1}-2u_i+u_{i-1}}{\Delta x^2}. \tag{12.88}$$

If one chooses $\beta = 1/2$, one recovers the discrete equation for first order upwind finite difference scheme. However, to *reduce* phase error, Raymond & Garder (1976) suggested using $\beta = 0.26$ and this value is used here too. For the discrete equation for SUPG method, as given in equation (12.86), the equivalent wave number that defines the spatial derivative is given by

$$\frac{k_{eq}}{k}=\left[\frac{6}{4+2\cos k\Delta x-i\beta\sin k\Delta x)}\frac{\sin k\Delta x-i\beta(1-\cos k\Delta x)}{k\Delta x}\right]. \tag{12.89}$$

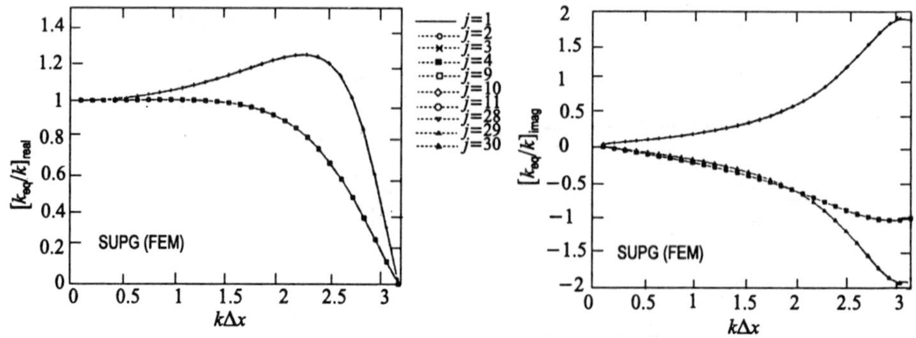

Fig. 12.16 *The spectral resolution and added numerical dissipation of SUPG formulation for the full domain.*

Note that the FEM uses an implicit time discretization scheme to improve the spectral accuracy. In the process the spatial discretization remains explicit and the boundary node problems, if any, do not percolate to the interior.

In Fig. 12.16 the real and imaginary part of (k_{eq}/k) is shown for all the points for the SUPG formulation. The spectral resolution for SUPG is identical to that of Galerkin FEM. However, the imaginary part shows that all the interior nodes are excessively dissipative. For $j = 1$ one has *anti-diffusion*, but it will be of no consequence for problems with Dirichlet boundary condition.

For the SUPG method the amplification factor is similarly calculated as

$$G = G_{\text{real}} + iG_{\text{imag}},$$

where

$$\left.\begin{aligned} G_{\text{real}} &= 1 - 6N_c\beta\left[\frac{4(1-\cos k\Delta x)(2+\cos k\Delta x) - \sin^2 k\Delta x}{4(2+\cos k\Delta x)^2 + \beta^2\sin^2 k\Delta x}\right] \\ G_{\text{imag}} &= -6N_c\left[\frac{\sin k\Delta x(4+2\cos k\Delta x) + 2\beta^2\sin k\Delta x(1-\cos k\Delta x)}{4(2+\cos k\Delta x)^2 + \beta^2\sin^2 k\Delta x}\right]\end{aligned}\right\}. \tag{12.90}$$

12.14 A Comparitive Study of FEM, FVM and FDM

We have alraedy seen the ability of these methods to resolve different length scales via the (k_{eq}/k) vs. $k\Delta x$ plots. The other useful criterion for comparison is the DRP property. In Fig. 12.17, the DRP patches in the $(k\Delta x - \omega\Delta t)$ plane are shown for the Galerkin and SUPG method along with the three FVS-FV methods (LI, QUICK and MUSCL schemes). The time integration is performed using Euler time marching.

Clearly the FE methods displays smallest range of wave numbers and frequencies for which DRP is satisfied and hence would be a poor candidate for DNS. In Fig. 12.18, the DRP patch for these methods are shown when Adams-Bashforth scheme is used for time integration. For the Galerkin formulation, using Equation (12.85) in Equation (12.81), one gets the ODE for each wave number components as

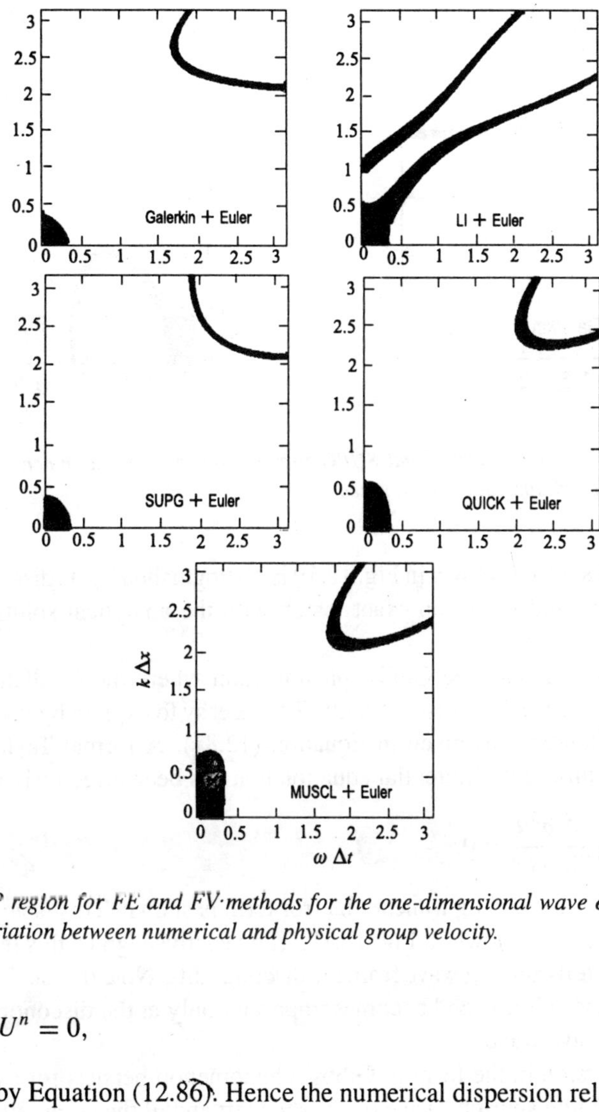

Fig. 12.17 *The DRP region for FE and FV methods for the one-dimensional wave equation shown here for a ±5% variation between numerical and physical group velocity.*

$$\frac{dU^n}{dt} + ik_{\text{eq}}cU^n = 0, \tag{12.91}$$

where k_{eq} is given by Equation (12.86). Hence the numerical dispersion relation is given by,

$$\frac{e^{-i\omega\Delta t} - 1}{\Delta t} + ik_{\text{eq}}c = 0. \tag{12.92}$$

This equation has been used to plot Fig. 12.18.

For all spatial discretization schemes, one notices a general degradation of DRP property with Adams-Bashforth schemes as compared to Euler time integration scheme. In Figs. 12.19 and 12.20, solution of 1D linear wave equation is shown for FE methods. Similar results were shown for higher order FD and compact methods in Figs. 11.29 to 11.34 with identical initial conditions. Thus, a direct comparision is possible among these methods.

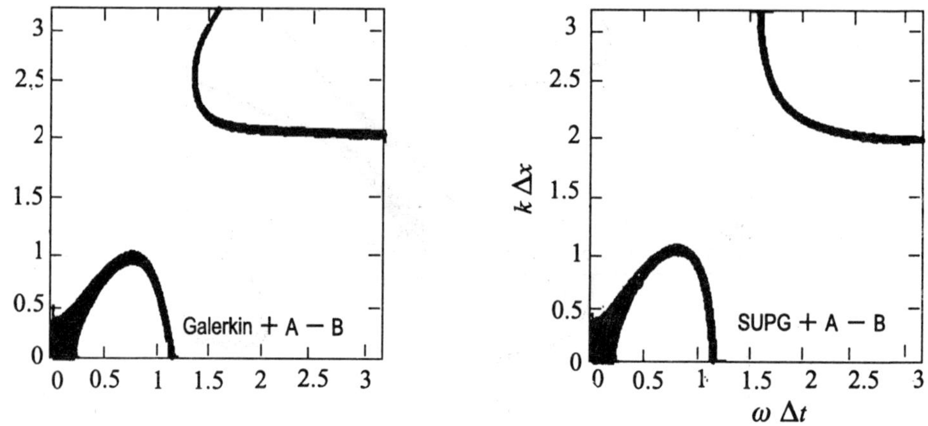

Fig. 12.18 *DRP region for the Galerkin and SUPG formulation for the one-dimensional wave equation for Adams-Bashforth time integration scheme.*

The Galerkin-FEM solution shown in Fig. 12.19 is distinguished by its dissipationless character of spatial discretization and shows an exact match with the analytical solution except near the discontinuity.

At the discontinuity one can see Gibbs' phenomenon where the oscillating spurious waves migrate to the left starting from the discontinuity. This is easy to explain by inspecting the discrete equation for this formulation, as given in Equation (12.81). A formal Taylor series expansion yields the following equivalent differential equation that has been effectively used.

$$\frac{\partial u}{\partial t} + c\frac{\partial u}{\partial x} - c\frac{\Delta x^4}{180}\frac{\partial^5 u}{\partial x^5} = 0. \tag{12.93}$$

In this equation only the leading truncation error term is shown. This is an odd derivative and would produce dispersion as shown in Fig. 12.19. The negative sign of this term implies that the dispersion will cause a left-running wave from the discontinuity. Note that such a term is negligible at the smooth part of the solution and becomes important only at the discontinuity, which acts as a source of such dispersive waves.

This spurious dispersion in the form of Gibbs' phenomenon persists for a long time. As time progresses this left-running spurious wave reaches the left end of the computing domain ($x = 0$). With passage of time, these streams of spurious disturbances add up. As the used scheme is completely non-dissipative, such disturbances continually increases with time, leading to solution breakdown.

In Fig. 12.20 the corresponding results are shown for SUPG formulation. Quite unlike the Galerkin formulation, SUPG formulation is a highly dissipative method. Results do not exhibit any Gibbs' phenomenon and this shows a totally attenuated solution without any match with the exact solution. One can proceed to show that the SUPG formulation given in Equation (12.88) equivalently solves the following equation:

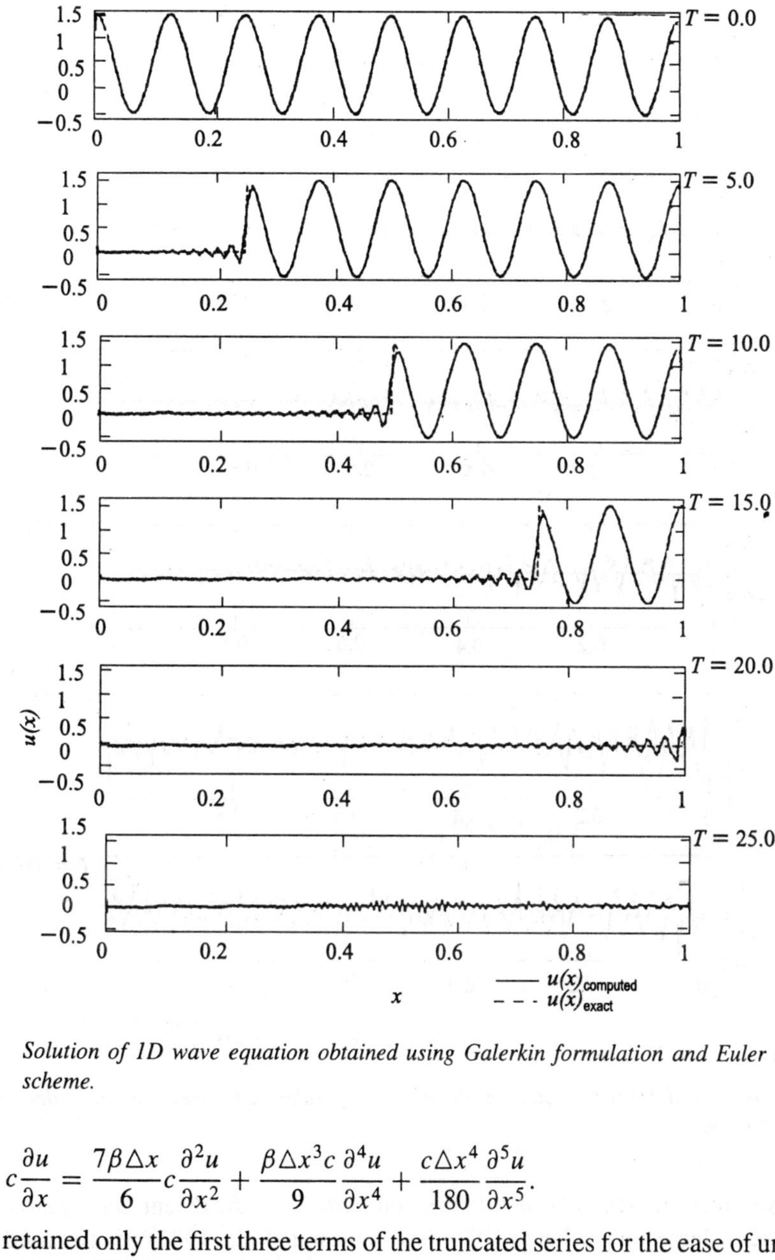

Fig. 12.19(a) *Solution of 1D wave equation obtained using Galerkin formulation and Euler time integration scheme.*

$$\frac{\partial u}{\partial t} + c\frac{\partial u}{\partial x} = \frac{7\beta \Delta x}{6} c \frac{\partial^2 u}{\partial x^2} + \frac{\beta \Delta x^3 c}{9} \frac{\partial^4 u}{\partial x^4} + \frac{c \Delta x^4}{180} \frac{\partial^5 u}{\partial x^5}. \tag{12.94}$$

We have retained only the first three terms of the truncated series for the ease of understanding. The poor results of SUPG formulation is mainly due to the presence of second derivative term. This term introduces numerical dissipation which is proportional to Δx. The fourth derivative term also introduces dissipation, but at higher wave numbers and by a smaller amount, as this is proportional to Δx^3. The last term on the right hand side of Equation (12.92) is a dispersion term and is the same one seen in Galerkin formulation. However, this term is a few orders of

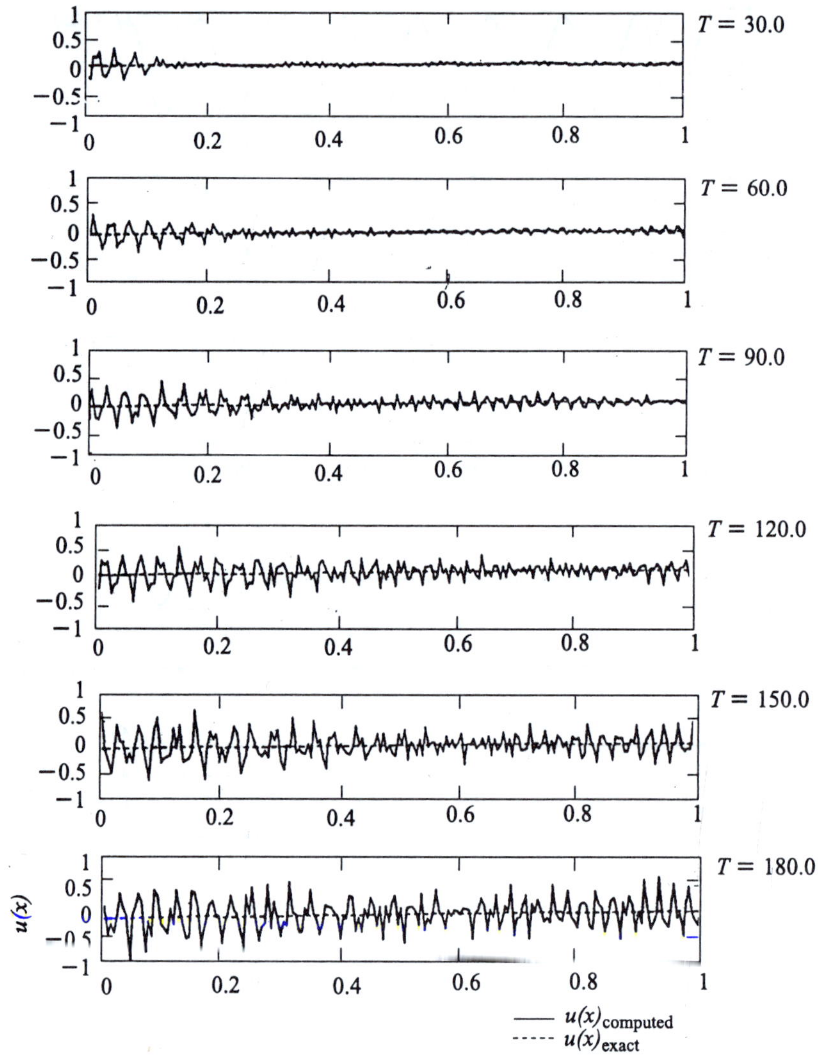

Fig. 12.19(b) *Solution of 1D wave equation obtained using Galerkin formulation and Euler time integration scheme.*

magnitude lower than the second order dissipation term. To some extent, this approach is similar to what is done in Jameson's scheme (JST) (see Jameson et al. (1981)) that is used for solving compressible flow problems with shocks in finite volume and finite difference framework. In JST scheme the second order dissipation term is switched on near the shock (discontinuity) based on some pressure based indicators. Also, in JST the fourth dissipation term is added at all points for numerical stability.

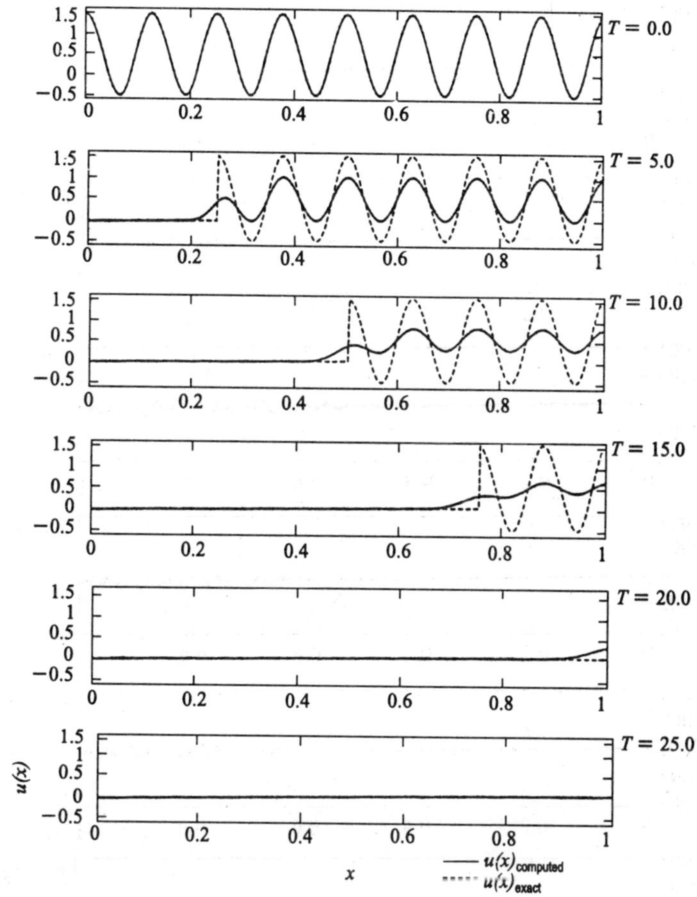

Fig. 12.20(a) *Solution of 1D wave equation obtained using SUPG formulation and Euler time integration scheme.*

However, despite adding massive dissipation everywhere, the presence of the dispersion term is visible through the Gibbs phenomenon at and beyond $t = 30$.

In contrast to the FEM, the FV methods did not show large dispersion generated Gibbs' phenomenon because the leading truncation terms have small dispersion error. Results were shown for the solution of one-dimensional wave equation by using LI, QUICK and MUSCL schemes in Figs. 12.11 to 12.13 respectively. The LI method shows large dissipation near the discontinuity and a general phase error at all points is visible at $t = 10$ onwards. Unlike the finite element methods, because of adequate dissipation and very low dispersion errors the results match well with exact solution beyond $t = 20$.

In Fig. 12.12 the results for the QUICK scheme is shown and the quality of result is better than the LI scheme results. This scheme has lesser dissipation and hence the mismatch at the discontinuity is lower. However, this scheme has larger dispersion error and it is visible in the solution.

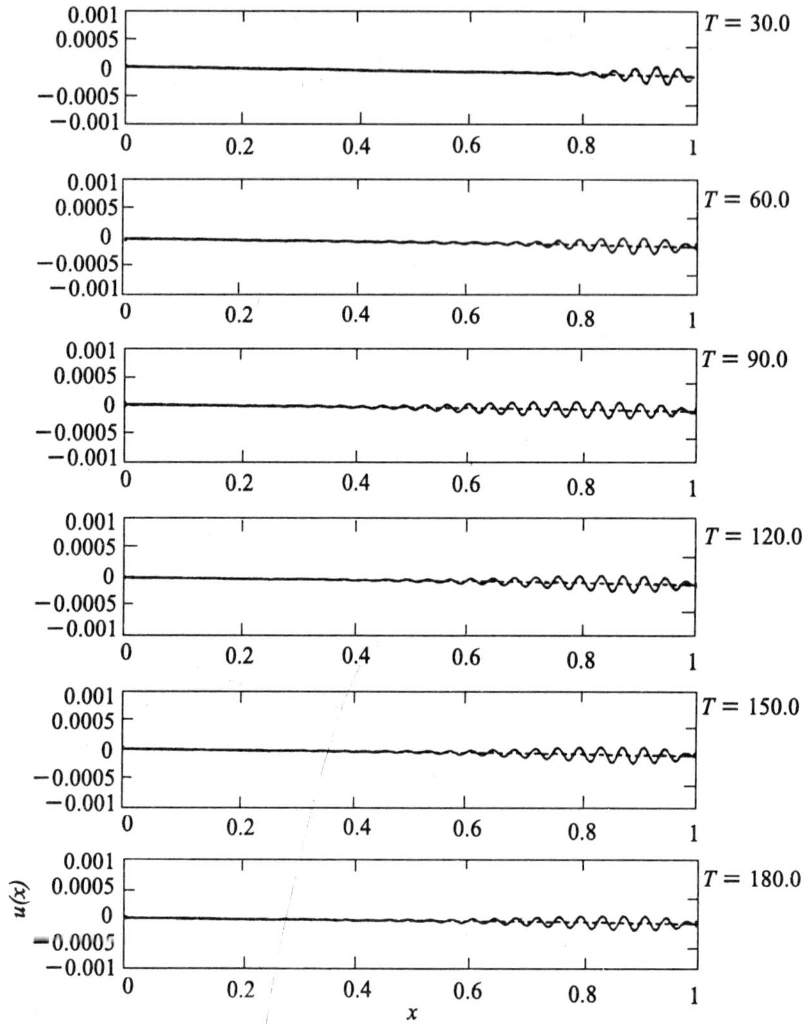

Fig. 12.20(b) *Solution of 1D wave equation obtained using SUPG formulation and Euler time integration scheme.*

The results in Fig. 12.13 are obtained using the MUSCL scheme and here the disperison error is even smaller. The scheme does not capture the discontinuity well, like the previous two methods. However, for incompressible flows this is not going to be a major issue.

The problem of Gibbs' phenomenon can be essentially explained through Equation (12.72). The equivalent equation which is solved by FVS method is given by this equation, and it is seen that the dissipation added is always present except when $\kappa = 1$, that is, when we use arithmatic averaging to calculate the cell fluxes, but the dispersion varies depending on the choice of κ. For example, for the MUSCL scheme ($\kappa = 1/3$) the third-derivative dispersion term is absent, but the fifth term indicates a right-propagating dispersion. For the QUICK scheme ($\kappa = 1/2$), the third derivative and fifth derivative terms indicate dispersion in the opposite directions which is visible at the discontinuity. Also, the coefficient of fifth derivative term is one order lower for FV methods as they are multipled by $(\Delta x)^5$ instead of $(\Delta x)^4$ for the FEM. Also, note that the sign of the fourth derivative term is opposite for QUICK scheme as opposed to that for MUSCL scheme. By and large, MUSCL is less dispersive than QUICK scheme.

Exercise

1. Derive the expression for the amplification factors of different $(\varepsilon - \kappa)$ schemes given in Equations (12.57) and (12.58).
2. Derive the amplification factors for the $(\varepsilon - \kappa)$ schemes for (a) Adams-Bashforth time integration scheme, (b) RK_2 scheme, (c) RK_3 scheme and (d) RK_4 scheme.
3. Consider the initial solution of 1D wave equation shown in the top panel of Fig. 12.11. Obtain the FFT of this solution and plot the spatial derivatives (as obtained by spectral method) upto 7th order and compare their relative importance in determining truncation errors at discontinuities. (Note: Take an extended domain while calculating the FFT of the initial solution).
4. Develop the equivalent differential equations that are solved by Galerkin and SUPG method (Equations (12.91) and (12.92)).

References

[1] ADAMS, N. A., Ph. D. Thesis submitted to Technische Univ. Munich; also availiable as DLR report DLR-FB 93–29, Göttingen (1993).

[2] BORIS, J. P. & BOOK, D. L., Solution of the continuity equation by the method of flux corrected transport, *J. Comp. Phys.*, **16**, 85–129 (1976).

[3] CAUSON, D. M., INGRAM, D. M. & YANG, G., On application of high resolution shock capturing methods to unsteady flows, in *Numerical Methods for wave propogation*, (eds. EF Toro, JF Clarke), Kluwer, Netherlands, 145–171 (1998).

[4] GOTTLIEB, S. & SHU, C.W., Total variation diminishing Runge-Kutta schemes, NASA ICASE report 96–50 (1996).

[5] GRESCHO, P.M. & SANI, R.L., *Incompressible Flow and The Finite Element Method*, Wiley, Chichester, England (1998).

[6] HARAS, Z. & TA'ASAN, S., Finite difference scheme for long time integration, *J. Comp. Phys.*, **114**, pp 265–279 (1994).

[7] HIRSCH, C., *Numerical Computation of Internal and External Flows. Vol I and II: Computational Methods for Inviscid and Viscous Flows*, Wiley (1990).

[8] JAMESON, A., SCHMIDT, W. & TURKEL, E., Numerical solution of the Euler equations by finite volume methods using Runge-Kutta time stepping schemes, *AIAA paper* 81–1259 (1981).

[9] KALLINDERIS, J. G. & BARON, J. R., The finite volume approach for the Navier-Stokes equations, in *Computational Methods in Viscous Aerodynamics*, Eds. T.K.S. Murthy & C. A. Brebbia, Elsevier (1990).

[10] LE, H. & MOIN, P., An improvement of fractional time methods for the incompressible Navier-Stokes equation, *J. Comp. Phys.*, **92**, 369–379 (1991).

[11] VAN LEE, B., Towards the ultimate conservative difference scheme I. The quest for monotonicity, *Lect. Notes in Physics*, **18**, 163–168. Springer Verlag, Berlin (1973).

[12] VAN LEER, B., Towards the ultimate conservative difference schemes 5. A second order sequel to Godunov's method, *J. Comp. Phys.*, **32**, 101–136 (1979).

[13] LEONARD, B.P., A stable and accurate convective modeling procedure based on quadratic upstream interpolation, *Comp. Methods App. Mech. & Eng.*, **19**, 59–98 (1979).

[14] RAYMOND, W.H. & GARDER, A., Selective damping in a Galerkin method for solving wave problems with variable grids, *Mon. Weather Rev.*, **104**, 1583–1590 (1976).

[15] SENGUPTA, T.K., SRIDAR D., SARKAR S. & DE, S., Spectral analysis of flux vector spllitting finite volume method, *Int. J. Num. Methods Fluids.*, **37**, 149–174 (2001).

[16] SHU, C.W. & OSHER, S., Efficient implementation of essentially non-oscillatory shock capturing schemes, *J. Comp. Phys.*, **77**, 439–471 (1988).

[17] SHU, C.W. & OSHER, S., Efficient implementation of essentially non-oscillatory shock capturing schemes II, *J. Comp. Phys.*, **83**, 32–78 (1989).

[18] VINOKUR, M., An analysis of finite-difference and finite volume formulations of conservations, *J. Comp. Phys.*, **81**, 1–52 (1989).

[19] ZHONG, X., High-order finite difference schemes for numerical simulattion of hypersonic boundary layer transition, *J. Comp. Phys.*, **144**, 622–709 (1998).

13 Solution of Navier-Stokes Equation

Having established all the tools of discretization of differential equations and methods of analysis of such discretized equations, we proceed to develop methods for solving the equations of steady and unsteady incompressible viscous flows given by Navier-Stokes equation. The various versions of Navier-Stokes equation in primitive and derived variables, expressed in inertial and non-inertial frames, were given in Chapter 2. Various versions of pressure Poisson equation were also derived in Chapter 2, which are usually required for the accurate evaluation of pressure field.

For unsteady flows, the Navier-Stokes equation can be viewed as evolution of vorticity, a primary physical quantity, for both laminar and turbulent flows. Hence the vorticity transport equation (Equation (2.48)) is central for the analysis and solution of Navier-Stokes equation. The attendant velocity field can be obtained from the solution of the stream function equation (for 2D flows) or vector potential equation (for 3D flows). Occasionally, Poisson equations relating velocity and vorticity field are solved for three-dimensional flows in *vorticity–velocity formulation*. As discussed in Chapter 2, the *vorticity–stream function (or vector potential formulation* is preferred over *vorticity–velocity formulation* whenever more accurate solution is desired. This is due to the former's ability to satisfy mass conservation more accurately. For incompressible flows, the velocity field obtained as solution of the latter formulation may not be divergence-free or solenoidal. Such inaccuracies may change dynamics of flows, especially for Direct Numerical Simulations (DNS) of turbulent and transitional flows.

There is another aspect of Navier-Stokes computations that should not be lost sight of. VTE being an evolution equation, all the analysis tool developed in Chapter 10 and 11 apply. Such solution processes are faster computationally. In contrast, the Poisson equation being an elliptic equation is slower and difficult to solve. In a typical unsteady simulation more than three quarters of time is usually spent in solving the stream function equation in (ψ–ω) formulation. This is one of the reasons which make vorticity vector potential formulation a formidable one to solve. One would be required to solve three Poisson equations—one for each components of vector potential. Additionally, of course, one would be required to solve another Poisson equation for the pressure field.

In contrast, the momentum equations in primitive variables are parabolic differential equations in time and should be able to utilize all the tools developed for convection–diffusion equation. However, the extension is not obvious due to the presence of the pressure gradient term. The presence of pressure gradient term – and not pressure – gives rise to what is known as pressure-velocity coupling problem. This is due to the fact that there is no evolution equation for pressure in explicit form for incompressible flows. In its place we have the condition of incompressibility to be satisfied. In primitive variable formulations this restriction is often expressed in terms of a Poisson equation for the pressure. This is an implicit equation for pressure. Such a problem does not exist for compressible flows where an evolution equation exists for density which can be related to pressure through an equation of state.

It is for the above reasons that we emphasize the role played by different formulations to solve Navier-Stokes equation for incompressible flows.

13.1 Stream Function–Vorticity Formulation (Two-Dimensional Flows)

The governing equations, in inertial frame of reference, are given here by

$$\nabla^2 \psi = -\omega, \tag{13.1}$$

$$\frac{D\omega}{Dt} = \frac{1}{Re}\nabla^2 \omega. \tag{13.2}$$

These equations are in non-dimensional form and $Re(= \frac{U_0 L}{\nu})$ is the Reynolds number based on velocity scale (U_0), length scale (L) and kinematic viscosity (ν). We will represent computational methods for these equations in an orthogonal grid (as defined in Chapter 9), via the transformations

$$\left.\begin{aligned} \xi &= \xi(x, y) \\ \eta &= \eta(x, y) \end{aligned}\right\}. \tag{13.3}$$

The scale factors of the orthogonal transformations are given by

$$h_1 = \left[\left(\frac{\partial x}{\partial \xi}\right)^2 + \left(\frac{\partial y}{\partial \xi}\right)^2\right]^{1/2} \quad h_2 = \left[\left(\frac{\partial x}{\partial \eta}\right)^2 + \left(\frac{\partial y}{\partial \eta}\right)^2\right]^{1/2}. \tag{13.4}$$

Equations (13.1) and (13.2), can be written in this frame of reference as

$$\frac{\partial}{\partial \xi}\left(\frac{h_2}{h_1}\frac{\partial \psi}{\partial \xi}\right) + \frac{\partial}{\partial \eta}\left(\frac{h_1}{h_2}\frac{\partial \psi}{\partial \eta}\right) = -h_1 h_2 \omega \tag{13.5}$$

and

$$h_1 h_2 \frac{\partial \omega}{\partial t} + h_2 u \frac{\partial \omega}{\partial \xi} + h_1 v \frac{\partial \omega}{\partial \eta} = \frac{1}{Re}\left[\frac{\partial}{\partial \xi}\left(\frac{h_2}{h_1}\frac{\partial \omega}{\partial \xi}\right) + \frac{\partial}{\partial \eta}\left(\frac{h_1}{h_2}\frac{\partial \omega}{\partial \eta}\right)\right]. \tag{13.6}$$

The contravariant components of velocity in the transformed plane are defined as

$$h_2 u = \frac{\partial \psi}{\partial \eta} \text{ and } h_1 v = -\frac{\partial \psi}{\partial \xi}. \tag{13.7}$$

Thus, Equation (13.6) can also be written as

$$h_1 h_2 \frac{\partial \omega}{\partial t} + \left(\frac{\partial \psi}{\partial \eta}\frac{\partial \omega}{\partial \xi} - \frac{\partial \psi}{\partial \xi}\frac{\partial \omega}{\partial \eta}\right) = \frac{1}{Re}\left[\frac{\partial}{\partial \xi}\left(\frac{h_2}{h_1}\frac{\partial \omega}{\partial \xi}\right) + \frac{\partial}{\partial \eta}\left(\frac{h_1}{h_2}\frac{\partial \omega}{\partial \eta}\right)\right], \tag{13.8}$$

where the quantity inside parenthesis on the left hand side is the Arakawa Jacobian

$$J(\psi, \omega) = \frac{\partial \psi}{\partial \eta}\frac{\partial \omega}{\partial \xi} - \frac{\partial \psi}{\partial \xi}\frac{\partial \omega}{\partial \eta}$$

defined in Equation (10.101). Thus the VTE can be written as

$$h_1 h_2 \frac{\partial \omega}{\partial t} + J(\psi, \omega) = \frac{1}{Re}\left[\frac{\partial}{\partial \xi}\left(\frac{h_2}{h_1}\frac{\partial \omega}{\partial \xi}\right) + \frac{\partial}{\partial \eta}\left(\frac{h_1}{h_2}\frac{\partial \omega}{\partial \eta}\right)\right]. \tag{13.9}$$

To calculate the pressure field, additionally a Poisson equation needs to be solved. Earlier, for the solution of Navier-Stokes equation using *stream function–vorticity formulation*, loads and moments used to be calculated from the surface pressure obtained by integrating the normal vorticity gradient at the surface. It has been conclusively established now, that this way of calculating surface pressure leads to inaccurate results. For accuracy, pressure Poisson equation (as discussed in Chapter 2) is to be solved. The following equation provides the most accurate results.

$$\nabla^2 P = \nabla \cdot (V \times \omega), \tag{13.10}$$

where P is the total pressure given by $P = p + \frac{1}{2}|V|^2$. This equation is particularly favoured because P is a good indicator of the total mechanical energy of the flow. This equation (Equation (2.63)) was derived in Chapter 2 by taking divergence of the Navier-Stokes equation in the rotational form (Equation (2.66)). This equation expressed in the orthogonal curvilinear coordinates is given by

$$\frac{\partial}{\partial \xi}\left(\frac{h_2}{h_1}\frac{\partial P}{\partial \xi}\right) + \frac{\partial}{\partial \eta}\left(\frac{h_1}{h_2}\frac{\partial P}{\partial \eta}\right) = \frac{\partial}{\partial \xi}(h_2 v \omega) - \frac{\partial}{\partial \eta}(h_1 u \omega). \tag{13.11}$$

Before we start discussing about the detailed steps about the solution procedure, it is worthwhile to show a flow chart which is used for solving Navier-Stokes equation for external flow past a body.

In the flow chart (Fig. 13.1) the solution procedure for unsteady flows is indicated. It is worthwhile to discuss the features of unsteady flows and what makes them unsteady. General unsteady flows can be classified based on whether: (a) their unsteady responses are due to dynamic disturbances or (b) the unsteadiness is self-generated and self-sustaining due to flow instabilities. In the latter case also, the presence of dynamic disturbances are necessary. These omnipresent disturbances have relatively smaller amplitudes, but they produce large effects due to flow receptivities and instabilities. So far CFD as a tool has been predominantly used to understand and explain the effects of large disturbances as classified under (a) above. For this class of problems the dynamic disturbances are either due to impressed translational and rotational acceleration/decelerations, or due to large perturbations inherent with the oncoming flow. When flow instabilities are not involved for the levels of imposed disturbances, one usually studies what is known as the *buffeting* problem.

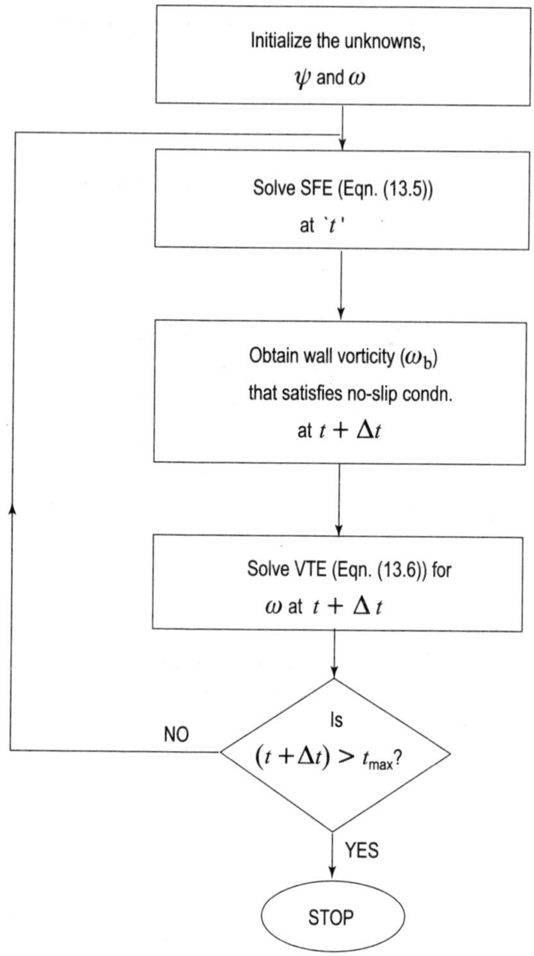

Fig. 13.1 *Flow chart for solving Navier-Stokes equation in* $(\psi\text{–}\omega)$ *formulation.*

Relatively fewer CFD efforts are recorded for the class of problems in (b). This is due to the following reasons. First and foremost, it is due to the inability of many numerical methods in use to resolve the excited spatial and temporal scales. This essentially requires higher resolution and accuracy for the solution process. Higher resolution needs larger computers with low accuracy schemes. On the contrary, development of higher accurate methods allows one to solve these problems. When more powerful computers with larger memory are available, one can solve geometrically challenging problems. In Section 13.9 we will see some results for a complicated physically problem, for which the geometry is simple, but requires very high level of accuracy.

For the class of problems in (a), unsteady flow past airfoils have been investigated quite often. We will use this problem to discuss some of the computational issues here. Various applications of this problem relevant to aircraft, helicopters and turbo-machineries have been discussed in

McCroskey (1982) and relates mostly to dynamic stall phenomenon, where an aerofoil executing pitching oscillation attains instantaneous angles of attack which are higher than the static stall angle for that aerofoil. There is another class of externally driven viscous flows, where unsteadiness is enhanced by the imposed translational accelerations/ decelerations. These problems are important, as they relate to (i) flow establishment in an experimental facility; (ii) in parachutes/ aerodynamic decelerator applications; and (iii) wave forces experienced by off-shore structures. The acceleration/ deceleration processes produce large transient loads, which are not understood completely.

For flow establishment in computations, the process is often started impulsively. By definition, impulsive start implies application of concentrated or distributed forces of large magnitude for a very short period of time. Unfortunately, true impulsive motion is neither feasible nor can be seen in natural environment.

Despite this, impulsive start is the basis of many theoretical and computational investigations. The adoption of impulsive start in analysis is prompted by the fact that in many cases this is a good approximation and also because it brings in significant simplification of the formulation for theoretical and computational investigations. The earliest study of typical impulsive viscous flow problem is due to Stokes (1851) and Rayleigh (1911) which involved motion of an infinite flat plate in its own plane. Similarly, Stuart (1963) and Schlichting (1979) have discussed a class of problems involving establishment of free stream speed with uniform acceleration.

With these basic remarks, we discuss the various modules of the flow chart shown in Fig. 13.1.

13.2 Start-up and Initial Condition

The various start-up cases used for stimulating unsteady flows discussed here are shown in Fig. 13.2. These are simple idealizations, and in real flows the initials conditions can take more complicated descriptions.

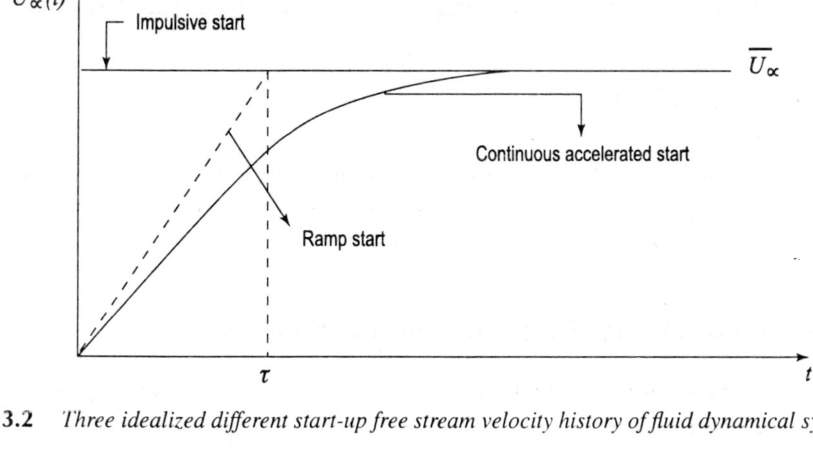

Fig. 13.2 *Three idealized different start-up free stream velocity history of fluid dynamical systems.*

Case I: The impulsive start case is given by

$$\left.\begin{aligned} U_\infty(t) &= 0 \text{ for } t < 0 \\ &= \bar{U}_\infty \text{ for } t \geq 0 \end{aligned}\right\}. \qquad (13.12)$$

The assumption of impulsive motion results in significant simplification, as boundary conditions become time independent. One applies the no-slip boundary condition for $t \geq 0$. In scope the initial condition is identical to that used for solving Stokes' first problem. Immediately after an impulsive start, the flow field is inviscid and irrotational everywhere and this can be used as the initial condition. The imposition of no-slip condition at $t = 0$, creates a shear layer of infinitesimal thickness on the no-slip wall. The role of VTE is then to diffuse and convect the generated vorticity at the wall. Theoretically, the incoming flow has singular acceleration. The role of diffusion and convection at early times is to moderate this singularity.

According to Telionis (1981): *"The early stages of impulsive viscous flows contain valuable information on flow properties that control the subsequent development of the phenomenon and a thorough understanding of its properties is essential in the study of unsteady flows"*. This issue will be discussed again in Section 13.10.

Non-singular acceleration cases shown in Fig. 13.2 are given by the following two cases.

Case II: The ramp-start case is given by

$$\left.\begin{aligned} U_\infty(t) &= \bar{U}_\infty t/\tau \text{ for } t \leq \tau \\ &= \bar{U}_\infty \text{ for } t > \tau \end{aligned}\right\}, \qquad (13.13)$$

where τ is the characteristic acceleration time. This type of accelerated start-up is occasionally studied experimentally with a view to study the effect of uniform acceleration. For example, Freymuth (1985) studied effects of constant acceleration where the free stream speed continuously accelerated during the experiment.

Case III: The continuous acceleration case is a typical of situation where the final free stream speed is achieved by a continuous changing acceleration. For example, in the water-Ludwig tunnel experiments performed by Morikawa & Grönig (1995), the free-stream speed was varied according to

$$U_\infty(t) = \bar{U}_\infty \tanh\left(\frac{t}{\tau}\right), \qquad (13.14)$$

where $\bar{U}_\infty$ is the final steady state free stream speed and τ is a characteristic acceleration time. In this representation the acceleration is uniform during the early stages followed by a duration when it is non-uniform and finally (beyond 3τ) it decays to zero smoothly.

13.3 Solution of Stream Function Equation (SFE)

The stream function equation given by Equation (13.5) is in the self-adjoint form and is discretized by second order central differencing, as given below.

$$\frac{\partial}{\partial \xi}\left(\frac{h_2}{h_1}\frac{\partial \psi}{\partial \xi}\right) = \frac{1}{\Delta \xi^2}\left[\left(\frac{h_2}{h_1}\right)_{i+1/2j}\psi_{i+1j} - \left\{\left(\frac{h_2}{h_1}\right)_{i+1/2j} + \left(\frac{h_2}{h_1}\right)_{i-1/2j}\right\}\psi_{ij} + \left(\frac{h_2}{h_1}\right)_{i-1/2j}\psi_{i-1j}\right] \tag{13.15}$$

and

$$\frac{\partial}{\partial \eta}\left(\frac{h_1}{h_2}\frac{\partial \psi}{\partial \eta}\right) = \frac{1}{\Delta \eta^2}\left[\left(\frac{h_1}{h_2}\right)_{ij+1/2}\psi_{ij+1} - \left\{\left(\frac{h_1}{h_2}\right)_{ij+1/2} + \left(\frac{h_1}{h_2}\right)_{ij-1/2}\right\}\psi_{ij} + \left(\frac{h_1}{h_2}\right)_{ij-1/2}\psi_{ij-1}\right]. \tag{13.16}$$

The stream function equation is an elliptic PDE of second order and requires only one boundary condition on every segment of the boundary. Let us consider the problem of solving the Navier-Stokes equation past an arbitrary 2D geometry as shown in Fig. 13.3.

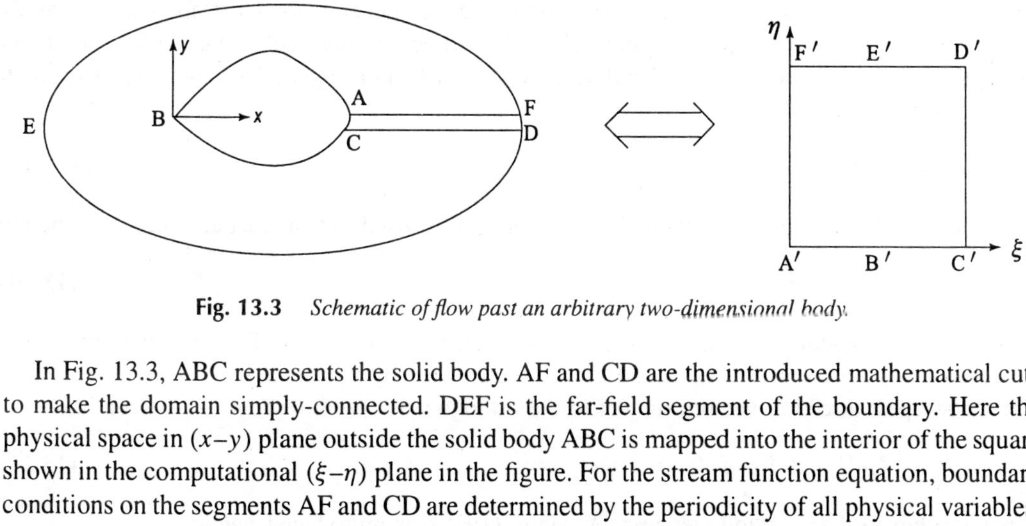

Fig. 13.3 *Schematic of flow past an arbitrary two-dimensional body.*

In Fig. 13.3, ABC represents the solid body. AF and CD are the introduced mathematical cuts to make the domain simply-connected. DEF is the far-field segment of the boundary. Here the physical space in $(x–y)$ plane outside the solid body ABC is mapped into the interior of the square shown in the computational $(\xi–\eta)$ plane in the figure. For the stream function equation, boundary conditions on the segments AF and CD are determined by the periodicity of all physical variables, i.e. $\psi_{AF} = \psi_{CD}$. The boundary condition on the far-field segment, DEF, is obtained by the value of stream function calculated from the velocity of the oncoming flow field. On the solid body ABC, there are two sets of boundary conditions given by no-slip conditions implemented by equating the contravariant components of fluid velocity to the prescribed velocity imposed by the body motion. For a body that is stationary, these corresponds to

$$\psi_{ABC} = \psi(\xi, \eta = 0) \equiv \text{ constant and } \frac{\partial \psi}{\partial \eta}(\xi, \eta = 0) = 0. \tag{13.17}$$

It is conceptually correct to use any one of these conditions in solving the SFE. However, both these conditions are needed to fix the wall vorticity value as will be discussed in the next section.

For the solution of SFE, the first condition given by Equation (13.17) is preferred, because this provides higher accuracy (satisfaction of Neumann condition computationally is an inexact procedure as compared to satisfying Dirichlet condition) and faster convergence during iteratively solving elliptic PDEs for solving SFE.

It was explained in Chapter 7 that the discretization of self-adjoint form creates a linear algebraic equation with associated diagonally dominant matrix. The discretized equation can be solved by any of the methods discussed in Chapter 7. Here we additionally describe a Conjugate Gradient (CG) method due to Jennings & Malik (1978) that works better for complex geometries as compared to ADI method or Strongly Implicit Procedure (SIP) described in Anderson, Tannehill & Pletcher (1984) and Nair & Sengupta (1997).

CG method uses the idea of steepest descent method. Hestenes & Stiefel (1952) have shown theoretically that the solution of

$$[A]\{X\} = \{b\} \tag{13.18}$$

by CG method can be obtained by no more than n iterations, where $[A]$ is a $(n \times n)$ matrix. However, due to round-off and aliasing errors the theoretical n-step convergence may not always be achieved. Despite these observations, the performance of CG method outstrips other methods as the *condition number* of the A matrix deteriorates. In general, for CFD applications (as for Equations (13.5)) $[A]$ is a sparse, symmetric and positive definite coefficient matrix and X and b are the unknowns and the right hand side, respectively. For any trial solution X, there is an associated residual

$$r = b - AX' \tag{13.19}$$

which is in general a non-null vector. This residual can be used to construct an objective function,

$$h = (r^T A^{-1} r)^{1/2}. \tag{13.20}$$

This objective function will attain a global minimum when $r \equiv 0$. This global minimum is searched along different directions and can be represented by vectors $p^{(0)}$, $p^{(1)} \ldots\ldots p^{(n)}$. These are the steepest direction vectors subject to the orthogonality (with respect to $[A]$) condition,

$$\{p^{(i)}\}^T Ap^{(j)} = 0 \ \text{ for } \ i \neq j. \tag{13.21}$$

An algorithm for CG method (Jennings & Malik (1978)) is reproduced below

$$\left.\begin{aligned}
u^{(k)} &= Ap^{(k)} \\
\alpha_k &= [r^{(k)}]^T r^{(k)} / [p^{(k)}]^T u^{(k)} \\
X^{(k+1)} &= X^{(k)} + \alpha_k p^{(k)} \\
r^{(k+1)} &= r^{(k)} - \alpha_k u^{(k)} \\
\beta_k &= [r^{(k+1)}]^T r^{(k+1)} / [r^{(k)}]^T [r^{(k)}] \\
p^{(k+1)} &= r^{(k+1)} + \beta_k p^{(k)}
\end{aligned}\right\} \tag{13.22}$$

that can be iterated with the initial conditions

$$p^{(0)} = r^{(0)} = b - AX^{(0)}. \tag{12.13}$$

The initial trial vector $X^{(0)}$ can be taken as the null vector.

The iteration given by Equation (13.22) is continued till $\|r^{(k)}\|/\|b\|$ is less than some prescribed tolerance. Note that the denominator in the above is chosen because

$$|r^{(0)}\| = \|b\|$$

for the initial trial vector $X^{(0)} \equiv 0$.

If the eigenvalues of the A matrix are ordered such that $\lambda_1 \leq \lambda_2 \leq \lambda_3 \ldots \leq \lambda_n$, then the CG method reduces error by at least a factor of ϵ in k iterations, given by

$$k \leq \frac{1}{2}\sqrt{\frac{\lambda_n}{\lambda_1}}\ln\left(\frac{2}{\epsilon}\right), \tag{13.24}$$

where λ_n/λ_1 is called the P-condition number. Larger this number, slower will be the convergence rate. Thus the convergence rate (as given by k) can be occasionally accelerated by symmetric scaling. If

$$[\bar{A}]\{\bar{X}\} = \{\bar{b}\} \tag{13.25}$$

is the linear algebraic equation to be solved with A as positive definite coefficient matrix, then one can rewrite Equation (13.25) as

$$(D\bar{A}D)(D^{-1}\bar{X}) = D\bar{b}.$$

Equations (13.18) and (13.26) are identical with $A = D\bar{A}D$, $b = D\bar{b}$ and $X = D^{-1}\bar{X}$. In the symmetric scaling method, D is chosen as the diagonal matrix with $d_i = \frac{1}{\sqrt{a_{ii}}}$, such that the $[A]$ matrix has unit diagonal elements. For such a A matrix, the Gerschgorin discs (as defined in theorems 1 and 2 of subsection 7.11) are concentric and from Equation (13.24) it is apparent that the rate of convergence improves significantly by symmetric scaling because the P-condition number reduces.

There are many other elegant techniques available to further accelerate the basic CG methods. Similarly one can use the variant of CG methods where $[A]$ is not a positive definite matrix. One can consult standard treatises on Linear Algebra for the same.

13.4 Wall--Vorticity Estimation

The solution of Navier-Stokes equations is dominated by vorticity dynamics. And in this process, the presence of the solid wall is vital because vorticity is generated at the solid walls. This is a consequence of the condition of no-slip that needs to be satisfied at the solid wall. The wall vorticity (ω_b) is calculated from the definition of vorticity (Equation (13.1)) along with the no-slip condition given by Equation (13.17). Thus,

$$\omega_b = -\nabla^2\psi\Big|_{\text{body}} = -\frac{1}{h_1h_2}\left[\frac{\partial}{\partial\xi}\left(\frac{h_2}{h_1}\frac{\partial\psi}{\partial\xi}\right) + \frac{\partial}{\partial\eta}\left(\frac{h_1}{h_2}\frac{\partial\psi}{\partial\eta}\right)\right]_{\text{body}}.$$

Using Equation (13.17),

$$\left(\frac{\partial\psi}{\partial\eta}\right)_{\text{body}} = 0 \text{ and } \left(\frac{\partial\psi}{\partial\xi}\right)_{\text{body}} = \left(\frac{\partial^2\psi}{\partial\xi^2}\right)_{\text{body}} = 0.$$

Therefore

$$\omega_b = \frac{-1}{h_1h_2}\left[\frac{h_1}{h_2}\frac{\partial^2\psi}{\partial\eta^2}\right]_{\text{body}} = -\frac{1}{h_2^2}\frac{\partial^2\psi}{\partial\eta^2}(\xi,0). \tag{13.27}$$

Thus one calculates the wall vorticity by numerically evaluating the second derivative on the right hand side of Equation (13.27). In deriving this second derivative, one needs to use the no-slip conditions. For example, in the following Taylor series expansion

$$\psi(\xi,\Delta\eta) = \psi(\xi,0) + \Delta\eta\frac{\partial\psi}{\partial\eta}(\xi,0) + \frac{\Delta\eta^2}{2}\frac{\partial^2\psi}{\partial\eta^2}(\xi,0),$$

when no-slip condition is used, one gets,

$$\frac{\partial^2\psi}{\partial\eta^2}(\xi,0) = \frac{2}{\Delta\eta^2}[\psi(\xi,\Delta\eta) - \psi(\xi,0)]. \tag{13.28}$$

Although this is formally a first order accurate expression, the leading truncation error term is proportional to $\frac{\partial^3\psi}{\partial\eta^3}$ which has a numerically stabilizing effect. This type of numerical estimate has an effect of under-predicting the wall vorticity value depending on the type of grid used. This effect is most pronounced at $t = 0$ for impulsive start problems. Theoretically, for impulsive start, the wall vorticity should be infinite at $t = 0$, while a numerical evaluation will severely 'filter' the calculated value. Does it affect the overall accuracy of solution? Without providing any proof, it can be stated that this does not produce serious effects for the implementation of impulsive start conditions. However, at subsequent times the effect of coarse grid will always underestimate the wall vorticity. A safe procedure is to use a very fine grid near the wall.

13.5 Solution of Vorticity Transport Equation (VTE)

For the solution of VTE as given in the orthogonal grid system by Equation (13.6), two accurate methods are indicated below. They are the explicit and implicit higher order upwind methods.

The upwinding is performed on the convection terms, there being no conceptual or physical requirement for doing so on the diffusion terms. Thus, diffusion terms can be discretized as in Equations (13.15) and (13.16).

13.5.1 Explicit Upwind Methods

In these method the convection term $u\frac{\partial f}{\partial x}$ is quasi-linearized first, i.e.,

$$\left[u\frac{\partial f}{\partial x}\right]^{(n+1)} \simeq u^{(n+1)}\frac{\partial f^{(n)}}{\partial x}. \tag{13.29}$$

Then depending on the sign of u^{n+1} the first derivative is appropriately up-or down-winded. The idea implicit in this process is to provide negative feedback (as discussed in Chapter 10). For example, in Equation (10.18) the third order upwind scheme was given by (for u negative),

$$\left(\frac{\partial f}{\partial x}\right)_{\text{UD}_3} = \left(\frac{\partial f}{\partial x}\right)_{\text{CD}_4} + \beta\frac{\Delta x^3}{4!}\frac{\partial^4 f}{\partial x^4}, \tag{13.30}$$

where $\left(\frac{\partial f}{\partial x}\right)_{\text{CD}_4}$ and $\frac{\partial^4 f}{\partial x^4}$ are given in Equations (10.16) and (10.17) respectively as obtained by central differencing.

For general $(2m-1)$th order upwind schemes Equation (13.29) is written as

$$\left[u\frac{\partial f}{\partial x}\right]_{\text{UD}_{2m-1}} = u\left[\frac{\partial f}{\partial x}\right]_{\text{CD}_{2m}} + \theta\frac{\Delta x^{2m-1}}{(2m)!}\frac{\partial^{2m} f}{\partial x^{2m}}|u|, \tag{13.31}$$

where θ takes an appropriate value and sign determined by the requirements of numerical stability and maximum DRP property.

A third order upwind scheme was used in Kawamura et al. (1985) and a fifth order upwind scheme was used in Rai & Moin (1991). Similarly in Sengupta & Sengupta (1994) and in Nair & Sengupta (1997), these methods have been used for calculating flow past bluff-bodies. In Kawamura et al. (1985), $\theta = 6$ was used. The rationale for explicit and implicit upwinding is discussed in Sengupta & Nair (1999), where it is shown that upwinding performs the multiple functions of numerical stabilization, altering numerical DRP property and controlling aliasing. This is discussed in some detail in Chapter 10. The real and imaginary part of the equivalent wave number of various upwind schemes have been shown in Figs. 10.2 and 10.3, which represent the spectral resolution and the added numerical dissipation respectively.

13.6 Implicit Upwind Methods

In this class of methods, we briefly mention the compact schemes that have shown extremely high spectral resolution and good numerical stability property. These compact schemes have been discussed in detail in Chapter 11. If we are solving Equations (13.5) and (13.6) in an O-grid topology (as indicated in Figure 13.3), then the ξ-direction will be the periodic direction. For the periodic direction, one can therefore use truly central schemes as given by Equation (11.29) or Equation (11.22) with coefficient values given by Equation (11.47) for the first derivative evaluation. These central schemes are non-dissipative. As there is no need for boundary closure due to periodicity, the induced numerical instability (as shown in Figure 11.1, 11.4 and 11.7) will not affect the solution due to this discretization. Similarly, one can also use the interior stencils of the upwind schemes given by Equations (11.70), (11.75) or (11.76).

In the η-direction, one needs to incorporate boundary and near-boundary stencils, as given in Chapter 11. It is to be pointed out that solution of Navier-Stokes equation requires upwind compact schemes and not simply central compact schemes. This is due to the fact that any numerical discretization scheme used for the physical dissipation term shows significant loss at high wave numbers. As mentioned in Chapter 4, the physical dissipation peak is always at high wave number where it is 'filtered' by numerical discretization. This loss can be augmented by suitably designed upwind compact schemes. Some representative results are shown in Section 13.9 to exhibit higher accuracy and speed-up of compact schemes. Specific problems that demand very high accuracy have been chosen, and in some cases the same can be provided by compact schemes and not by higher order explicit upwind schemes.

13.7 Solution of Navier-Stokes Equation using Pressure Velocity Formulation

In this section, a couple of numerical methods for solving Navier-Stokes equation, using primitive variables for constant density, and constant viscosity flows, are discussed. The Navier Stokes equation is rewritten as

$$\frac{\partial \vec{V}}{\partial t} = N(\vec{V}) - \frac{\nabla p}{\rho} + \nu \nabla^2 \vec{V}, \tag{13.32}$$

where

$$N(\vec{V}) = -\vec{V}.\nabla\vec{V} = -\nabla.(\vec{V}\vec{V}),$$

The MAC method of Harlow & Welch (1965)

The Marker and Cell (MAC) method developed by Harlow & Welch (1965) is one of the often used accurate and stable finite difference techniques. Its accuracy is mainly due to the way the mass conservation is satisfied via the solution of Poisson equation for pressure, as discussed below. The numerical stability is ensured by using a staggered grid when second order central difference is used. If higher order upwind is used, then there is no need to use staggered grid; instead one can use collocated grid.

The finite difference form of Equation (13.32) is obtained after quasi-linearization and time advancing the solution by Euler scheme, as

$$V^{n+1} = V^n + \Delta t \left[N(V^n) + \nu\nabla^2(V^n) - \frac{\nabla p^n}{\rho} \right]. \tag{13.33}$$

We will not use the vector symbol for the velocity hereafter, until and unless there is any scope for confusion. To obtain p^n in Equation (13.33), we take divergence of Equation (13.32) to get

$$\frac{\partial}{\partial t}(\nabla.V) = \nabla.N(V) - \frac{1}{\rho}\nabla^2 p + \nu\nabla.\nabla^2 V.$$

When this equation is discretized, using the notation $D = \nabla.V$, we get

$$\frac{1}{\Delta t}[D^{n+1} - D^n] = \nabla.N(V^n) - \frac{1}{\rho}\nabla^2 p^n + \nu\nabla^2 D^n. \tag{13.34}$$

In MAC method one imposes $D^{n+1} \equiv 0$ to get the Poisson equation for pressure

$$\nabla^2 p^n = \frac{\rho}{\Delta t} D^n + \rho\nabla.N(V^n) + \mu\nabla^2 D^n. \tag{13.35}$$

Note that in Equation (13.35), the first and third terms are theoretically zero, but are retained to prevent the onset of numerical stability. It is for this purpose, that this equation is termed as the consistent pressure Poisson equation in literature. The following steps are involved in setting up a MAC method solver for unsteady flow simulations:

(*a*) One starts with a solenoidal velocity field ($\nabla.V \equiv 0$) as the initial condition, which satisfies the prescribed boundary condition.

(*b*) The pressure field is obtained by solving Equation (13.35) subject to the Neumann boundary condition given by Equation (2.64).

(*c*) The velocity field is time advanced by solving Equation (13.33). In the original MAC method, all the spatial derivatives are evaluated by using central differences. Since this is an explicit method, the time steps are kept very small so that the resultant scheme is *dispersion relation preserving*. This is one step of solution-advancement in time and the process can be continued by solving first the PPE, followed by solving the momentum equation repeatedly.

Harlow & Welch (1965) used purely central differencing for spatial derivatives, and implemented the method on a staggered grid. In such a grid the velocity components are defined at locations displaced by half a grid space from the cell center, on cell faces over which the component is perpendicular. Thus, the horizontal component of velocity is defined at the center of vertical cell faces and vice versa. The pressure is defined at the center of each cell. It can be shown that such a staggering of variables is equivalent to adding numerical dissipation. It is for this reason that Kawamura et al, (1985) solved incompressible Navier-Stokes equation in a collocated grid using MAC method while discretizing the convection terms by third order upwind scheme defined by Equation (13.30) with $\beta = 6$.

13.8 Operator-Splitting Projection Methods

In these methods the Navier-Stokes equation is split based on various physical processes and solved in a sequence in time. In the governing equation, since pressure gradient term is decoupled with the velocity terms, the Navier-Stokes equation is solved in the sequence

$$\frac{\partial V}{\partial t} = N(V) + \nu\nabla^2 V, \tag{13.36}$$

and

$$\frac{\partial V}{\partial t} = -\frac{1}{\rho}\nabla p, \tag{13.37}$$

subject to the condition that the velocity field is solenoidal, i.e., the velocity field is divergence-free. The implication of the operator splitting is such that the solution of Equation (13.36) produces a velocity field V^* that is not guaranteed to be solenoidal. Equation (13.37) will be solved in such a way that the resultant velocity field will satisfy the divergence-free condition. It is to be realized that the right hand side is strictly not the pressure gradient term, instead it denotes $-\frac{1}{\rho}\nabla\phi$, as shown in Grescho (1990), where ϕ is termed as the projection function.

For illustration purposes, we discuss a method that was used by Kim & Moin (1985) in which three time level Adams-Bashforth time integration scheme was used, while the viscous term was discretized via Crank-Nicholson procedure.

Thus, Equations (13.36) and (13.37) are discretized in the following manner going from the nth to $(n+1)$th time level:

$$\frac{V^* - V^n}{\Delta t} = \frac{1}{2}[3N(V^n) - N(V^{n-1})] + \frac{\nu}{2}[\nabla^2 V^* + \nabla^2 V^n] \tag{13.38}$$

and

$$\frac{V^{n+1} - V^*}{\Delta t} = -\frac{1}{\rho}\nabla\phi^{n+1}. \tag{13.39}$$

Taking the divergence of Equation (13.39) and demanding $\nabla.V^{n+1} = 0$, one gets

$$\nabla^2\phi^{n+1} = \frac{\rho}{\Delta t}\nabla.V^*. \tag{13.40}$$

One can relate ϕ and p by using Equation (13.39) to obtain V^* from

$$V^* = V^{n+1} + \frac{\Delta t}{\rho}\nabla\phi^{n+1}.$$

Substituting this in Equation (13.38) one gets,

$$\frac{V^{n+1}-V^n}{\Delta t} = -\frac{1}{\rho}\nabla\left[\phi^{n+1} + \frac{\nu}{2}\Delta t\nabla^2\phi^{n+1}\right]$$

$$+\frac{1}{2}(3N(V^n) - N(V^{n-1})) + \frac{\nu}{2}[\nabla^2 V^n + \nabla^2 V^{n+1}]. \tag{13.41}$$

Comparing Equations (13.32) and (13.41), it is easy to see that

$$p = \phi + \frac{\nu}{2}\Delta t\nabla^2\phi.$$

Kim & Moin (1985) solved Equation (13.38) by the following split operator form

$$\left(I - \frac{\nu\Delta t}{2}D_x^2\right)\left(I - \frac{\nu\Delta t}{2}D_y^2\right)\left(I - \frac{\nu\Delta t}{2}D_z^2\right)(V^* - V^n)$$

$$= \frac{\Delta t}{2}[3N(V^n) - N(V^{n-1})] + \Delta t\nu\nabla^2 V^n. \tag{13.42}$$

Thus the solution of Equation (13.42) requires solving three sets of tridiagonal systems. It is assumed that V^* satisfies the same boundary conditions as that for V^{n+1}, and that ϕ satisfied the homogeneous Neumann boundary condition $\nabla\phi.\hat{n} = 0$. The far-field boundary condition for ϕ is same as that for p.

Overall, one can conclude that the accuracy of solution for Navier-Stokes equation using derived variables is superior to the solution obtained using primitive variables. This is due to fact that the latter formulation has many possible error sources. Also, it was shown in Chapter 10 and 11 that adequate care must be exercised which choosing a method for time integration, so that the numerical instability is avoided while preserving DRP. For example it is shown that the second order Adams-Bashforth scheme used in Kim & Moin (1985) has a spurious computational mode that will smooth out physical instabilities. Also, such a scheme has poor DRP property despite the fact that the scheme was adopted to achieve higher accuracy. Once again we encounter the case where higher order scheme does not ensure higher accuracy.

13.9 Solution of Navier-Stokes Equation: A Comparative Study of Different Formulations

Here we show the solution of Navier-Stokes equation for a difficult flow phenomenon. This is the *vortex-induced eruptions* inside a boundary layer, as excited by a convecting vortex far outside the shear layer. A related phenomenon of unsteady erupting flow over an infinite plane excited by a two-dimensional vortex in motion above the wall for very large Reynolds number was studied in Peridier et al. (1991). The authors studied this problem by looking at the boundary layer and interactive boundary layer equations in Lagrangian coordinates. It was mentioned by the authors that this problem was not amenable to solutions when posed in the Eulerian frame work followed in this book.

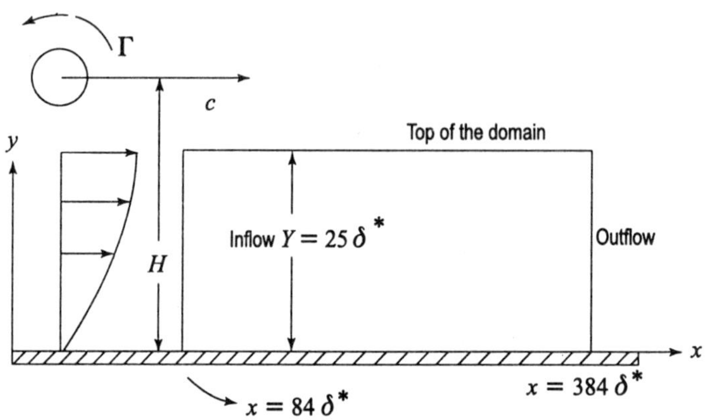

Fig. 13.4 *Schematic of the solved vortex-induced instability problem. The vortex (of strength Γ) convects from left to right at a speed c, moving at a constant height, $H = 32\delta^*$.*

Here the related problem is solved for the physical problem sketched in Fig. 13.4 by third order upwind schemes using pressure–velocity and (ψ–ω) formulation in identical grid structure in the wall normal direction. For the pressure–velocity formulation, MAC method is used. For the stream function–vorticity formulation, a forward in-time and centered-space discretization for the diffusion terms are used. For the convection terms third order upwinding of Kawamura et al. (1985). and Sengupta & Sengupta (1994) is used. Since the computations are performed in the transformed plane, a smoothly varying AP grid (whose beneficial effects were discussed in Chapter 10 for higher order upwind schemes) is used in the wall normal direction to resolve the shear layer and the perturbed erupting flow field.

To ensure accuracy of the solutions for both the formulations, the wall resolution is used as $1.75 \times 10^{-3}\ \delta^*$, where δ^* is the displacement thickness at the inflow. (In the computational domain indicated in Fig. 13.4, 176 points are used in η-direction for both the formulations.) In the streamwise direction a uniform grid is used for the computations.

The initial condition is obtained over the full computational domain by using the Blasius boundary layer solution everywhere and adding the inviscid perturbation that is created by the convecting vortex and its image. The image vortex ensures zero normal velocity at the wall for all times. The time dependent boundary condition at the top lid is analytically computed and used. For the (p–V) formulation, at the outflow a fully developed condition, $\frac{\partial u}{\partial x} = 0$, for velocity is used. Therefore for the (ψ–ω) formulation, the corresponding condition applied to the wall normal velocity, $\dfrac{\partial^2 \psi}{\partial x^2} = 0$, is used for ψ. For the vorticity, the fully developed condition $\dfrac{\partial \omega}{\partial x} = 0$ is used. Neumann condition for pressure is obtained at the solid wall from y-momentum equation. At the inflow, the shear layer solution plus the perturbed inviscid flow field due to the vortex and its image is used. It is to be pointed out that for the MAC method the solution was obtained in the collocated grid, and because of the added dissipation in the numerical scheme, staggering of grid was not necessary.

The streamline contours compared for the two formulation are shown in Fig. 13.5. This shows a close agreement of the number and speed of the coherent erupting vortices, initially.

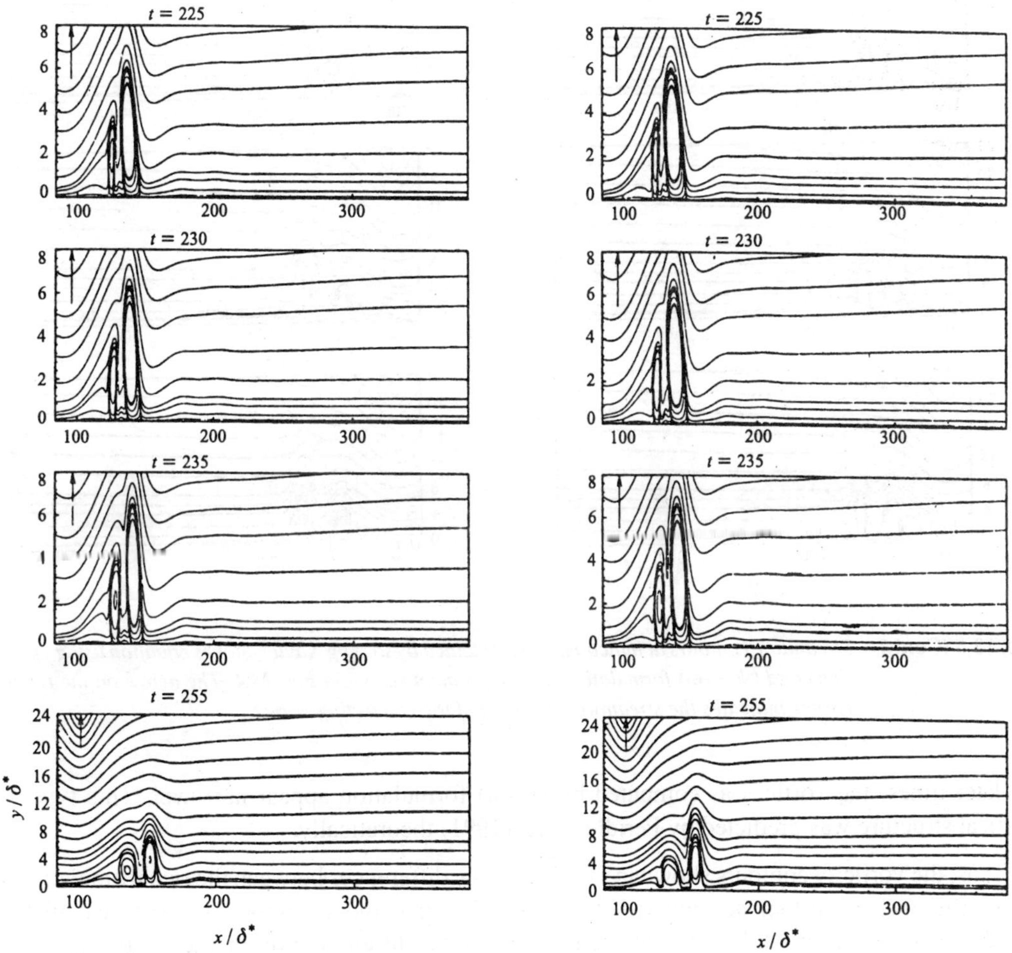

Fig. 13.5 *Streamline contours at indicated times obtained by solving Navier-Stokes equation using $(p - V)$ (left) and $(\psi$–$\omega)$ formulation for the problem shown in Fig. 13.4. The arrow on the top left corner indicates the streamwise location of the convecting vortex. (Cont.)*

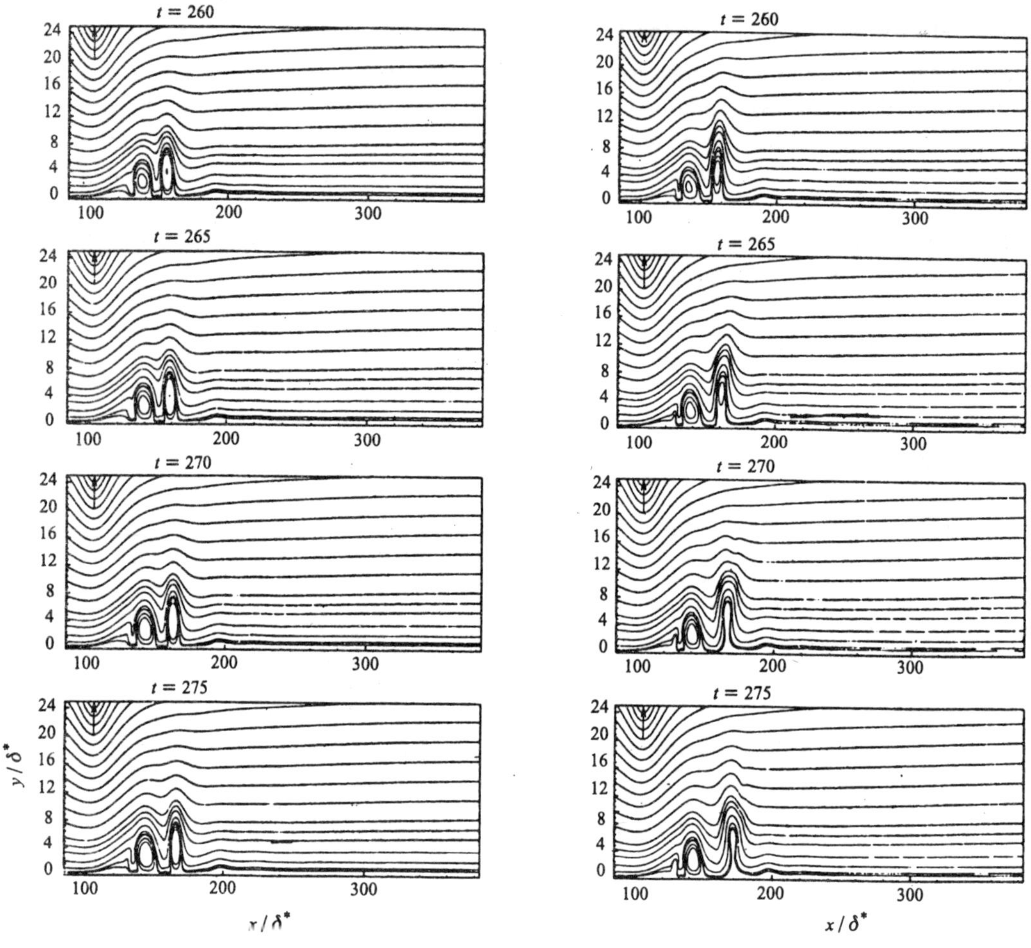

Fig. 13.5 (Cont.) *Streamline contours at indicated times obtained by solving Navier-Stokes equation using (p–V) (left) and (ψ– –ω) formulation for the problem shown in Fig. 13.4. The arrow on the top left corner indicates the streamwise location of the convecting vortex.*

At later times, the vortices as computed by (ψ–ω) formulation appear narrower. This type of vortical structure was predicted by Peridier et al. (1991) theoretically.

Thus, the vorticity contours shown in Fig. 13.6 show differences between the two formulations. This difference is further accentuated due to the fact that the vorticity is directly calculated in (ψ–ω) formulation while it is obtained by numerical differentiation in (p–V) formulation, which introduces error. To show a proper comparison in all the plots, identical contour values are plotted.

A quantitative comparison between the two formulations is provided in Fig. 13.7 where skin friction variation is shown. Overall, the differences in results between the two formulations also

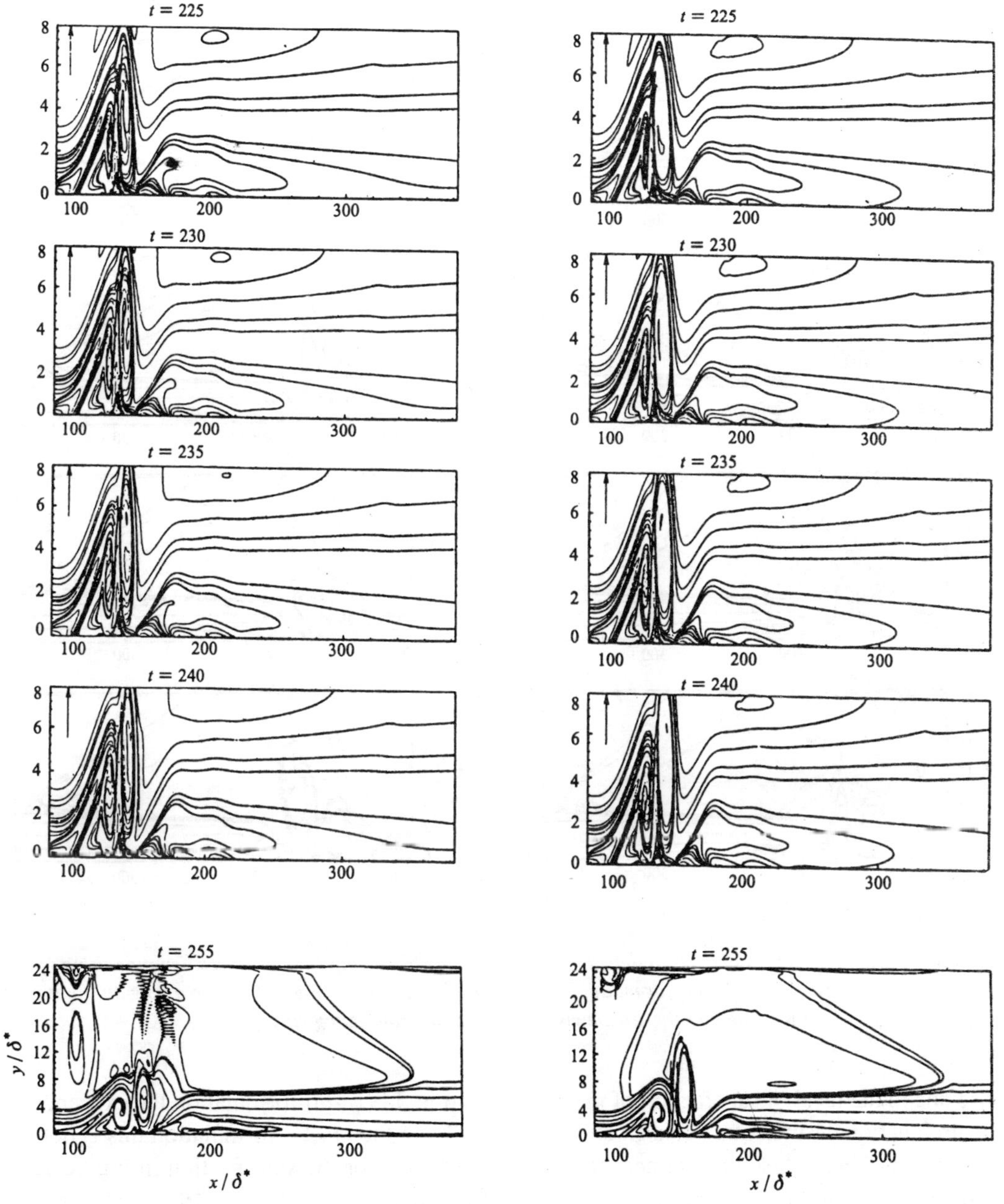

Fig. 13.6 *Vorticity contours at indicated times obtained by solving Navier-Stokes equation using* $(p\text{–}V)$ *(left) and* $(\psi\text{–}\omega)$ *formulation for the problem shown in Fig. 13.4. The arrow on the top left corner indicates the streamwise location of the convecting vortex. (Cont.)*

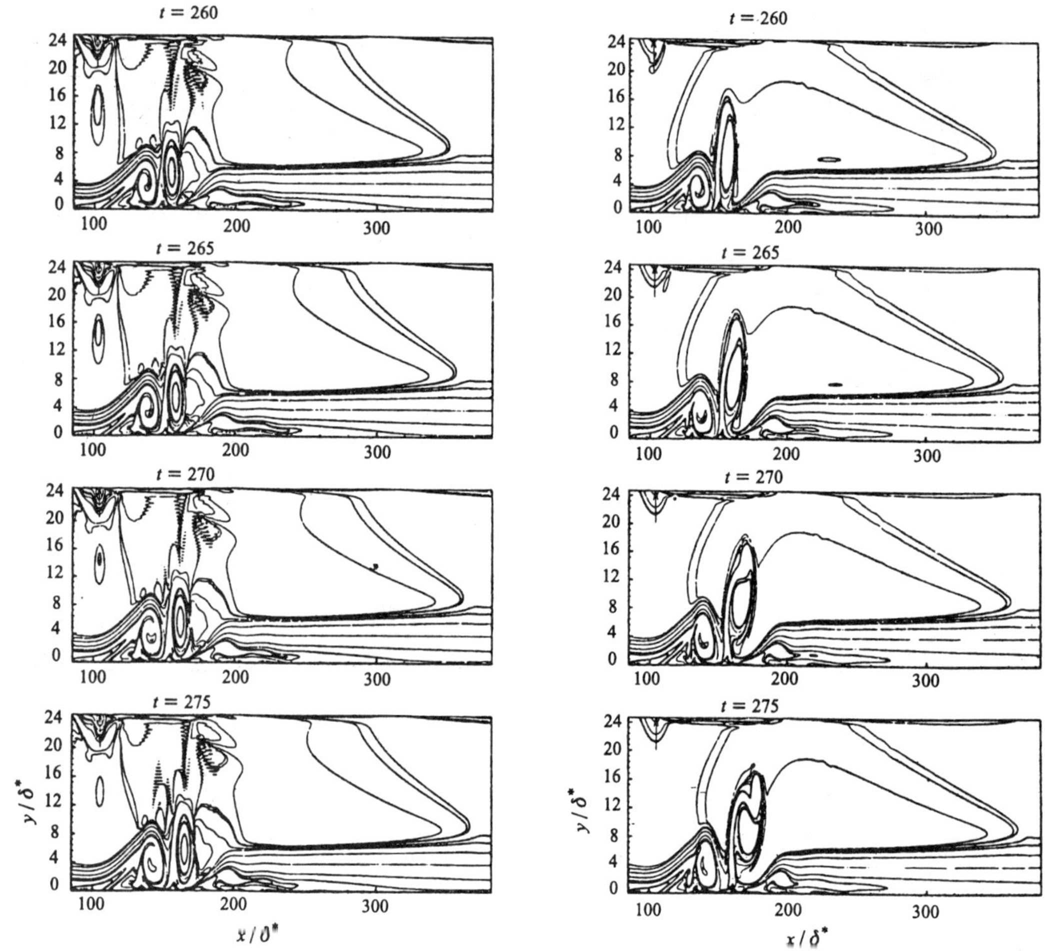

Fig. 13.6 (Cont.) *Vorticity contours at indicated times obtained by solving Navier-Stokes equation using (p–V) (left) and (ψ–ω) formulation for the problem shown in Fig. 13.4. The arrow on the top left corner indicates the streamwise location of the convecting vortex.*

arise due to the level of satisfaction of mass conservation. In $(\psi$–$\omega)$ formulation the existence of stream function ensures mass conservation numerically. In the $(p$–$V)$ formulation this is not the case and one notices oscillations near the inflow due to this for the skin friction in Figure 13.7. Numerical instabilities were reported for $(p$–$V)$ formulations in Williams (1969). Specifically Chorin (1968) has reported numerical instability using $(p$–$V)$ formulations for transient flows and he attributed this to non-satisfaction of mass conservation. Figure 13.7 also displays that the $(\psi$–$\omega)$ formulations captures detailed small scale variations without numerical oscillations whatsoever.

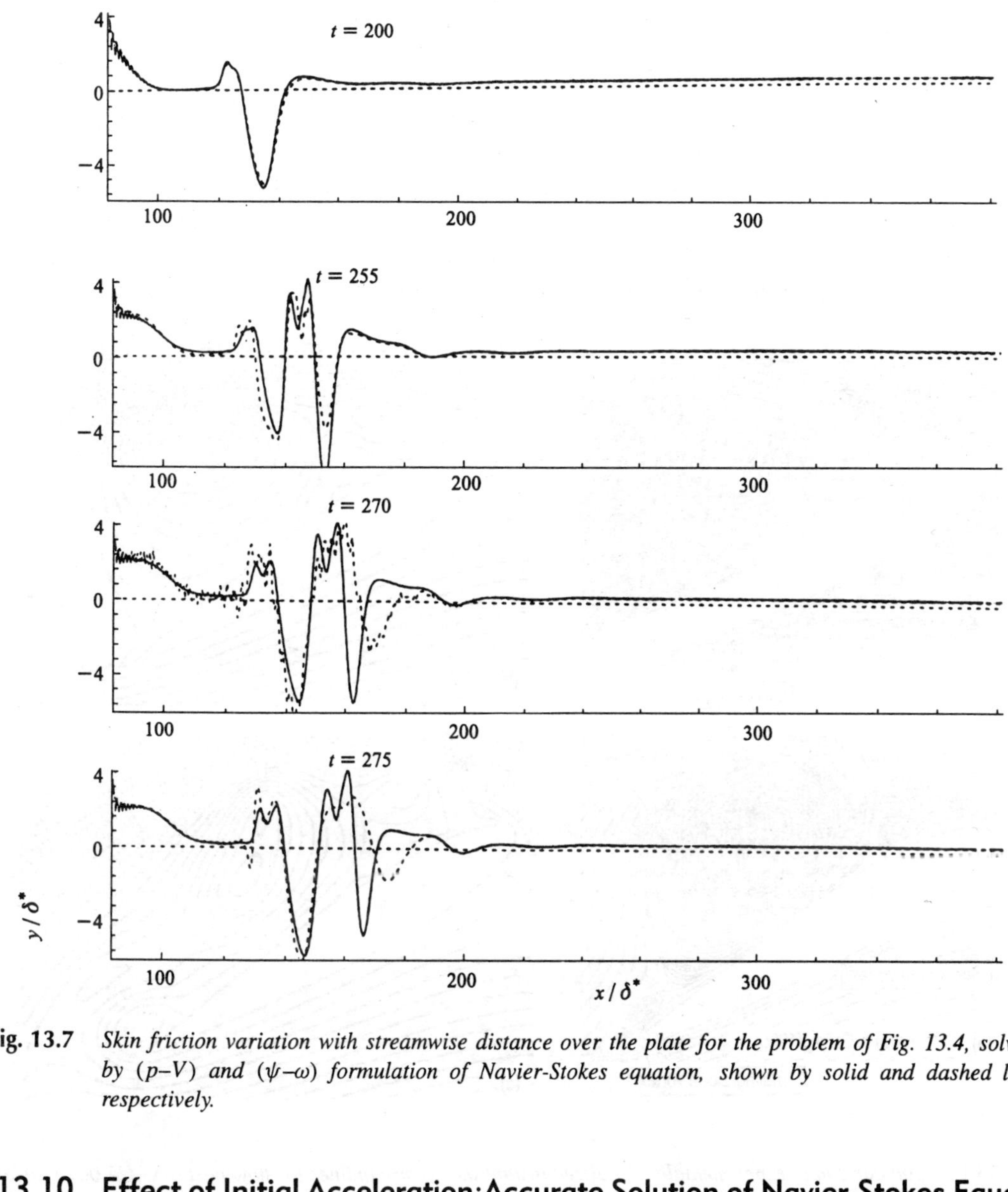

Fig. 13.7 *Skin friction variation with streamwise distance over the plate for the problem of Fig. 13.4, solved by (p–V) and (ψ–ω) formulation of Navier-Stokes equation, shown by solid and dashed line respectively.*

13.10 Effect of Initial Acceleration:Accurate Solution of Navier-Stokes Equation

For a general fluid dynamical system, it is not known *a priori* whether the numerical solution would display a steady state or not. In general, one can obtain a steady state solution for viscous flow by solving the full unsteady equation. If a steady state exists, it will be obtained as a long time solution of the full Navier-Stokes equation. However, the problem of starting the unsteady solution has not received its due importance and attention. In this context, impulsive start is the basis of many computational studies. Since this is an idealization, it is necessary to follow the effect of initial accelerations for the cases depicted in Fig. 13.2 and starting condition given by Equations (13.12)-(13.14). The calculated transient loads for flows with initial acceleration is totally different from the impulsive start case. To study these effects, the flow past NACA0015 airfoil at high angles of attack (30°) at moderate Reynolds number (35,000) is reported here, for which some experimental results exist. This particular case was considered so that the experimental results of Morikawa & Grönig (1995) can be used to validate airfoil computations. An orthogonal O-type grid with 233 points in the azimuthal direction and 401 points in the wall normal direction

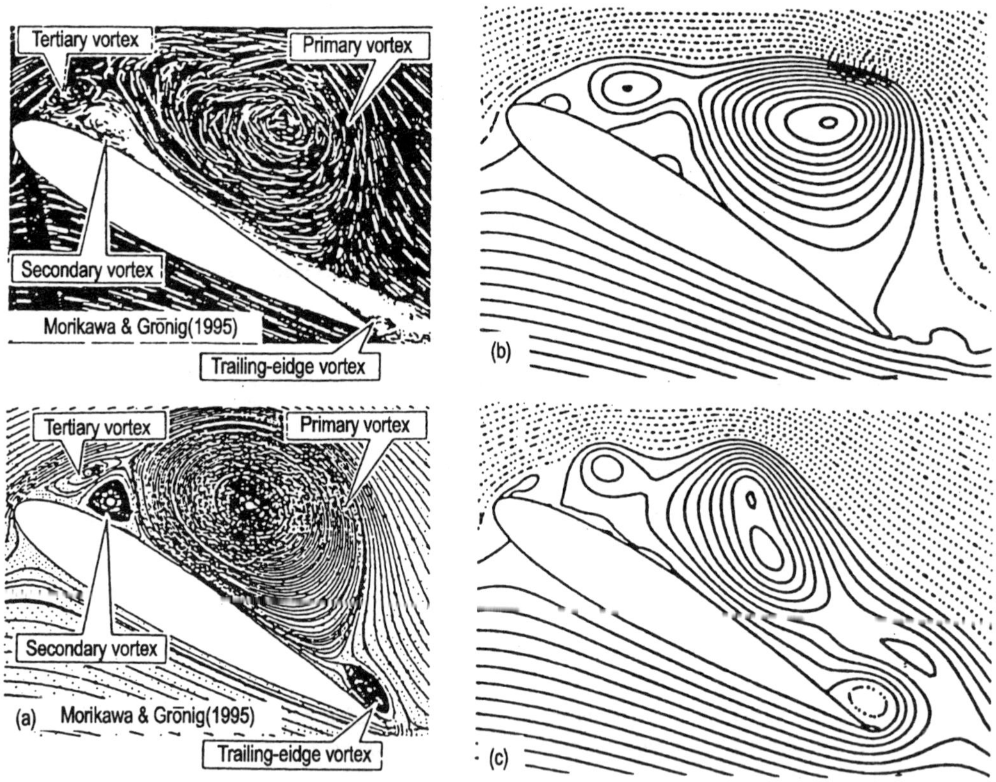

Fig. 13.8 *Comparison of experimentally visualized instantaneous streamlines for flow past a NACA 0015 airfoil for* $Re = 35000$ *and* $\alpha = 30°$, *at* $t = 2.903$ *(top left) by Morikawa & Gronig (1995) and the computed solutions: (a) The computed solution by Morikawa & Gronig (1995), shown in the bottom leftside, using* $(p–V)$ *formulation with actual start-up condition (Case-III in Fig. 13.2); (b) results from* $(\psi–\omega)$ *formulation using start-up condition of Case-III (top right); and (c) results from* $(\psi–\omega)$ *formulation using start-up condition of Case-II in Fig. 13.2. The computed results match the best with (b). Note that the primary bubble is flat on top, due to the presence of another weak vortex of opposite sign on top of the primary bubble, which can be seen in vorticity contour plots (not shown here).*

is used for this purpose, following the method described in Chapter 9, with regard to work reported in Nair & Sengupta (1998).

In the experiments, the free stream speed was varied according to Equation (13.14) with $\tau = 0.6$. This relation describes the mean motion adequately and no effort is made to model the high frequency fluctuations observed in the water tunnel speed record. High frequency oscillations noticed for the tunnel speed is a typical characteristic of empty tunnel free-stream turbulence.

The reported results were obtained by solving Equations (13.6) using the third order upwind scheme, which was also used in Sengupta & Sengupta (1994) and Nair & Sengupta (1997). The stream function equation (Equation (13.5)) is solved by the CG method described by Equation (13.22). In Fig. 13.8 the streamline plots of all the three start-up cases (as indicated in Fig. 13.2), are shown for $t = 2.903$ along with the experimental record. In this figure the computed results due to Morikawa & Grönig (1995) is also shown, who used a $(p\text{–}V)$ formulation. This figure clearly reveals that the best match between experimental and the computed results was achieved for the case when the actual start up condition given by Equation (13.14) was used for the $(\psi\text{–}\omega)$ formulation. However, the results of $(p\text{–}V)$ formulation with correct start-up condition produced by Morikawa & Grönig (1995) failed to provide the correct results that surprisingly resembled the solution obtained using $(\psi\text{–}\omega)$ formulation with impulsive start condition Equation (13.12). The noticeable feature of the experimental results is the generation of severely strained primary bubble. The straining of the bubble causes it to shrink in the lateral direction. The ramp- start case results show that the primary bubble has burst due to *alleyway* formation associated with a strong trailing edge bubble. The impulsive start case exhibits neither the straining of the primary bubble nor its bursting.

Although an impulsive start implies infinite acceleration at $t = 0$, but because of dominant diffusion and negligible convection of the generated vorticity at the wall during $t = 0^+$, the acceleration effects are least for the impulsive start, among the three cases shown here.

The loads and the pitching moments are compared for the Case-II and Case-III in Fig. 13.9. For these two cases the vorticity and acceleration levels associated with the free stream speed are identical up to $t = 0.125$ and again from $t = 2$ onwards. Despite this, the calculated instantaneous loads and the pitching moment vary significantly.

Case-II exhibits a larger lift and drag (on average) up to the displayed time of computation ($t \simeq 30$). This observation is also supported by the experimental results of Ham & Garelick (1982) who investigated the effects of constant pitch-acceleration of NACA 0012 airfoils. It was demonstrated that the lift coefficient and lift curve slope are different for the accelerating cases as compared to the corresponding steady state case. However, Gendrich et al. (1995) and Koochesfahani & Smiljanovski (1991) have concluded, based on their investigations for pitching oscillations of airfoils, that the unsteady effects are significant only for a short time after the acceleration is over. These observation may appear to contradict the finding of persistent effects of the start-up process shown here.

This apparent contradiction can be explained in terms of the most significant parameter that determines the unsteady effects of different start-ups, namely the time (T_B) to form the first bubble after the starting vortex is shed. Here (T_B) can be estimated via an impulsive start-up computations. In any start-up processes if the acceleration phase is over before (T_B), then the impulsive start

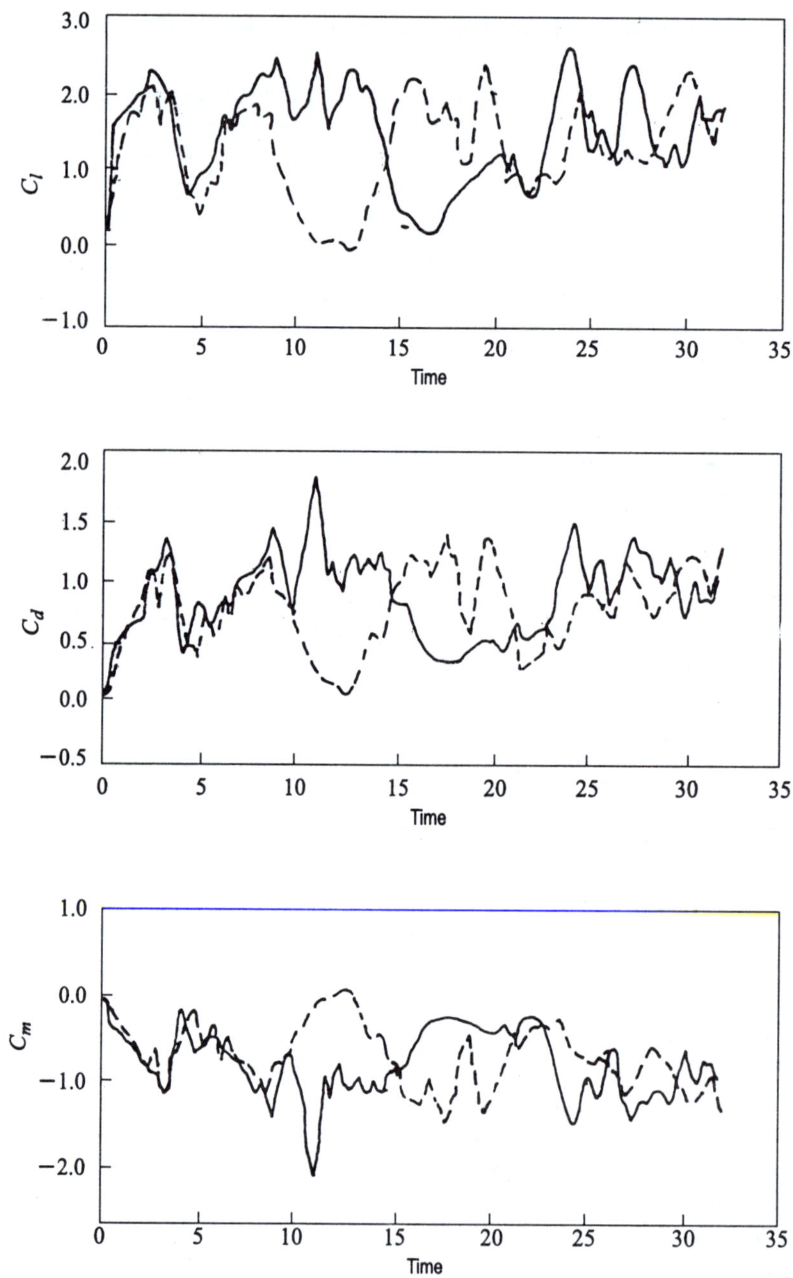

Fig. 13.9 *The load and moment calculated for flow past NACA 0015 airfoil for Re = 35000 and α = 30° for the cases (b) and (c) of Fig. 13.8, shown by dotted and solid lines respectively.*

assumption is quite good. It is for this reason, that many computations done with an impulsive start matches with the experimental results which are strictly not impulsively started.

The above discussions ought to convince the reader that the initial condition and start-up process can leave behind its long-time effects for laminar flows. In comparison, turbulent flows have been studied and understood even less. It is worth recording from George (1999) who states "I was surprised to learn that some years ago the DNS isotropic turbulence simulators usually throw away all the data prior to the peak in the derivative skewness since the rapid rise is believed to be part of the starting transient." Therefore some more systematic studies are required for turbulent flows using DNS to understand the effects of start-up.

13.11 Application of Compact Schemes in Solving Navier-Stokes Equation

The advantage of compact schemes over higher order explicit scheme is shown here by solving a problem that investigates the receptivity of a boundary layer to convecting vortex in free stream, as depicted in Fig. 13.4. The stream function equation (Equation (13.5)) is again solved by CG method, also describe by the algorithm of Equation (13.22). The convection terms of the VTE given by Equation (13.6) are treated differently by three methods that are compared here. In the first method, we adopt once again the third order upwind scheme due to Kawamura et al. (1985) and is given by Equation (13.31) with $m = 2$ and $\theta = 6$. This is referred to here in the discussion as K-scheme. The other two are the compact schemes given by,

(*a*) Adams (1993) via Equations (11.27)-(11.29). This is refereed to here as A-scheme, and

(*b*) the scheme due to Haras & Ta'asan (1994) modified for non-periodic boundary condition problem, called as the OUCS-3 scheme here. The Haras & Ta'asan (1994) scheme, given in Equation (11.47) for periodic problems, is used here for the interior nodes. The modification are due to (i) the usage of boundary closure scheme is given by Equations (11.73) and (11.74), and (ii) the explicit addition of higher order dissipation at the interior nodes to achive negative feedback for numerical stablization. Thus, this is similar to the OUCS-3 scheme described in Chapter 11.

These results are also reported in Sengupta et al. (2003).

For the problem shown in Fig. 13.4, uniform grid is used in the streamwise direction using 801 points for the K-scheme and 401 points for the compact schemes. In the wall-normal direction a non-uniform grid that stretches slowly in arithmetic progression is used for all the cases. For the K-scheme 175 points in this direction are taken with wall resolution of 0.00175 δ^* where δ^* is the displacement thickness of the shear layer at the inflow of the domain. The Reynolds number based on this length scale is 250 at the inflow. For the compact schemes, only 100 points are taken in the wall-normal direction. The inflow of the domain is located at the $x/\delta^* = 84$ and the convecting vortex is located at the $x = 50\delta^*$ at $t = 0$. The vortex subsequently convects to the right at a speed of $c = 0.20U_\infty$.

In discretizing the diffusion operators central differencing is used. It is to be noted that the compact schemes are usually defined with one type of convection velocity in the whole domain.

However, for a general problem the convection velocity can take any sign in the interior, which is not known *a priori*. The A-scheme is a central scheme for the interior nodes with fixed boundary stencils, and does not require the velocity direction information. The OUCS-3 scheme used here adds explicitly fourth order dissipation to provide negative feedback, and this information is not required once again. Application of truly upwind schemes would therefore pose problems.

The streamline contour plots are shown in Fig. 13.10 for $t = 190$ and $t = 198$. After $t = 198$ the A-scheme blows up due to numerical instability. The reason for this numerical instability is due

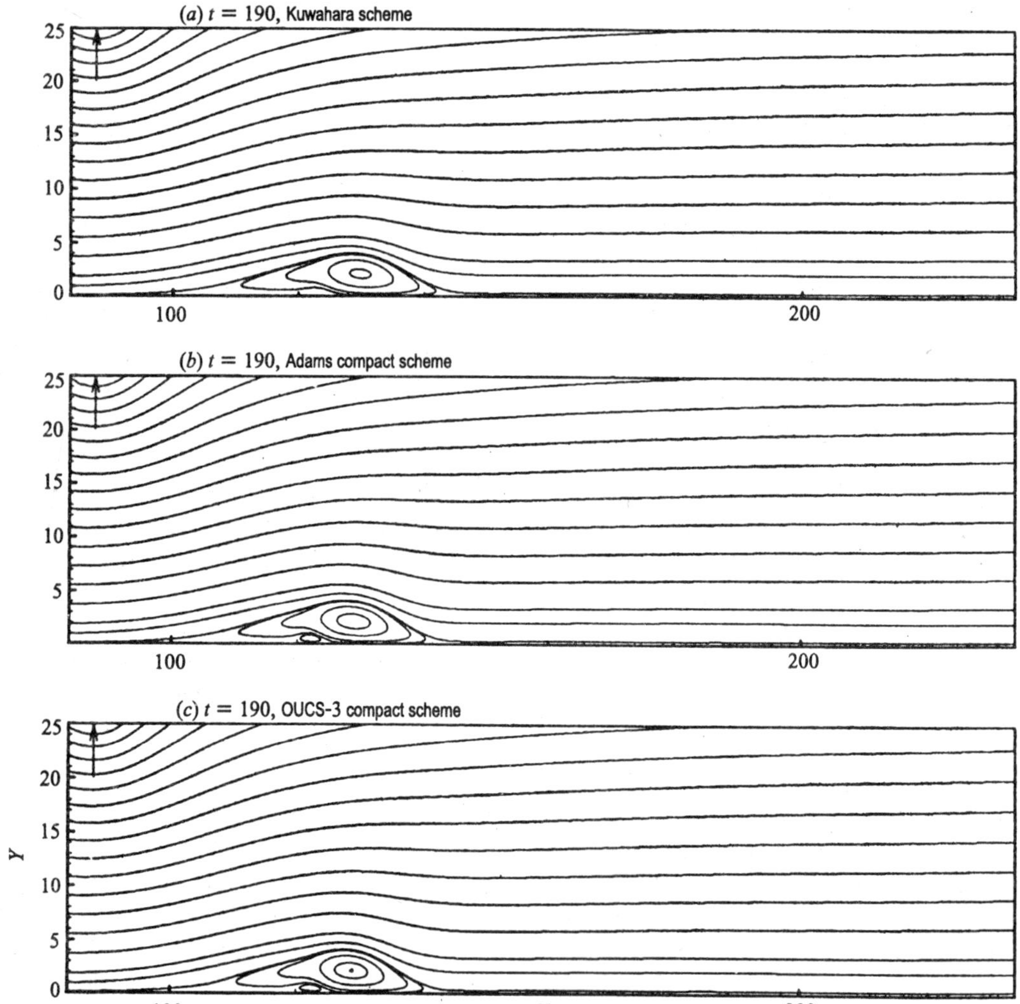

Fig. 13.10(a) *Streamline contours at the indicated time for (a) K-, (b) A- and (c) OUCS-3 schemes as obtained by solving Navier-Stokes equation using (ψ–ω) formulation. The arrow on the top left corner indicates the streamwise location of the convecting vortex. The flow is depicted in Fig. 13.4.*

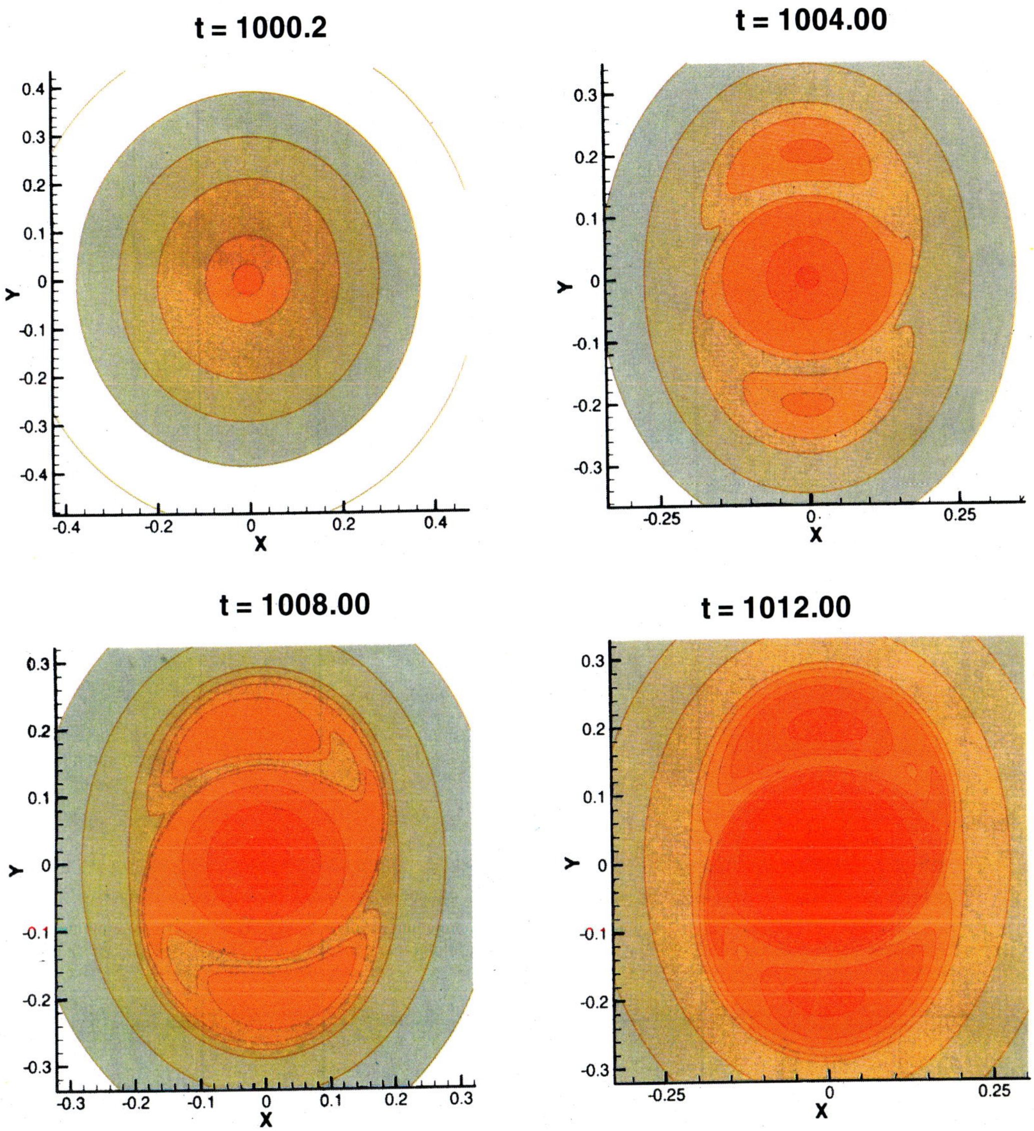

Vorticity contours for a problem of an Oseen-Lamb vortex interacting with uniform background shear. Here the value of background shear is low (2.5 per second) and the interaction begins at $t = 1000$ *for the Oseen-Lamb vortex of strength 50 per second. The Navier-Stokes solution is performed in a polar grid using OUCS-3 compact scheme for spatial discretization and* RK_4 *time integration strategy. Compare this interaction with the results shown overleaf.*

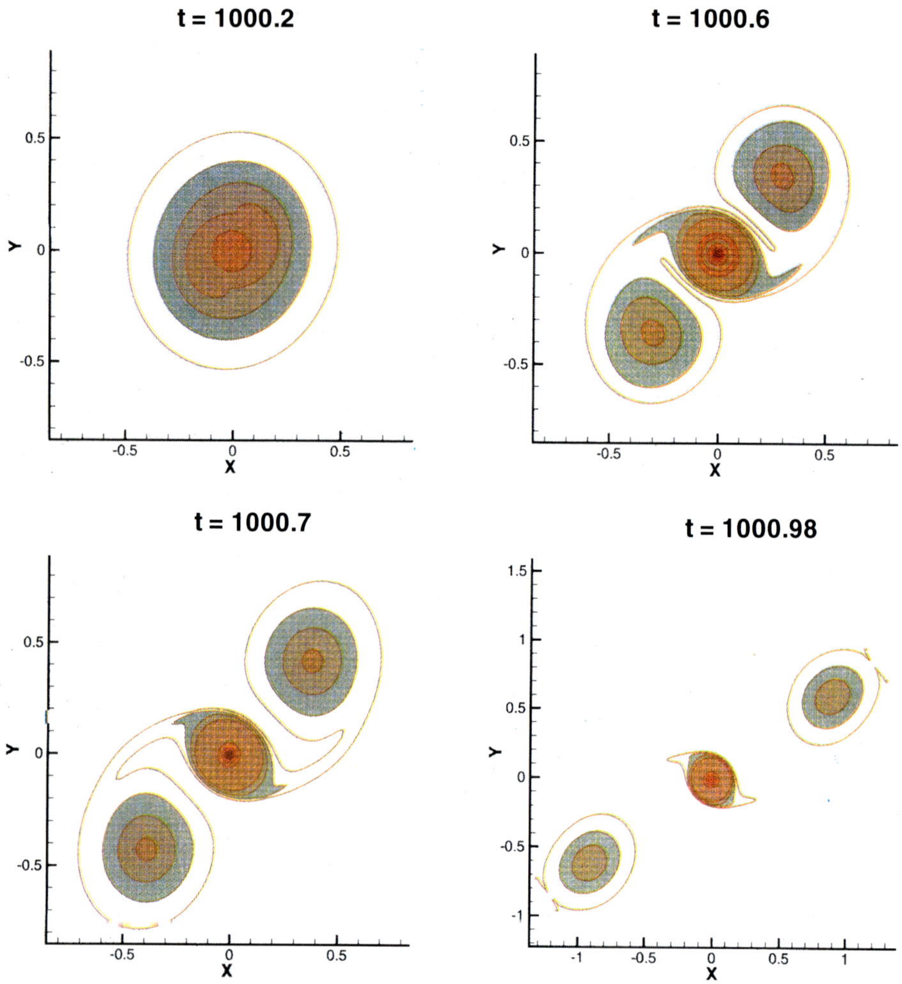

Vorticity contours of a Oseen-Lamb vortex interacting with uniform background shear. Here the value of shear moderate (5.0 per second) and the interaction begins at $t = 1000$ sec for the Oseen-Lamb vortex of strength 50 per sec. The Navier-Stokes solution is performed in a polar grid using high accuracy compact scheme for spatial discretization and RK_4 time integration strategy.

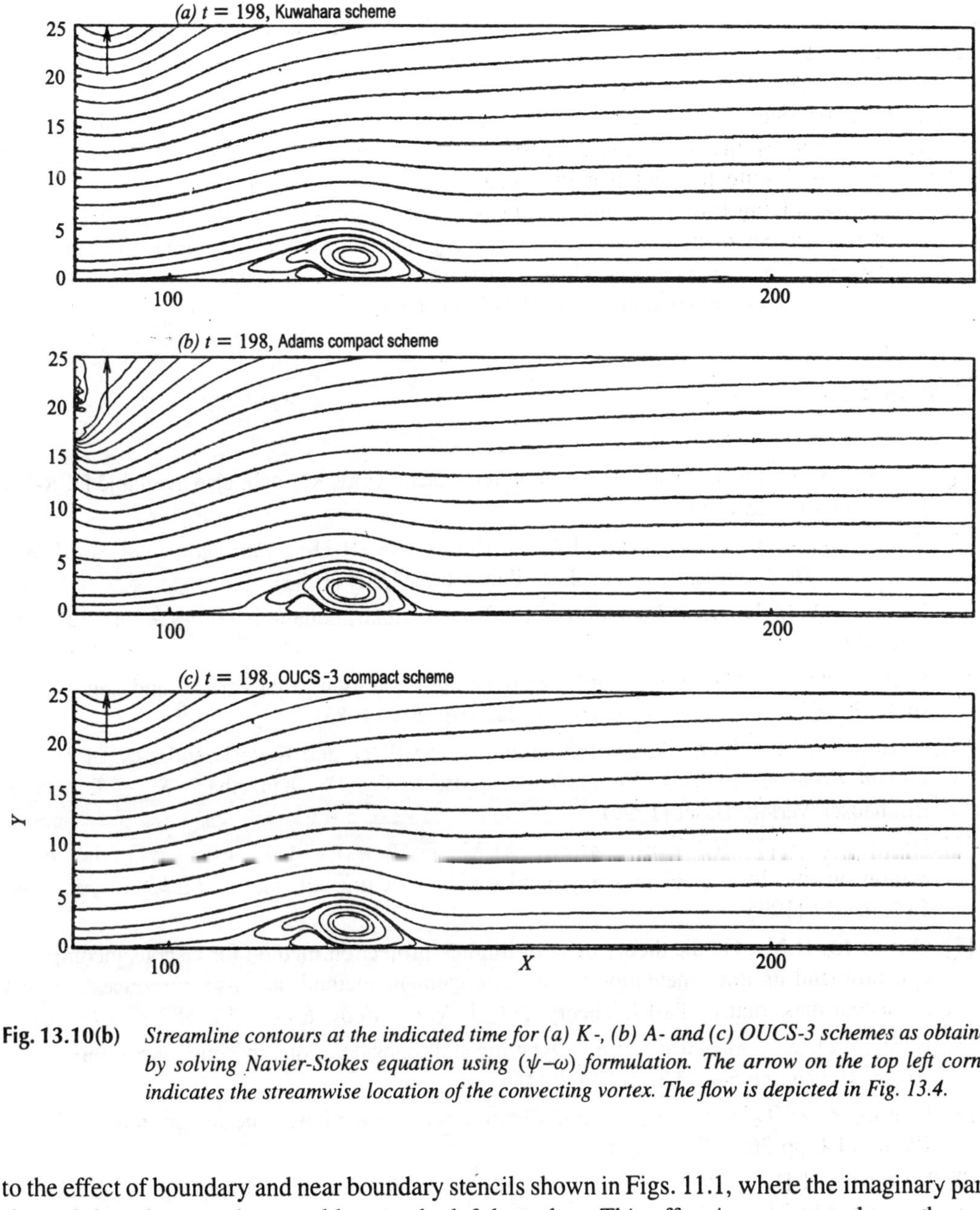

Fig. 13.10(b) *Streamline contours at the indicated time for (a) K-, (b) A- and (c) OUCS-3 schemes as obtained by solving Navier-Stokes equation using (ψ–ω) formulation. The arrow on the top left corner indicates the streamwise location of the convecting vortex. The flow is depicted in Fig. 13.4.*

to the effect of boundary and near boundary stencils shown in Figs. 11.1, where the imaginary part showed the scheme to be unstable near the left boundary. This effect is accentuated near the top left corner of the domain where the imposed disturbances by the convecting free-stream vortex is strongest at around $t \simeq 200$ when the streamwise location of the vortex is at $x = 90\delta^*$. However, another cause that contributed to the instability is the fact that A scheme is dissipationless in the interior, which makes it incapable of controlling numerical instability and aliasing error. In contrast to the A-scheme, the boundary stencils for the OUCS-3 and K-schemes are stable. Additionally the interior stencils are stabilized against accumulation of aliasing error by the added higher order numerical dissipation. Both these schemes can be integrated indefinitely.

The results obtained from K- and OUCS-3 schemes are virtually the same. But, the computational time taken by the K-scheme is an order of magnitude more compared to the compact schemes. This is despite the fact that the evaluation of the convection term for the compact scheme requires solving tridiagonal matrix equation to evaluate $\frac{\partial\omega}{\partial\xi}$ and $\frac{\partial\omega}{\partial\eta}$. This is entirely due to fact that the number of points required by the compact schemes are 3.5 times less than that for the K-scheme. Thus the compact schemes provide powerful and faster alternatives for three-dimensional flow computations required for DNS and LES.

References

[1] ADAMS, N.A., Ph.D Thesis submitted to Technische Univ. Munich, also available DLR-FB 93–29, Göttimgen (1993).

[2] ANDERSON D.A., TANNEHILL, J.C. & PLETCHER, R.H., *Computational Fluid Mechanics & Heat Transfer*, Hemisphere Pub. Corp., USA (1984).

[3] CHORIN, A.J., Numerical solution of the Navier-Stokes equation, *Math Comp.*, **22**, 745–762 (1968).

[4] FREYMUTH, P., The vortex pattern of dynamic separation: a parametric and comparative study, *Progress in Aerospace Sciences*, **22**, 161–208 (1985).

[5] GEORGE, W.K., Some thought on similarity, the POD, and finite boundaries, in *Fundamental Problematic Issues In Turbulence*(eds. A. Gyr, W. Kinzelbach & A. Tsinober), Birkhäuser Verlag, Basel (1999).

[6] GENDRICH, S.C., KOOCHHESFAHANI, M.M. & VISBAL, M.R., Effect of initial acceleration on the flow development around rapidly pitching airfoils, *J. Fluids Engineering*, **117**, 45–49 (1995)

[7] GRESCHO, P.M., On the theory of semi-implicit projection method for viscous incompressible flow and its implementation via a finite element method that also introduces a nearly consistent mass matrix, Part I: Theory, *Int. J. Num. Meth. Fluids*, **11**, 587–620 (1990).

[8] HAM, N.D. & GARELICK, M.S., Dynamic stall considerations in helicopter rotors, *J. of Amer. Helicopter Soc.*, **13**, 2–12 (1982).

[9] HARAS, Z. & TA'ASAN, S., Finite difference scheme for long time integration, *J. Comp. Phys.*, **114**, pp 265–279 (1994).

[10] HARLOW, H.H. & WELTCH, J.E., Numerical calculation of time-dependent viscous incompressible flow of fluid with free surface, *Phys. Fluids*, **8**, 2182–2189 (1965).

[11] HESTENES, M.R. & STIEFEL, E., Methods of conjugate gradients for solving linear systems, *J. Res. Nat.Bur. Stand.*, **49**, 409–436 (1952).

[12] JENNINGS, A. & MALIK, G.M., The solution of sparse linear equations by the conjugate gradient method, *Int. J. for Num. Meth. in Engineering.*, **12**, 141–158 (1978).

[13] KAWAMURA, T., TAKAMI, H. & KUWAHARA, K., A new higher-order upwind scheme for incompressible Navier-Stokes equations, *Fluid Dynamic Res.*, **1**, 145–162 (1985).

[14] KIM, J. & MOIN, P., Application of a fractional-step method to incompressible Navier-Stokes equation, *J. Comp. Phys.* **59**, 308–323 (1985).

[15] KOOCHHESFAHANI, M.M. & SMILJANOVSKI, V., Initial acceleration effects on flow evolution around airfoils pitching to high angles of attack, *AIAA Journal*, **31(8)**, 1529–1531 (1991).

[16] MCCROSKEY, W.J., Unsteady Airfoils, *Annual Rev. Fluid Mech.*, **14**, 285–311 (1982).

[17] MORIKAWA, K. & GRÖNIG, H., Formation of structure of vortex systems around a translation and oscillating airfoil, Z. Fluwiss, Weltraumforsch, **19**, 391–396 (1995).

[18] NAIR, M.T. & SENGUPTA, T.K., Unsteady flow past elliptic cylinders, *J. Fluid & Structures*, **11**. 555–595 (1997).

[19] NAIR, M.T. & SENGUPTA, T.K., Orthogonal grid generation for Navier-Stokes computation, *Int. J. Num. Meth. Fluids*, **28**, 215–224 (1998).

[20] NAIR, M.T. & SENGUPTA, T.K., Upwind schemes and large eddy simulation, *Int. J. for Num. Methods in Fluids*, **31**, 897–889 (1999).

[21] PERIDIER, V.J., SMITH, F.T. & WALKER, J.D.A., Vortex induced boundary layer separation. Part-2. Unsteady interacting boundary layer theory, *J. Fluid Mech.*, **232**, 133–165 (1991).

[22] RAI, M.M. & MOIN, P., Direct simulation for turbulent flows using finite difference schemes, *J. Comp. Phys.*, **96**, 15–53 (1991).

[23] RAYLIEGH, L., On the motion of solid bodies through viscous liquids, *Philosophical Magazine*, **21(6)**, 697–711 (1911).

[24] SCHLICHTING, H., *Boundary Layer Theory*, 7th edn., McGrawhill Book Company (1979).

[25] SENGUPTA, T.K. & SENGUPTA, R., Flow past an impulsively started circular cylinder at high Reynolds number, *Comput. Mech.*, **14(4)**, 298–310 (1994).

[26] SENGUPTA, T.K. & NAIR, M.T., Upwind schemes and lagge eddy simulation, *Int. J. Num. Meth. Fluids*, **31**, 879–889 (1999).

[27] SENGUPTA, T.K., GANERIWAL, G. & DE, S., Analysis of central and upwind compact schemes, *J. Comp. Phys.*, **192(2)**, 677–694 (2003).

[28] STOKES, G.G., On the effect of the internal friction of fluids on the motion of pendulums, *Transactions of Cambridge Philosophical Society*, **9**, 8–106 (1851).

[29] STUART, J.T., Unsteady boundary layers in *Laminar Boundary Layers*(ed. L. Rosenhead), Oxford Univ.. Press, Chap. VII, 285–311 (1963).

[30] TELIONIS, D.P., *Unsteady Viscous Flows*, Springer Verlag, New York (1981).

[31] WILLIAMS, G.P., Numerical integration of the three dimensional Navier-Stokes equation for incompressible flow, *J. Fluid Mech.*, **37**, 727–750 (1969).

Appendix A: Brief Review of Vector Analysis

(a) **Vector Algebra**

a, b are scalar constants.
$\vec{A}, \vec{B}$ are vector constants.
f, g, h are scalar variables; $f(x, y, z)$, etc.
$\vec{F}, \vec{G}, \vec{H}, \vec{I}$ are vector variables, $\vec{F}(x, y, z)$, etc.

(i) $\vec{F} = \vec{i}F_x + \vec{j}F_y + \vec{k}F_z = \sum \vec{e}_i F_i$ ← Summational convection.

(ii) $F = |\vec{F}| = \sqrt{F_x^2 + F_y^2 + F_z^2}$ ← Magnitude of F.

(iii) $\frac{\vec{F}}{F}$ ← Unit vector in the direction of $\vec{F}$.

(iv) $\vec{F} + \vec{G} = \vec{i}(F_x + G_x) + \vec{j}(F_y + G_y) + \vec{k}(F_z + G_z)$ ← Summation.

(v) $f\vec{F} = \vec{i}(fF_x) + \vec{j}(fF_y) + \vec{k}(fF_z)$ ← Multiplication by scalar.

(vi) $\vec{F}.\vec{G} = F_xG_x + F_yG_y + F_zG_z = \sum F_iG_i$ ← Scalar product.

(vii) $\vec{F}.\vec{F} = F^2$

(viii) $\frac{\vec{F}}{F}.\vec{G}$ ← Component of $\vec{G}$ in the direction of $\vec{F}$.

(ix) $\vec{F} \times \vec{G} = \vec{i}(F_yG_z - F_zG_y) + \vec{j}(F_zG_x - F_xG_z) + \vec{k}(F_xG_y - F_yG_x)$

$$= -\vec{G} \times \vec{F} = \begin{vmatrix} \vec{i} & \vec{j} & \vec{k} \\ F_x & F_y & F_z \\ G_x & G_y & G_z \end{vmatrix}$$ ← Vector product.

(x) $\frac{\vec{F}}{F} \times \vec{G}$ ← Component of $\vec{G}$ in the direction normal to $\vec{F}$.

(xi) $\vec{F} \times \vec{F} = 0$

(xii) $\vec{F}.(\vec{G} \times \vec{H}) = (\vec{F} \times \vec{G}).\vec{H}$ ← Scaler triple product, volume of parallelepiped with sides $\vec{F}$, $\vec{G}$, and $\vec{H}$.

(xiii) $\vec{F}.(\vec{F} \times \vec{G}) = 0$

(xiv) $\vec{F} \times (\vec{G} \times \vec{H}) = \vec{G}(\vec{F}.\vec{H}) - \vec{H}(\vec{F}.\vec{G})$ ← Vector triple product.

(xv) $\vec{F} \times (\vec{G} \times \vec{H}) + \vec{G} \times (\vec{H} \times \vec{F}) + \vec{H} \times (\vec{F} \times \vec{G}) = 0$

(xvi) $(\vec{F} \times \vec{G}).(\vec{H} \times \vec{I}) = \vec{F}.[\vec{G} \times (\vec{H} \times \vec{I})] = (\vec{F}.\vec{H})(\vec{G}.\vec{I}) - (\vec{F}.\vec{I}).(\vec{G}.\vec{H})$

(xvii) $(\vec{F} \times \vec{G}) \times (\vec{H} \times \vec{I}) = [(\vec{F} \times \vec{G}).\vec{I}]\vec{H} - [(\vec{F} \times \vec{G}).\vec{H}]\vec{I}$

(xviii) $[(\vec{F} \times \vec{G}) \times \vec{H}].\vec{I} = [(\vec{G} \times \vec{F}) \times \vec{I}].\vec{H} = [(\vec{H} \times \vec{I}) \times \vec{F}].\vec{G} = [(\vec{I} \times \vec{H}) \times \vec{G}].\vec{F}$

(b) **Vector Differential Calculus** .

(i) $\vec{r} = \vec{i}x + \vec{j}y + \vec{k}z$ ← Position vector.

(ii) $df = \frac{\partial f}{\partial x}dx + \frac{\partial f}{\partial y}dy + \frac{\partial f}{\partial z}dz$

(iii) $d\vec{r} = \vec{i}dx + \vec{j}dy + \vec{k}dz.$

(iv) $\vec{\nabla} = \vec{i}\frac{\partial}{\partial x} + \vec{j}\frac{\partial}{\partial y} + \vec{k}\frac{\partial}{\partial z}$ ← Del operator, ready to differentiate.

(v) $\vec{\nabla} f = \vec{i}\frac{\partial f}{\partial x} + \vec{j}\frac{\partial f}{\partial y} + \vec{k}\frac{\partial f}{\partial z}$ ← Gradient of f.

$df = \vec{\nabla} f.d\vec{r}$ ← Change of f in the $\vec{r}$ direction.

$\frac{df}{dr} = \vec{\nabla} f.\frac{d\vec{r}}{dr}$ ← Rate of change of f in the $\vec{r}$ direction.

$\frac{df}{dr}$ is a maximum if $d\vec{r}$ is in the direction of $\vec{\nabla} f$.

$\vec{\nabla} f$ gives the direction of maximum rate of change of f.

The maximum rate is $\vec{\nabla} f$.

(vi) $\vec{\nabla}.\vec{F} = \frac{\partial F_x}{\partial x} + \frac{\partial F_y}{\partial y} + \frac{\partial F_z}{\partial z}$ ← Divergence of $\vec{F}$, net outward flux of $\vec{F}$ per unit volume.

(vii) $\vec{\nabla}\times\vec{F} = \vec{i}\left(\frac{\partial F_z}{\partial y} - \frac{\partial F_y}{\partial z}\right) + \vec{j}\left(\frac{\partial F_x}{\partial z} - \frac{\partial F_z}{\partial x}\right) + \vec{k}\left(\frac{\partial F_y}{\partial x} - \frac{\partial F_x}{\partial y}\right)$ ← Curl of $\vec{F}$, component of $\vec{\nabla} \times \vec{F}$ normal to s is the circulation per unit area of s.

(viii) $\vec{\nabla}.(\vec{\nabla} f) = \nabla^2 f = \frac{\partial^2 f}{\partial x^2} + \frac{\partial^2 f}{\partial y^2} + \frac{\partial^2 f}{\partial z^2} = \frac{\partial^2 f}{\partial x_i \partial x_i}$ ← Laplacian of f.

(ix) $\nabla^2 \vec{F} = (\vec{\nabla}.\vec{\nabla})\vec{F} = \vec{i}\nabla^2 F_x + \vec{j}\nabla^2 F_y + \vec{k}\nabla^2 F_z = \vec{\nabla}(\vec{\nabla}.\vec{F}) - \vec{\nabla} \times (\nabla \times \vec{F})$

(x) $\vec{\nabla} \times (\vec{\nabla} f) = 0$

(xi) $\vec{\nabla}.(\vec{\nabla} \times \vec{F}) = 0$

(xii) $\vec{\nabla}(fg) = f\vec{\nabla} g + g\vec{\nabla} f$

(xiii) $\vec{\nabla}.(f\vec{F}) = f\vec{\nabla}.\vec{F} + \vec{F}.\vec{\nabla} f$

(xiv) $\vec{\nabla} \times (f\vec{F}) = f\vec{\nabla} \times \vec{F} - \vec{F} \times \vec{\nabla} f$

(xv) $\vec{\nabla}.(\vec{F} \times \vec{G}) = \vec{G}.(\vec{\nabla} \times \vec{F}) - \vec{F}.(\vec{\nabla} \times \vec{G})$

(xvi) $\vec{\nabla} \times (\vec{F} \times \vec{G}) = \vec{F}(\vec{\nabla}.\vec{G}) - (\vec{F}.\vec{\nabla})\vec{G} - \vec{G}(\vec{\nabla}.\vec{F}) + (\vec{G}.\vec{\nabla})\vec{F}$

(xvii) $\vec{\nabla}(\vec{F}.\vec{G}) = \vec{F} \times (\vec{\nabla} \times \vec{G}) + (\vec{F}.\vec{\nabla})\vec{G} + \vec{G} \times (\vec{\nabla} \times \vec{F}) + (\vec{G}.\vec{\nabla})\vec{F}$

(xviii) $\vec{\nabla}.\vec{r} = 3$

(xix) $\vec{\nabla} \times \vec{r} = 0$

(xx) $\vec{\nabla} r = \frac{\vec{r}}{r}$ ← Unit vector in the r direction.

(xxi) $\vec{\nabla}(1/r) = -\frac{\vec{r}}{r^3}$

(xxii) $\vec{\nabla}.\left(\frac{\vec{r}}{r^3}\right) = -\nabla^2(1/r) = 0$ if $r \neq 0$

(xxiii) $\vec{\nabla}.(\vec{A}/r) = \vec{A}.\vec{\nabla}(1/r) = -\frac{\vec{A}.\vec{r}}{r^3}$

(xxiv) $\vec{\nabla} \times (\vec{A} \times \vec{r}/r^3) = -\nabla(\vec{A}.\vec{r}/r^3) = -(\vec{A}.\vec{\nabla})(\vec{r}/r^3)$ if $r \neq 0$

(xxv) $\nabla^2(A/r) = A\nabla^2(1/r) = 0$ if $r \neq 0$

(xxvi) $\vec{\nabla} \times (\vec{A} \times \vec{F}) = \vec{A}(\vec{\nabla}.\vec{F}) - (\vec{A}.\vec{\nabla})\vec{F} = \vec{A}(\vec{\nabla}.\vec{F}) - \vec{\nabla}(\vec{A}.\vec{F}) + \vec{A} \times (\vec{\nabla} \times \vec{F})$

(xxvii) If $\vec{\nabla} \times \vec{F} = 0$, then there is a scalar potential ψ such that $\vec{\nabla}\psi = \vec{F}$.

(xxviii) If $\vec{\nabla}.\vec{F} = 0$, then there is a vector potential $\vec{G}$ such that $\vec{\nabla} \times \vec{G} = \vec{F}$.

(c) **Vector Integral Calculus**

Curve P, element of curve dr
Surface S, element of surface ds
Volume V, element of volume dv
Unit vector normal to S is $\vec{n}$
$\int_s \vec{F}.\vec{n}ds$ = Flux of $\vec{F}$ through surface
$\oint_p \vec{F}.d\vec{r}$ = Circulation of $\vec{F}$ around closed curve P.

Theorem I: Line integral of $\vec{\nabla} f$ is independent of path.

$$\int_a^b (\vec{\nabla} f).d\vec{r} = f(b) - f(a)$$

Theorem II: Gauss' or Divergence Theorem

$$\oint_s \vec{F}.\vec{n}ds = \int_v (\vec{\nabla}.\vec{F})dv$$

V is the volume enclosed by the closed surface S.
n is directed outward from the volume.

Corollary I:

$$\oint_s f\vec{n}ds = \int_v (\nabla f)dv$$

Corollary II:

$$\oint_s \vec{F} \times \vec{n} ds = -\int_v (\vec{\nabla} \times \vec{F}) dv$$

Theorem III: Stokes' Theorem

$$\oint_p \vec{F}.d\vec{r} = \int_s (\vec{\nabla} \times \vec{F}).\vec{n} ds$$

S is any surface bounded by the closed curve P. The direction of $\vec{n}$ is given by the direction of integration around P following the right hand rule.

Corollary III:

$$\oint f d\vec{r} = -\int_s (\vec{\nabla} f) \times \vec{n} ds$$

Appendix B: Finite Difference Approximations to $\frac{\partial u}{\partial t} = \frac{\partial^2 u}{\partial x^2}$

(a) Explicit method:

$$\frac{u_j^{n+1} - u_j^n}{\Delta t} = \frac{(\delta_x^2 u)_j^n}{(\Delta x)^2}$$

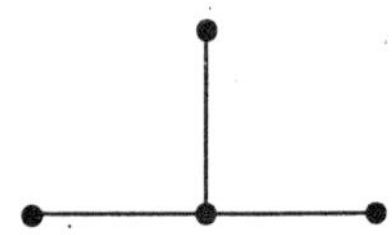

where the truncation error (T.E.) is of the order of $O(\Delta t) + O[(\Delta x)^2]$.

This is stable only if $\Delta t/(\Delta x)^2 = \text{Const.} \leq \frac{1}{2}$

as $\Delta t, \Delta x \to 0$,

where, $\delta^2 u_j^n = u_{j+1}^n - 2u_j^n + u_{j-1}^n$.

(b) Crank-Nicolson method: An implicit method

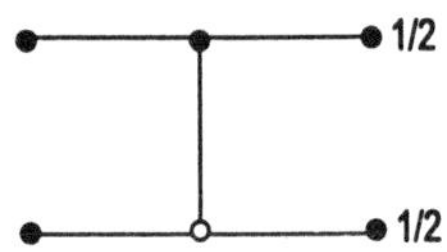

$$\frac{u_j^{n+1} - u_j^n}{\Delta t} = \frac{(\delta^2 u)_j^n + (\delta^2 u)_j^{n+1}}{2(\Delta x)^2}$$

T.E.: $O(\Delta t^2) + O[(\Delta x)^2]$

This is an unconditionally stable method.

(c) Fully implicit method:

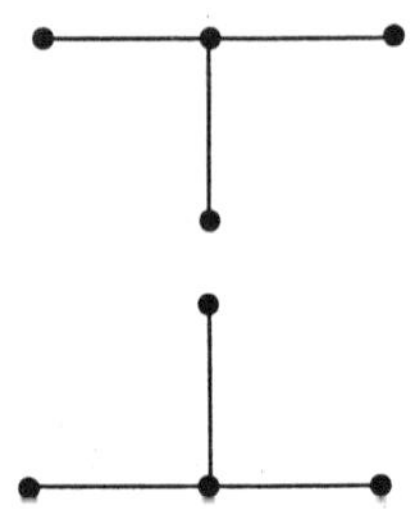

(Special)

$$\frac{u_j^{n+1} - u_j^n}{\Delta t} = \frac{(\delta^2 u)_j^{n+1}}{(\Delta x)^2}$$

T.E. : $O(\Delta t) + O[(\Delta x)^2]$

Another unconditionally stable method.

(d) Milne's method: Same as (a), but with $\dfrac{\Delta t}{(\Delta x)^2} = \dfrac{1}{6}$

T.E.: $O[(\Delta t)^2] = O[(\Delta x)^4]$

special case of (a)- a stable method.

(e) Crandall's method: $\dfrac{u_j^{n+1} - u_j^n}{\Delta t} = \dfrac{\theta(\delta^2 u)_j^{n+1} + (1-\theta)(\delta^2 u)_j^n}{(\Delta x)^2}$

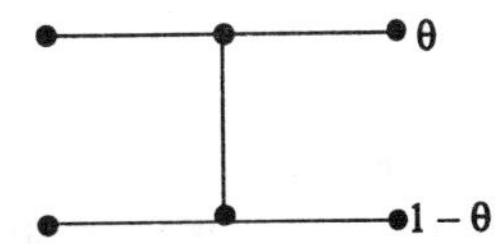

where $\theta =$ const., $0 \le \theta \le 1$.

T.E.: $0(\Delta t) + 0[(\Delta x)^2]$

for $0 \le \theta < \frac{1}{2}$, stable if $\Delta t/(\Delta x)^2 =$ Const. $\le 1/(2 - 4\theta)$; for $\frac{1}{2} \le \theta \le 1$, always stable, includes the above as special cases.

(f) Higher order implicit method: Same as (e), but with $\theta = \frac{1}{2} - (\Delta x)^2/12\Delta t$

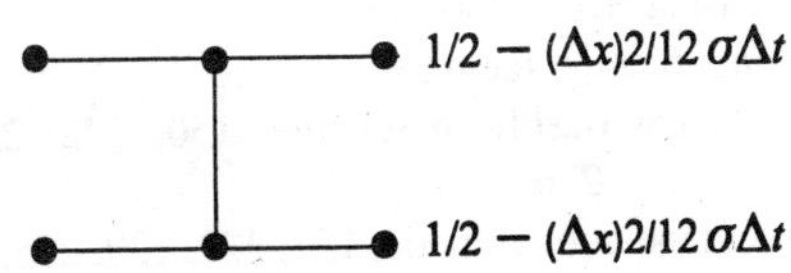

T.E.: $O[(\Delta t)^2] = O[(\Delta x)^4]$

-A stable method.

(g) Richardson's method: $\dfrac{u_j^{n+1} - u_j^{n-1}}{2\Delta t} = \dfrac{(\delta^2 u)_j^n}{(\Delta x)^2}$

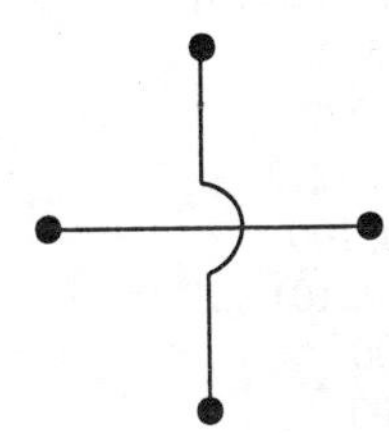

This method is always unstable.

(h) Dufort-Frankel's method:

$$\frac{u_j^{n+1} - u_j^{n-1}}{2\Delta t} = \frac{u_{j+1}^n - u_j^{n+1} - u_j^{n-1} + u_{j-1}^n}{(\Delta x)^2}$$

where $\Delta t/\Delta x \to 0$ as Δt, $\Delta x \to 0$.

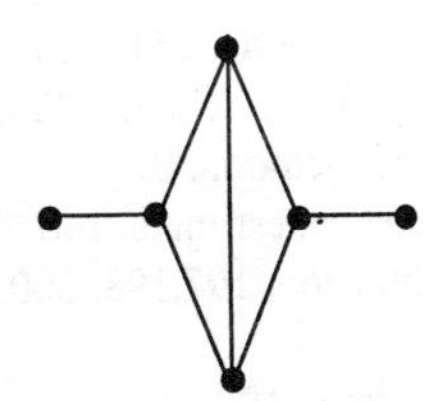

T.E.: $O(\Delta t)^2 + O[(\Delta x)^2] + O\left(\frac{\Delta t}{\Delta x}\right)^2$

An explicit, stable but *inconsistent method.*

(i) Adam-Bashforth method: $\dfrac{3}{2}\dfrac{u_j^{n+1} - u_j^n}{\Delta t} - \dfrac{1}{2}\dfrac{u_j^n - u_j^{n-1}}{\Delta t} = \dfrac{(\delta^2 u)_j^{n+1}}{(\Delta x)^2}$

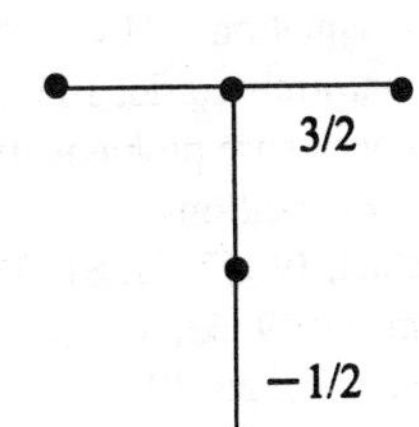

T.E.: $0[(\Delta t)]^2 + 0[(\Delta x)^2]$

Always stable, but has a spurious computational mode.

Index